Collection and Analysis of Text Data in Sensory and Consumer Science

Michel Visalli • Mara Virginia Galmarini
Editors

Collection and Analysis of Text Data in Sensory and Consumer Science

Springer

Editors
Michel Visalli
Centre for Taste and Feeding Behavior (CSGA)
National Research Institute for Agriculture, Food and the Environment (INRAE)
Dijon, France

Mara Virginia Galmarini
CONICET - UTDT
Buenos Aires, Argentina

ISBN 978-3-032-04850-9 ISBN 978-3-032-04851-6 (eBook)
https://doi.org/10.1007/978-3-032-04851-6

This Springer imprint is published by the registered company Springer Nature Switzerland AG
The registered company address is: Gewerbestrasse 11, 6330 Cham, Switzerland

Foreword

Sensory and consumer science has always relied on language to collect responses from test participants, principally through well-defined vocabulary and structured question formats. Utilizing explicit questions, normed attributes, labeled scales, and anchors, researchers in the field have developed most refined tools to measure perception and quantify sensations.

By contrast, open-ended questions were the poor cousin. They were (they still are) typically used in the hope they will provide additional insights into consumers' opinions and thoughts about the products. Yet, with limited means of analysis, I suspect that most of us kept open-ended questions in our surveys with the feeling that we could be missing important points and that our standardized questionnaires were not enough to capture consumer experience. Occasionally, analysis of those open comments would indeed unveil key consumer feedback and were helpful to raise warnings about aspects of the product that were not fully covered in the rest of the test.

It is only with the recent—and exponential—development in natural language processing techniques and with the power of AI that the real value of open-ended questions has emerged, allowing us to analyze comments at light speed, to make sense of larger corpus, and to quantify percepts and arrange concepts hierarchically.

However, we would be mistaken to see this evolution in the techniques only as a matter of efficiency and analysis power. With this book, Michel Visalli and Mara Galmarini highlight how open-ended questions allow for deeper, more nuanced insights into how people perceive products, make choices, and experience the world. Letting participants use their own words also opens the door to different ways of interviewing them and offers a more natural way to respond. It allows sensory and consumer scientists to capture nuances, as well as the richness and diversity of sensory experiences. This evolution of data collection and analysis techniques reflects a broader shift in our field: from measuring the "sensory properties" of a product to measuring people's perception, with the recognition that every product experience is unique.

The coming years will no doubt bring further advances in natural language processing, but what is already clear is that open-ended comments will no longer be the "poor cousin" of sensory evaluation. Instead, they will take their rightful place at the center of our understanding of consumer perception. This book is an important step on that path, and a timely contribution. Bringing together multidisciplinary perspectives, this book provides both the conceptual grounding and the practical guidance to make the most of this evolution, and I trust it will become a reference for sensory and consumer scientists who wish to fully integrate open comments and social media content into their research.

Julien Delarue

Preface

Open-ended questions and text data have become indispensable tools in sensory and consumer science. Far from being a simple methodological choice, their use reflects a growing recognition of the complexity and richness of human perception. Unlike closed questions, which limit responses to predefined options, open-ended questions invite individuals to express their thoughts in their own words. This allows for deeper, more nuanced insights into how people perceive products, make choices, and experience the world around them.

Historically, open-ended questions were first introduced in marketing research, where they were often dismissed as inefficient and no more insightful than check-box questions. However, the landscape has changed dramatically. The rise of natural language processing (NLP) and advances in technology have transformed how we collect, analyse, and interpret qualitative text data. Today, open-ended responses can be explored at scale, using sophisticated tools that extract meaning, patterns, and emotional depth. Beyond traditional survey formats, the availability of large, unstructured text corpora, ranging from online reviews to social media content, has opened new doors for research in sensory and consumer science. In the era of big data, understanding consumer experiences through the lens of language is not only possible but essential.

This book brings together a multidisciplinary perspective on this evolving field. It covers foundational knowledge in language and perception, experimental considerations for data collection and analysis, and emerging opportunities to harness web-based text data. Contributions span from neuroscience and psychology to linguistics and data science, reflecting the inherently interdisciplinary nature of sensory and consumer research. Real-world case studies illustrate how text data can be used to generate actionable insights in both academic and industry settings.

The primary audience for this book includes professionals in sensory, marketing, and consumer sciences, particularly within the food sector, whether working in academic research or industrial applications. A secondary audience includes professionals in non-food domains such as cosmetics, textiles, and gastronomy, who can also benefit from a deeper understanding of how consumers describe and relate to products. Additionally, this book serves as a valuable complementary resource for university-level programmes in sensory and consumer science. It is accessible to undergraduate students—especially those studying food science, product development, and consumer behaviour—as well as to educators in these fields. We hope this book will serve as both a practical guide and a source of inspiration for anyone seeking to better understand consumer-product interactions and the broader consumer experience through the power of words.

Dijon, France Michel Visalli
Buenos Aires, Argentina Mara Virginia Galmarini

Contents

Contributors

Gastón Ares Department of Sensometrics & Consumer Science, Instituto Polo Tecnológico de Pando, Facultad de Química, Universidad de la República, Canelones, Uruguay

Þóranna Bender Department of Applied Mathematics and Computer Science, Technical University of Denmark, Kongens Lyngby, Denmark

Jing Cao Department of Statistics and Data Science, Southern Methodist University, Dallas, TX, USA

Sok L. Chheang The New Zealand Institute for Plant and Food Research Limited, Auckland, New Zealand

Danièle Dubois French National Centre for Scientific Research (CNRS), Paris, France

Magalie Dubois School of Wine and Spirits Business, Burgundy School of Business, Dijon, France

Michał Folwarczny Discipline of Marketing, J.E. Cairnes School of Business and Economics, University of Galway, Galway, Ireland

Mara Virginia Galmarini Escuela de Negocios. Centro de Inteligencia Artificial y Neurociencia (CIAN). Laboratorio de Neurociencia, Consejo Nacional de Investigaciones Científicas y Técnicas (CONICET) and Universidad Torcuato Di Tella, Buenos Aires, Argentina

Enrique García-Marco Departamento de Psicología Clínica y Experimental, Universidad de Huelva, Huelva, Spain

Carlos Gómez-Corona Human Insight Department - TTH, dsm-firmenich, Neuilly-sur-Seine, France

Clémentine Hugol-Gential CIMEOS Laboratory, Université Bourgogne Europe, Dijon, France

Sara R. Jaeger Department of Food Science, Aarhus University, Aarhus, Denmark

Benjamin Mahieu Oniris, INRAE, StatSC, Nantes, France

Katia Rolán Laboratorio de Linguaxe e Cognición, Universidade de Vigo, Vigo, Spain

Instituto de Investigación Lingua, Universidade de Vigo, Vigo, Spain

Departamento de Análise e Intervención Socioeducativa, Universidade de Vigo, Vigo, Spain

Cátedra Unesco Educación Transformadora: Ciencia Comunicación y Sociedad, Universidade de Vigo, Vigo, Spain

Grupo Language Variation and Textual Caterogorisation (LVTC), Universidade de Vigo, Vigo, Spain

Ronan Symoneaux GRAPPE, ESA, USC 1422 INRAE, SensoVeg, SFR 4207 QUASAV, Angers, France

Amparo Tárrega Instituto de Agroquímica y Tecnología de los Alimentos (IATA-CSIC), Paterna, Spain

Michel Visalli Centre for Taste and Feeding Behavior (CSGA), National Research Institute for Agriculture, Food and the Environment (INRAE), Dijon, France

Magalie Weber BIA, National Research Institute for Agriculture, Food and the Environment (INRAE), Nantes, France

Part I
Introduction

Chapter 1
Introduction

Mara Virginia Galmarini

1 Sensory and Consumer Science: A Discipline in Constant Evolution

The field of Sensory and Consumer Science has undergone major transformations together with continuous expansion throughout its history of over 100 years (Jaeger et al., 2025; Meiselman et al., 2022). What began as a discipline focused on understanding the chemical senses and quantifying physiological responses to stimuli, with a solid orientation to be used by the industry as a complementary tool for product description—evolved into a comprehensive science interested in consumer–product interaction which goes beyond foods and beverages (Lahne, 2016).

Today, it encompasses not only the evaluation and description of the perception of food products, but also a wide range of products and experiences, including the evaluation of consumer behaviour in relation to complex, multisensory experiences. This expansion has brought about methodological changes as well as the interaction and adaptation of methods from other (social) sciences such as anthropology, behavioural psychology, and sociology.

M. V. Galmarini (✉)
Escuela de Negocios. Centro de Inteligencia Artificial y Neurociencia (CIAN). Laboratorio de Neurociencia, Consejo Nacional de Investigaciones Científicas y Técnicas (CONICET) and Universidad Torcuato Di Tella, Buenos Aires, Argentina
e-mail: mara.galmarini@utdt.edu

M. Visalli, M. V. Galmarini (eds.), *Collection and Analysis of Text Data in Sensory and Consumer Science*, https://doi.org/10.1007/978-3-032-04851-6_1

2 Early Understanding of the Consumer

During many years, in the field of sensory science, asking consumers to explain their choices (e.g., why they liked a certain product, what they disliked of an experience) was considered controversial and even frowned upon. According to a classic textbook of sensory science (Lawless & Heymann, 1998) consumers respond to the product as a whole, even when sometimes one or two salient characteristics may drive their decisions. In addition, it was said that these characteristics may cause other attributes to be viewed as more positive (an example of what is known as "halo effect" where liking a product implies liking all its characteristics). These are some of the reasons why in early 2000 it was said that it was difficult to get consumers to explain the basis of their choices and that asking them for the reason for their preferences could be a risky practice (Symoneaux et al., 2012).

Consumers' preferences and choices were understood by relating hedonic values obtained from rating liking on a 9–point hedonic scale (Peryam & Pilgrim, 1957) (which could also be adapted to 7 or 5 points depending on the aim of the experiment) to a detailed product description which was separately obtained by a trained panel using methods such as Quantitative Descriptive Analysis (Stone & Sidel, 1985) consisting of objectively rating on scales the intensity of several previously defined and reported attributes. This correlation of data resulted in *preference maps* which were used (specially by the industry) to understand the drivers of liking or disliking of their products.

These maps, defined as Internal Preference Mapping (IPM), were based solely on consumer liking data, with consumers and products projected into a perceptual space; or External Preference Mapping (EPM) which incorporate sensory data—from trained panels who quantifies the intensity of different attributes—to model the relationship between product characteristics and consumer liking. Both have long been used in sensory and consumer science to understand and visualize consumer preferences in relation to product attributes (Greenhoff & MacFie, 1994). These techniques were (and are still) particularly useful for identifying preference segments, uncovering sensory drivers of liking, consumer clusters, and informing product optimization. When the goal is to detect general liking trends or market segments within a product category, both internal and external preference mapping still provide valuable, reliable insights (McEwan, 1996).

However, these methods also have some limitations and drawbacks. Their implementation requires a trained panel that can describe products using the mentioned techniques. This can be time consuming and not always in line with the costs available for everyday tests (not every business can afford to have a trained panel and—here comes the good part of the new methods—they might not need one). Moreover, methods grounded in quantitative data reduce consumer responses to numerical values, often assuming linear relationships between liking and product attributes. As a result, they are limited in their capacity to capture the richness and multidimensionality of consumer perception (Meiselman, 2015).

Also, using the perception of trained panellists (people who were probably screened on the basis of uniform perception) it is usually implied that consumers perceive the product in the same way as people whose work is to perceive (and quantify) certain characteristics of the product and that, if given the possibility, they would use the same words to describe their experience. However, this assumption overlooks the fact that perception is not universal and that consumers often interpret and describe their experiences through the lens of their own personal, cultural, and contextual backgrounds. Unlike trained panellists, whose responses are shaped by a standardized sensory vocabulary and calibration procedures, consumers rely on their individual memories, emotions, and expectations, which can lead to different—and equally valid—ways of perceiving and expressing sensory experiences. In addition, traditional preference mapping does not capture the *why* behind consumer responses. Emotional, contextual, and cultural aspects—key components of the consumer experience—are absent from the analysis (Cardello, 2003; Köster, 2009). Also, hedonic scores are influenced by the artificiality of the test environment and the structured nature of the task, where consumers are rarely allowed or encouraged to elaborate on their responses (Dacremont & Valentin, 2020). Furthermore, outliers or minority opinions are often underrepresented in preference mapping. The methods tend to prioritize the dominant trends in the dataset, potentially overlooking small but important consumer subgroups with different expectations or usage contexts.

Therefore, with the goal of understanding consumer perception in a broader, more holistic sense, and with a modern context that changed how consumers' opinions are perceived, it was time for sensory and consumer science to integrate new methodology "borrowed" from other social sciences.

3 The Consumer in the Center of the Scene

The rise of social media platforms together with the design of mobile devices equipped with front-facing cameras, has profoundly changed the way we communicate, express ourselves, and narrate experiences. For consumers, this shift has meant that the act of evaluating or reviewing a product is no longer confined to private opinions or structured surveys; instead, it becomes a public, even performative act, often staged in the first person. Platforms like *Instagram*, *X*, or *TripAdvisor* have amplified the individual voices, allowing consumers to produce and circulate their own accounts of consumption to potentially massive audiences.

This transformation aligns with what scholars describe as the *self-branding logic of networked culture* (Marwick, 2013), where individuals become both subjects and broadcasters of their own lived experiences. At the same time, it reflects broader trends in contemporary society where personal narratives, authenticity, and emotional resonance are privileged over traditional expertise or institutional authority.

In this consumer-centric landscape, the experience as told by the user—not the company nor a few experts—holds increasing power. A negative review on Amazon can significantly alter a product's success, and peer-to-peer commentary has made it harder for companies to obscure flaws in products, as seen in user-generated reviews of platforms like *Shein*. Today, consumers appropriate the marketplace through storytelling, turning the first-person voice into both a source of truth and a mechanism of influence (Chen et al., 2024).

In this new context, underestimating consumers' comments and opinions is not an option. Considering this broadening of the scope of study, traditional quantitative methods result insufficient in the face of this bigger challenge which is to capture the richness and complexity of the complete consumer experience in a vast world of consumer generated text data. Quantitative methods, while useful for measuring liking or attribute intensity, tend to reduce responses to a single dimension and limit the possibility of uncovering underlying meanings, expectations, and emotional nuances. The same applies to classical descriptive analysis and closed-ended questionnaires, which may constrain participant responses and fail to reflect the way people spontaneously express their perceptions and experiences in real life (Dacremont & Valentin, 2020). Even rapid methods like CATA (Check All That Applies, Ares & Jaeger, 2013) or RATA (Rate All That Applies, Vidal et al., 2018) lack the richness of using the vocabulary generated by the own consumers which, if well-handled and analysed, can provide a lot of information to improve products, services and experiences.

Moreover, the explosion of consumer-generated content—ranging from social media posts and online reviews to voice notes and hashtags—has created a new territory for researchers. For the first time, we have access to large-scale, unsolicited, spontaneous verbalizations of consumer experiences. These texts are rich, messy, emotional, and diverse. They reflect not only individual perceptions but also collective imaginaries and cultural narratives (Kozinets, 2019; Kleres, 2011).

At the same time, advances in computational linguistics and artificial intelligence have made it possible to process and interpret massive, unstructured text corpora in ways that were previously unimaginable (Cambria et al., 2013; Feldman, 2013). These tools allow researchers to extract not only semantic content but also emotional tone, intensity, and evaluative meaning from consumer-generated text, offering powerful insights into perceptions, preferences, and affective responses giving way to sentiment analysis, topic modelling, and large-scale language modelling (Visalli et al., 2020). What was once discarded as "anecdotal" or "unstructured" can now be visualized, clustered, mapped, and interpreted with scientific rigor (Thelwall, 2016).

In this digitalized era with so much consumer-generated information, it is the perfect moment to evaluate the available methods for analysing free text, comments, and open-ended questions and use them not as a complement, but as a standalone method (Symoneaux et al., 2012; Visalli et al., 2023) to analyse consumer decisions and behaviour.

4 The Challenge of Consumer Centered Methods

The words people choose when describing foods or beverages, whether in a lab setting or on a social media platform, are rarely neutral. Language is shaped by cultural norms, personal memories, and emotional states. For example, describing a flavour as "comforting" versus "rich" reveals more than intensity: it signals familiarity, emotion, and even social meaning. This spontaneous discourse is different from what can be obtained with a fixed list of words, which is the basis for many of the most used sensory descriptive methods (Symoneaux & Galmarini, 2014).

The words people use to describe foods are not just vehicles of communication; they are part of the sensory experience itself. Language shapes how we remember a product, how we compare it with others, and how we make sense of what we consume. Calling a texture "creamy" or "slimy", or a smell "fresh" versus "chemical", involves complex layers of interpretation that go beyond physical properties. These verbal choices are influenced by culture, emotion, memory, and social norms. Therefore, understanding consumer language means understanding consumption itself.

In sensory and consumer science, the shift from measuring responses to interpreting narratives reflects a deeper epistemological transformation: from treating perception as objective data to recognizing it as socially and cognitively constructed. Language is not merely descriptive, it is generative. It brings to life experiences that are multisensory, contextual, and affective. The challenge, then, is not just to *record* what people say, but to *understand* what those words mean, why they were chosen, and what they reveal.

But this change of paradigm which supposes a modification in approach from a purely quantitative framework designed by chemists, physiologists and psychophysicist in the early years of sensory science, to one that increasingly incorporates observational and qualitative perspectives, can also bring about certain methodological pitfalls. In many cases, qualitative data has been approached with a similar perspective to that used for quantitative data, which is not necessarily appropriate.

As noted by Ares and Jaeger (2013) while open-ended responses (Symoneaux et al., 2012) and free-text data offer valuable insights into consumer perception, their use often lacks standardized protocols for data collection and analysis. In many cases, techniques developed for structured data are applied uncritically to these more complex forms of expression, leading to inconsistencies and potential biases. Without clear methodological rigor, we risk misinterpreting the meaning of consumer discourse or giving disproportionate weight to certain types of responses. It is therefore essential to take a deeper look at how we obtain, analyse and interpret this kind of information, to ensure fair and valid conclusions and to properly assess the role of the consumer's voice in sensory and consumer science.

In order to properly approach this challenge, we believe it is essential to move beyond a purely technical view of open-ended questions and text data. Understanding how people describe their sensory experiences requires not only suitable methodological tools, but also a deeper appreciation of how language is produced, processed, and shaped by cognitive, emotional, and social mechanisms. For this reason, the book is structured into five interconnected parts, presenting some complementary perspectives from authors from different backgrounds (which can be sometimes controversial) that guide the reader through both the theoretical and practical dimensions of working with text in sensory and consumer science.

Part II offers foundational knowledge from cognitive neuroscience and linguistics, highlighting how the brain processes language, how meaning is constructed, and how the act of naming and describing sensory experiences is far from neutral or universal. These perspectives help us frame consumer responses not as isolated data points but as situated, embodied expressions.

Part III moves into the realm of data handling. It explores how spontaneous text responses (whether collected in consumer tests or mined from online platforms) can be pre-processed, categorized, and analysed using both manual content analysis and computational techniques from natural language processing (NLP). This section serves as a technical bridge for researchers looking to apply qualitative and quantitative methods in a systematic and rigorous way.

Part IV focuses on experimental and methodological considerations when collecting data from open-ended questions in practical cases. It revisits key practices such as the formulation of questions, response constraints, and the integration of free text with hedonic and behavioural measures. It also introduces more recent tools such as sentiment analysis and text highlighting and discusses how these can be used to enrich our understanding of consumer attitudes and decision-making.

Finally, Part V brings together a series of applied case studies that illustrate the power of text data in action. From co-occurrence network analysis of online comments to netnographic studies and digital wine reviews, this section demonstrates the richness of consumer-generated language in both academic and industrial contexts. These examples show how words can map experiences, shape product meaning, and even reveal cultural shifts in how we relate to food, consumption, and identity.

Altogether, this book brings a multidisciplinary approach to a growing area of inquiry. By integrating insights from sensory science, consumer behaviour, linguistics, data science, and digital culture, we aim to equip researchers and practitioners with tools to listen to consumers in their own words. At a time when personal narratives, peer opinions, and unfiltered expressions carry more weight than ever, understanding what people say—and how they say it—may be one of the most powerful ways to innovate, empathize, and grow.

References

Ares, G., & Jaeger, S. R. (2013). Check-all-that-apply questions: Influence of attribute order on sensory product characterization. *Food Quality and Preference, 28*(1), 141–153.

Cambria, E., Schuller, B., Xia, Y., & Havasi, C. (2013). New avenues in opinion mining and sentiment analysis. *IEEE Intelligent Systems, 28*(2), 15–21.

Cardello, A. V. (2003). Consumer concerns and expectations about novel food processing technologies: Effects on product liking. *Appetite, 40*, 217.

Chen, T., Fan, X., He, J., Fan, J., & Chen, W. (2024). When "I" or "S/He" uses the product: The impact of narrative perspective on consumers' brand attitudes in storytelling ads. *Frontiers in Psychology, 15*, 1338249.

Dacremont, C., & Valentin, D. (2020). Free descriptions and projective methods in sensory and consumer research. In *Novel techniques in sensory characterization and consumer profiling*. CRC Press.

Feldman, R. (2013). Techniques and applications for sentiment analysis. *Communications of the ACM, 56*(4), 82–89.

Greenhoff, K., & MacFie, H. J. H. (1994). Preference mapping in practice. *Measurement of Food Preferences*, 137–166.

Jaeger, S. R., Meiselman, H. L., & Giacalone, D. (2025). Sensory and consumer science: A complex, expanding, and interdisciplinary field of science. *Food Quality and Preference, 122*, 105298.

Kleres, J. (2011). Emotions and narrative analysis: A methodological approach. *Journal for the Theory of Social Behaviour, 41*(2), 182–202.

Köster, E. P. (2009). Diversity in the determinants of food choice: A psychological perspective. *Food Quality and Preference, 20*(2), 70–82.

Kozinets, R. (2019). *Netnography: The essential guide to qualitative social media research*. SAGE.

Lahne, J. (2016). Sensory science, the food industry, and the objectification of taste. *Anthropology of Food, 10*.

Lawless, H. T., & Heymann, H. (1998). *Sensory evaluation of food: Principles and practices*. Chapmann and Hall.

Marwick, A. E. (2013). *Status update: Celebrity, publicity, and branding in the social media age*. Yale University Press.

McEwan, J. A. (1996). Preference mapping for product optimization. In *Data handling in science and technology* (Vol. 16, pp. 71–102). Elsevier.

Meiselman, H. L. (2015). A review of the current state of emotion research in product development. *Food Research International, 76*, 192.

Meiselman, H. L., Jaeger, S. R., Carr, B. T., & Churchill, A. (2022). Approaching 100 years of sensory and consumer science: Developments and ongoing issues. *Food Quality and Preference, 100*, 104614.

Peryam, D. R., & Pilgrim, F. J. (1957). Hedonic scale method of measuring food preferences. *Food Technology, 11*, 9–14.

Stone, H., & Sidel, J. L. (1985). Sensory evaluation practices: Relating sensory, instrumental and consumer data. *Journal of Food Quality, 7*(4), 223–233.

Symoneaux, R., & Galmarini, M. V. (2014). 12 open-ended questions. In *Novel techniques in sensory characterization and consumer profiling* (p. 307). CRC Press.

Symoneaux, R., Galmarini, M. V., & Mehinagic, E. (2012). Comment analysis of consumer's likes and dislikes as an alternative tool to preference mapping. A case study on apples. *Food Quality and Preference, 24*(1), 59–66.

Thelwall, M. (2016). The heart and soul of the web? Sentiment strength detection in the social web with SentiStrength. In *Cyberemotions: Collective emotions in cyberspace* (pp. 119–134). Springer International Publishing.

Vidal, L., Ares, G., Hedderley, D. I., Meyners, M., & Jaeger, S. R. (2018). Comparison of rate-all-that-apply (RATA) and check-all-that-apply (CATA) questions across seven consumer studies. *Food Quality and Preference, 67*, 49–58.

Visalli, M., Mahieu, B., Thomas, A., & Schlich, P. (2020). Automated sentiment analysis of free comment: An indirect liking measurement? *Food Quality and Preference, 82*, 103888.

Visalli, M., Dubois, M., Schlich, P., Ric, F., Cardebat, J.-M., Georgantzis, N. (2023). Relevance of free-comment to describe wine temporal sensory perception: An application with panels varying in culture and expertise. *Food Quality and Preference, 105*, 104785. https://doi.org/10.1016/j.foodqual.2022.104785

Part II
Fundamentals on Natural Language and Food Perception

Understanding free comment responses in sensory and consumer science requires more than appropriate linguistic or statistical tools—it also demands an appreciation of how the human brain processes and produces language. Therefore, before diving into the practical applications of free comment and text analysis, this section begins with two chapters that offer insights from cognitive neuroscience (Chapter 2) and psycholinguistics (Chapter 3), shedding light on the mechanisms underlying the generation of spontaneous verbal responses in evaluation contexts.

By examining how language is represented and activated in the brain, and how meaning is constructed during verbal expression, this section offers a foundation for interpreting free comments and text responses beyond surface-level content. These perspectives may help practitioners and researchers better understand the cognitive constraints and potentials that shape the way people describe their sensory experiences.

Chapter 2
Neuroscience and Language Processing

Enrique García-Marco and Katia Rolán

1 Introduction. Exploring the Intersection of Language and the Mind

"Human language production is caused by neuronal activity, and any speech signal necessarily activates neurons in the brains of listeners when being perceived." (Pulvermüller, 1999, p. 253)

Through the brain's functioning, humans can communicate in many ways, from everyday conversations to scientific discussions and storytelling, exploring both real and imaginary worlds. Language allows us to share information precisely, communicate and express our emotions, check if we maintain the communicative bond, influence others and create beauty by means of poetry. We can even use language to discuss language itself. In this chapter we will focus on this latter function using cognitive neuroscience. The reader will find out that despite the complexity of language, theories and our current understanding of brain function, neuroscientists,

E. García-Marco
Departamento de Psicología Clínica y Experimental, Universidad de Huelva, Huelva, Spain
e-mail: enrique.garcia@dpces.uhu.es

K. Rolán (✉)
Laboratorio de Linguaxe e Cognición, Universidade de Vigo, Vigo, Pontevedra, Spain

Instituto de Investigación Lingua, Universidade de Vigo, Vigo, Spain

Departamento de Análise e Intervención Socioeducativa, Universidade de Vigo, Vigo, Spain

Cátedra Unesco Educación Transformadora: Ciencia Comunicación y Sociedad, Universidade de Vigo, Pontevedra, Spain

Grupo Language Variation and Textual Caterogorisation (LVTC), Universidade de Vigo, Vigo, Spain
e-mail: katiarolan@uvigo.gal

M. Visalli, M. V. Galmarini (eds.), *Collection and Analysis of Text Data in Sensory and Consumer Science*, https://doi.org/10.1007/978-3-032-04851-6_2

psychologists, and linguists are still working on fully deciphering the processes of language and its underlying mechanisms. While a pure physiological approach is far from explaining the vast world of linguistic phenomena in terms of neurons, it is timely to tackle specific, fundamental questions about language and seek answers that sooner or later will end up being based on biological principles.

In this chapter, we will explore key aspects of the neuroscience of language, starting with the historical approaches and major theoretical and neurobiological models. The evolution of language theories and the methods used to study them, such as the lesional model, EEG and MRI will be discussed. Finally, we will expose the role of artificial intelligence in language processing. The chapter concludes by summarizing our findings.

1.1 Measuring the Brain

In the dynamic field of cognitive neuroscience, we (including you, dear reader) aim to understand how the human brain's structure relates to its function. Along with this, the methodologies utilized to explore this relationship have evolved considerably since the 90s decade, particularly with the trend from neuropsychological studies from brain lesions to functional neuroimaging. The main purpose is establishing correlations about (psycho)linguistic and neurobiological data to determine which areas of the brain support specific language functions and/or how various linguistic and cognitive representations are processed at the interneuronal level (García, 2012).

Functional neuroimaging has become the central method, rapidly surpassing lesion studies, which traditionally involved analysing brain function following specific brain damages. While both approaches offer valuable insights, they are inherently different in nature and provide complementary, rather than competitive, evidence. Lesion studies can suggest a brain area's necessity for specific cognitive functions by observing what happens when that area is damaged. Conversely, functional imaging, which often measures changes in blood flow to infer brain activity, provides correlational data that cannot establish causality but can suggest associations between brain regions and cognitive tasks.

However, cognitive neuroscience continues to evolve with technological advancements, in the search of a more nuanced understanding of the brain. The shift towards functional imaging is also underscored by broader changes, such as the digitization of scientific literature, which provides resources to conduct systematic reviews and assess the comparative impact of different research methods. Interestingly, but not surprisingly, the choice of certain methods influences the impact and citation frequency of research within the scientific community. Recognizing the strengths and weaknesses of these methods, cognitive neuroscience now emphasizes a convergent approach. This approach advocates using multiple methodologies to offset the limitations of one method with the strengths of another. This holistic view is essential for a more accurate understanding of the brain's intricate workings (Fellows et al., 2005).

1.2 Theoretical Approaches in Cognitive Neuroscience

For a better understanding of the methodologies used in cognitive neuroscience we will use Caplan's (1987) distinctions among arguments. These distinctions were developed to outline different approaches to reasoning about the relationship between cognitive theories and empirical data, each serving as a way to illustrate how researchers draw conclusions from data about cognitive processes. By considering these forms of reasoning, we gain a clearer framework for understanding how theoretical models and empirical findings interact in the study of cognition.

- The first form of argument in neurolinguistics involves utilizing linguistic and psycholinguistic models to infer and describe corresponding neural structures. This approach assumes that specific features of language structure and psychology suggest particular brain structures and operations.
- The second form of argument adopts the reverse approach of the first form. Starting from known neurophysiological structures and theories, this method uses neurobiological insights to hypothesize about language functions and the organization of language processing in the brain.
- The third, and logical, approach in neurolinguistics synthesizes elements of the first two methods, exploring the mutual impact of neural structures on language function and vice versa. This integrative method highlights the parallels between language features and brain structures, proposing a bidirectional relationship where each informs the understanding of the other.
- The fourth argument centers on analysing how naturally occurring brain diseases influence language functions. The fact that is listed as the last argument is historically inaccurate, as this approach was the first method used for systematic scientific studies of the language-brain relationship. Precisely identifying the locations and characteristics of brain diseases has significantly advanced our understanding of how specific brain regions and their pathologies impact language.

The reader will notice that research in the neuroscience of language often employs both top-down and bottom-up approaches. This justifies the inclusion of sections in the chapter that address both cognitive theories and data-driven research methods as well as the study of the damaged brain.

2 First Neurobiological Models of Language

> *"(...) the facts assembled by science so far allow to state, as I said before, that the great regions of the mind correspond to the great regions of the brain. [(...) comme je l'ai dit ailleurs, que les grandes régions de l'esprit correspondent aux grandes régions du cerveau.]"*
> (Broca, 1861, p. 336)

If we could pinpoint a starting date for an experimental approach to the timeless interest in understanding how language is organized in the brain, it would be the description of patients with an aphasia condition. This first approach leads to a localizationist model of the language. Aphasia is a disorder that disrupts both the understanding and expression of language, arising from dysfunctions in specific brain regions. Damasio (1992) explains this disorder as a failure in the two-way translation process that typically maps thoughts to language. If we can imagine thoughts as a mind-related non- verbal representation, we normally can ground it in language by means of linguistic symbols and the syntaxis. On aphasic patients this conversion is compromised. Aphasia refers to the (in)ability to use the written and spoken forms of any language, including auditory-based languages like English, ideograms like many Asian languages and based on visual-motor systems, such as sign languages.

Aphasia fundamentally disrupts the processes that facilitate the translation of thought into language, rooted not in perceptual or motor deficits but in neurological damage affecting language processing. The impact of aphasia can differ greatly, potentially influencing multiple language components, including syntax (sentence structure), lexicon (vocabulary), and morphology (the creation of words from phonemes, which are the smallest sound units in speech). More than one of these aspects can be compromised in the same patient. This latter phenomenon, in addition to the large number of people affected by focal brain damage, mostly in older ages, constitutes an important corpus of study of the cerebral basis of language.

2.1 *Broca's Aphasia*

Louis Victor Leborgne, known as "Tan" or "Tan-tan" was a patient of Paul Broca and one of the most famous patients in psychology history. He suffered from very early years of brain damage, epilepsy fits. In 1839 having lost his ability to speak and thinking it was a temporary condition, he delayed seeking help. He entered at the age of 30 in hospital Bicetre. When admitted it was notorious he had difficulty speaking, he could only say the word "Tan" (usually twice). He was first admitted to the hospital's psychiatry division and was later placed under Broca's care towards the end of his life. Both his life previous and after entering the hospital are documented (e.g., Domanski, 2013). Broca, besides being a great neurosurgeon, was a fortunate person, to whom a patient with a very specific lesion came knocking at his door. Both instances culminated in the identification and designation of a novel region within the brain, now known as Broca's area. Following Broca's initial assessments, Broca's area was initially considered a speech production region because patients had articulation difficulties. Later research on aphasic patients suggested that Broca's area is responsible for syntactic processing or the grammatical system, due to the agrammatism observed in Broca's aphasics along with difficulties processing complex sentences, such as the passive form.

Typically, in Broca's aphasia, there is extensive damage not only to Broca's area, located in the inferior left frontal gyrus and involving Brodmann's areas 44 and 45, but also to the surrounding frontal fields, including the external aspect of area 6 and areas 8, 9, 10, and 46, also underlying white matter and basal ganglia (Damasio, 1992). Monsieur Leborgne's condition is often referred to as being caused by a particularly precise lesion. More recent studies, however, have suggested a dissociation between Broca's aphasics' difficulties with syntactic processing and comprehension and lesions in Broca's area (Embick et al., 2000). However, it led Paul Broca to establish this case as a milestone in the study of the brain basis of language production and processing. But this case was not the only one.

2.2 *Wernicke's Aphasia*

The reader might also refer to Wernicke's aphasia. Named after Carl Wernicke, a German neurologist and psychiatrist. He hypothesized a model of functional neuroanatomy responsible for the repetition, production, and comprehension of spoken words. In his initial model in 1874, he discussed the aphasic symptoms of ten patients, four of them also had an autopsy, to base the location of brain injuries. Wernicke proposed a connection between the left posterior section of the superior temporal gyrus and the fiber tracts linking sensory and motor areas, which associate the sensory and motor images of spoken words (Roelofs, 2024). Unlike Broca's aphasia, which is marked by effortful speech, Wernicke's aphasia allows patients to speak fluidly but with little to no meaning, indicating a disconnection between speech production and comprehension capabilities.

Wernicke's aphasia impairs language comprehension, making it challenging for patients to understand spoken words and select appropriate words when speaking. Patients with Wernicke's aphasia can speak at normal or even accelerated rates, which is why it can be referred to as fluid aphasia. Their speech often includes errors in phoneme and word choice, leading to unintelligible or irrelevant content and semantic paraphasia, where patients substitute an incorrect word for the intended one. Additionally, they struggle with auditory comprehension and are often unaware of their speaking errors.

2.3 *An Aphasia-Based Linguistic Model*

The then prevailing ideas by Franz Joseph Gall that language can be measured indirectly from the shape of the skull were overcome by the studies of Broca and Wernicke. By the beginning of the twentieth century the anatomical basis of language was solidly described and integrative models of neurobiology of language speech and production started to arise. Please note this section is not intended to be a comprehensive review of aphasiology, but rather to illustrate the theoretical development of the field.

One of the earliest models proposed was developed by Joseph Jules Dejerine and his wife Augusta Dejerine-Klumpke between 1895 and 1902. Their model described language as arising from an intricate network of cortical language centers and associated fibers, primarily located in the left hemisphere. According to their framework, language functions stem from a network, implying that aphasia, or the loss of language abilities, results from lesions anywhere within this "language zone". The extent of the language impairment varies depending on the specific location of these lesions. Although the Dejerine model is considered a localizationist approach, it recognized the interconnectedness of various regions in language processing (Krestel et al., 2013).

Norman Geschwind, in 1965, emphasized the role of disconnection syndromes, where aphasia results not just from damage to specific brain areas but from disruptions in the connections between them. This model underscored the importance of neural pathways, such as the arcuate fasciculus, in linking key language areas like Broca's and Wernicke's areas (Anderson et al., 1999).

Luria's neuropsychological model was developed after the 1960s and 1970s. His model integrates speech as an interconnected system involving multiple components, which requires the synchronized activity of several brain areas. If any of these areas are compromised, it disrupts the whole system of speech processing. However, the impact of this impairment is significant: it varies depending on the function associated with the affected brain region, the secondary systemic impairments triggered by the initial damage, and the functional adaptations that occur in response.

The examination of injury-based approaches carries significant theoretical, educational, and clinical implications. However, modelling based on pathological foundations presents significant limitations, as it is not always feasible to accurately represent the linguistic system of a healthy brain based on the characterization of a neurocognitive dysfunction.

3 Symbolic Theories of Language

> *"What a neurosurgeon sees when he removes a portion of skull does not look at all like what an engineer sees when he opens the cover of an IBM 701 or a Univac (…)."* (Newell et al., 1957, on the computer metaphor)

Besides notable contributions to the science of behaviour and cognition, like Psychoanalysis and the Gestalt school, among many others, until the 1950s, conductism was the prevailing paradigm in psychological study. Conductism focused on predicting behaviour, theorizing about the learning process, and detecting any observable behaviour, at the expense of ignoring internal processes, which were even referred to as a 'black box'. The cognitive revolution emerged during the 1950s and 1960s in a context of the necessity of a novel paradigm which englobes new epistemological, social and scientific inquiries; the necessity of studying the human being, a development on the methods and technological advances, and the development of a more refined study in memory and attention. Additionally, the 1960s were marked by the development of computers, leading to the widespread use of computer-brain analogies in a trend known as 'the computer metaphor.' Within the cognitive revolution, it became common to define models with boxes and arrows, representing cognitive processes and information flow respectively. The cognitive approach still prevails today.

In the application of cognitive psychology to language study, symbolist theories emerged. According to the 'symbolists', understanding language involves activating and computing mental symbols, which are processed by specialized neural modules, each designed to solve particular information processing challenges. These symbols are arbitrary, bearing no resemblance to their referents, and are amodal, thus independent of sensory modalities. The symbolist proposal suggests that meaning itself forms a mental language, composed of such symbols. Similar to natural language, symbols in this mental language follow syntactic patterns. Symbolists infrequently develop specific hypotheses or conduct research on the neurological bases of these symbols (de Vega, 1984).

To further explain the methods and theory behind these approaches, we will show some of the first computational models in the second half of the twentieth century. In them, we will assume that the mind operates with abstract symbols applying syntactic rules, analogous to the algorithms used by computers to process information.

Proposed by Allen Newell and Herbert A. Simon in 1976, the physical symbol system hypothesis is a foundational principle in computer science and cognitive psychology. It posits that a system can exhibit intelligent behaviour if, and only if, it is a physical symbol system. In this context, "symbols" are any type of patterns upon which operations can be performed, with some patterns functioning as commands or instructions. Newel & Simon proposed logic, numbers, and chess as examples of material permeable to being treated as symbols. The hypothesis suggests that both computers and human brains process information through physical symbol systems, enabling them to perform complex and intelligent tasks. Such a system is capable of inputting, storing, modifying, and outputting symbolic structures, and performing some of these actions in response to the symbols themselves (Simon, 1990).

Technically, the hypothesis implies that any system, like a computer or the human brain, that can manipulate symbols according to a set of rules could theoretically perform any cognitive task that a human can, provided the correct symbols and rules are in place. In other words, a hypothesis that necessarily follows is that a computer could achieve intelligence. Thus, this hypothesis not only underpins much research in artificial intelligence, where computers are designed to mimic human thinking, but also suggests that human understanding of thought and cognition can be fundamentally mechanistic and symbolic. Additionally, it bridges computer science and cognitive psychology by suggesting that both fields study variants of the physical symbol system and Newel & Simon extensively worked on this issue.

4 Models of Semantic Memory

"You shall know a word by the company it keeps." (Firth, 1957, p. 179)

Semantic memory is a type of declarative memory proposed by Tulving in 1972, which stores general knowledge about the world and the meanings of verbal symbols (i.e., words). According to the proposals of models of semantic memory, it requires an amodal store of information, this means it is not linked to any specific sensory modality and encompasses knowledge about concepts, facts, and the properties of objects. In contrast, episodic memory involves memories of specific experiences linked to particular times and places. Despite debates about the strong separation between the two, Tulving's distinction between episodic and semantic memory has been supported by neuropsychological studies, showing that patients could acquire new semantic knowledge without remembering the context in which they learned it (see patient KC cited in Rosenbaum et al., 2005).

Research into semantic memory has led to the development of various models to understand its structure and organization. These theories are based on the assumption that the meaning of words is represented in a hypothetical mental dictionary or lexicon, and associative network models (de Vega, 2002). In these models, words are represented as nodes in a network connected by edges that reflect their relatedness in meaning; distributional models, which focus on patterns of co-occurrence between words; and feature-based models, which represent words based on their constituent features. The reader would still easily recognize the metaphor of the computer when reading the premises of the following models.

4.1 The Hierarchical Network Model

The Hierarchical Network Model proposes semantic memory is organized based on a system of hierarchical networks and concepts are represented as nodes. In this model, more general concepts like "bird" are positioned higher in the hierarchy, while more specific concepts like "canary" are placed lower. Additionally, the model includes properties or characteristics of concepts, which are placed at the highest possible level of the hierarchy to optimize memory storage. Following their original example, if "canary" is stored with the description "a yellow bird that can sing", there would be pointers to "bird" (the category of canary) and the properties "yellow" and "can sing". General information about birds, such as their ability to fly and their features (wings and feathers), is not stored with each specific bird type but can be inferred from the general category. Exceptions, like "an ostrich cannot fly", are stored explicitly to prevent incorrect inferences. The model is sustained on behavioural experiments, based on response times on yes-no response on a verification task. The authors claimed that this is an efficient organization which minimizes the storage space required. This organizational system applies computer storage models to human memory and Collins and Loftus (1975) addressed that criticism.

4.2 The Spreading Activation Theory of Semantic Processing

Collins and Loftus (1975) proposed a more flexible model avoiding the constraints of a purely computerized vision. A concept can be represented as a node within a network, with its attributes linked to other concept nodes. The connections between nodes are typically bidirectional and with a numerical value indicating how crucial that link is to define the concept. The importance of these links can vary in each part of the connection; typewriter is solidly connected to the concept of machine but is not as critical for the concept of a machine to include a typewriter.

When a concept is specified, activation spreads from the initial node to all linked nodes and continues outward. Each node leaves an activation tag noting the starting node and predecessor. An intersection is found when a tag from another starting node is encountered, allowing the path to be traced back and reconstructed. The path is then evaluated for syntactic and contextual accuracy. If it fails, other paths are considered. Other studies on memory are also covered, for instance, Collins & Loftus pay attention to the priming effect. When a concept is primed, activation tags spread through the network to a certain depth. When another concept is presented, it connects with these tags to find intersections. Priming affects both nodes and links, as links are treated as concepts. Thus, priming a concept like "red" also primes related links, enhancing the context mechanism in the network.

4.3 Distributional Semantic Models

The meaning of words from the linguistic environment on semantic memory indicates that words referring to concepts such as "boat" and "ship" typically occur in contexts with related words like "sea" and "sailor". Under this simple assumption, words in these models are represented as vectors in a high-dimensional space, where proximity between vectors indicates semantic similarity.

A dimension is a direction in which one can move or a characteristic that can be measured. In a three-dimensional space, for example, we have three dimensions: length, width, and height. In contrast, in a high-dimensional space, there are many more dimensions, which can be difficult to visualize intuitively. The vector components can represent not only individual words but also larger linguistic units like paragraphs or documents, or more complex representations such as n-grams, which is every possible n number of words that occur together in a text (Kumar, 2021).

Notable models like Latent Semantic Analysis (LSA) and Hyperspace Analogue to Language (HAL) derive meanings and automatically compute semantic similarity from large document collections. The primary advantage of these methods is the ease with which word meaning similarity can be measured mathematically by the distance in the vector space or the cosine of the angle between vectors. For instance, in a literary corpus, the words "king" and "castle" have a high co-occurrence, resulting in the cosine of the angle between their vectors being close to 1, whereas "cell" and "castle" do not show any significant co-occurrence, leading to the cosine of the angle between their vectors being close to 0.

4.4 Challenges of Symbolist Models and the Grounding Problem

Semantic space models have been successful in simulating a wide range of psycholinguistic and linguistic phenomena. Their very clear premises about the organization of linguistics symbols permit creating sophisticated software's and normative data (e.g. de Deyne et al., 2019). Yet, their ability to handle the combination of meanings at more complex levels remains an area of active development. However, semantic models typically represent words in isolation and do not adequately address the compositionality of language. While simple methods like vector averaging can combine meanings, they do not account for syntactic structure and word order, leading to inaccurate representations for complete phrases or sentences. The following is an example from Landauer et al. (1997). Sentences (a) and (b) contain exactly the same set of words, but their meanings are entirely different.

(a) It was not the sales manager who hit the bottle that day, but the office worker with the serious drinking problem.
(b) That day the office manager, who was drinking, hit the problem sales worker with the bottle, but it was not serious.

This argument is analogous to Searle's well-known experiment, the "Chinese Room" (1980) (cited in Glenberg & Kaschak, 2002). In this experiment, a hypothetical individual is locked in a room where he receives messages written in Chinese through a slot and has access to very precise instruction manuals that allow him to generate response messages also in Chinese. However, this ability to manipulate symbols does not enable him to understand the meaning of the messages, as he cannot relate the Chinese messages to real-world referents. The claim proposes "*computational manipulation of abstract symbols merely produces more abstract symbols, not meaning*" (Glenberg & Robertson, 2000).

Traditional theories fail to explain how cognition interfaces with perception and action. Understanding terms requires something qualitatively different from the application of formal rules to symbols; it requires uncovering the sensorimotor referents. The "grounding problem" addresses the challenge of how a formal symbol system can have intrinsic semantic interpretation, rather than relying solely on the meanings we attribute to the symbols in our minds.

A proposed solution involves grounding symbolic representations in two types of nonsymbolic representations: iconic representations, which are analogs of sensory projections of objects and events, and categorical representations, which are feature detectors identifying invariant features of object and event categories from their sensory projections. To achieve this, a computer would need a certain sense of sensorimotor knowledge of its environment.

5 Embodied Cognition Theories

Grounded cognition theories suggest that our knowledge and thought processes are rooted in, and emerge from, our sensory and motor interactions with the world. Unlike traditional models of cognition, grounded cognition posits that our brain utilizes multimodal representations, including visual, auditory, haptic, olfactory, and gustatory perceptions, as well as motor, proprioceptive, and interoceptive sensations we have experienced (Barsalou, 2008). These representations are activated or "simulated" when we need to access knowledge or face new situations, functioning as a tacit reactivation of previously experienced states in contact with specific objects and scenarios. Conversely, cognition is influenced and shaped by the body, action, and the social and physical environment in which it unfolds. Thus, physical experiences and social interactions play a critical role in how we process and remember information, suggesting that our cognitive capabilities are not only rooted in the brain but also in our bodies and surroundings (Barsalou, 2008).

For instance, the concept of "affordance", introduced by James J. Gibson in his work on visual perception and ecological psychology, refers to the actionable properties between the world and an actor (an organism or a person). These properties are not just the physical characteristics of an object but also involve the relationship between the object and the capabilities of the actor (Greeno, 1994). For example, a chair affords sitting because it has a flat, horizontal surface at a height that accommodates the human body's ability to bend and rest. The concept of affordance is quite flexible: intrapersonally, if we want to get up to an unreachable position over a wall, the chair affords a step; interpersonally, if we are toddlers, the chair could afford a desk. Gibson's theory laid the foundation for a corpus of studies placing the body's interactions with the environment at the center of cognitive processes.

5.1 The Indexical Hypothesis

Glenberg and Robertson (2000) published an empirical study that is highly critical of models based on arbitrary symbols, such as those used in LSA. These models cannot capture the richness of human understanding because they lack a direct connection to sensory and perceptual experiences. In their study, they presented participants with sentences controlled for semantic indexes by means of LSA, but these sentences were clearly not sensible.

Context-setting: Classical conditioning is a procedure that can induce a type of learning.
Alternative 1: The procedure is based on pairing two types of stimuli.
Alternative 2: The procedure is based on pairing two types of Pavlov.

Readers familiar with psychology would understand that stimuli and Pavlov are semantically connected among them and to the context-setting. However, this illustrates the problem; while abstract symbols might recognize connections, they fail to capture meaningful comprehension. The study presents a theoretical argument that arbitrary symbols cannot be combined in a way that captures the meaning humans derive from word combinations. Human cognition relies on the ability to simulate and imagine objects and their interactions, which abstract symbols cannot achieve.

The index hypothesis proposes, first, words or phrases are indexed or associated with perceptual symbols. These symbols are modal in nature and represent the neural activation states underlying perceptual processes (Barsalou, 1999). Second, affordances are derived for the formed perceptual symbols based on the context. Finally, the different affordances are combined or integrated according to the syntactic construction of the sentence, resulting in a coherent simulation of the message's content. In this way, the theory proposes that language acquires meaning through the cognitive simulation of the actions described in the sentences.

The authors connect this proposal with Gibson's theory, as the construction of the meaning of a phrase, object, or situation depends on the possible actions an individual can perform in a given context. Thus, a theory grounded in perceptual and embodied cognition, where meaning is tied to real-world experiences and sensory information, is posited as a more accurate model of how humans derive meaning from language. This approach grounds meaning in the interaction between action and environment. Additionally, representations built from perceptual symbols can be compared to perceptual experiences, as these symbols and experiences are related. For example, a cd can be perceived by an individual as useful for scratching their back, even though this object was never designed for that purpose and even if the individual had never used it in that way or learned about this possible use in any prior context.

5.2 The Immersed Experiencer

To understand the theory, we may refer to the model of situation. Zwaan et al.'s (1995) model propose that readers comprehend narratives by indexing events along five dimensions: time, space, causation, motivation and protagonist. As the reader processes the narrative, they construct an initial situation model and continuously update it. For instance, when reading "*Peter took the elevator to the fifth floor*", the reader establishes a temporal framework (past) and a spatial setting (building) for Peter. Moving on to "*He went to talk to his professor*", the reader maintains temporal and spatial continuity and constructs a goal dimension (Peter's intent to talk to the professor). This ongoing integration process ensures that new information is appropriately indexed within the existing model.

The Immersed Experiencer Framework (Zwaan, 2004) explains language comprehension as a process where words serve as cues for the interlocutor to construct an experiential simulation of the described situation. This model proposes that understanding language involves simulating the perceptual and motor experiences related to the words'referents. For example, when reading "*The ranger saw the eagle in the sky*", the reader visualizes an eagle with its wings opened in the sky. Conversely, "*The ranger saw the eagle in the nest*" activates a mental image of an eagle resting.

Zwaan (2004) distinguishes three key processes of language comprehension: activation, construal, and integration. Words activate relevant perceptual and motor experiences associated with their referents. Construal operates at the clause level, integrating the activated experiences to form a coherent representation of the described narrative. Integration functions at the discourse level, linking the current clause-level representation with prior knowledge and context. These processes ensure that the comprehension of language is an immersive experience, once again, grounded in perceptual and motor simulations.

6 Other Proposals

The hub-and-spoke theory of semantic representation provides a comprehensive framework for understanding how conceptual knowledge is formed and organized in the brain. This theory integrates two key ideas to explain the development and neural basis of concepts. It builds on classical proposals from Wernicke, as well as contemporary embodied theories, by positing that multimodal experiences—both verbal and non-verbal—form the basic elements for constructing concepts. These elements are encoded in modality-specific cortices distributed throughout the brain. For instance, sensory experiences such as sights, sounds, and actions are processed in their respective perceptual and motor areas.

The ATL-Hub (anterior temporal lobe hub) theory posits that the bilateral anterior temporal lobes serve as a central, transmodal region crucial for integrating semantic knowledge across various modalities and conceptual domains. As we anticipated, this concept is supported by cognitive neuropsychology; particularly studies of semantic dementia (SD). Neuroanatomical evidence shows that SD is associated with atrophy and hypometabolism specifically in the anterior ventral and polar temporal regions. Individuals with SD exhibit widespread semantic impairments affecting all modalities and types of knowledge, suggesting the ATL region's central role in semantic processing. The consistent patterns of deficits in SD, such as better performance with familiar and typical items and worse performance with highly specific knowledge tasks, indicate that this region functions as a domain-general hub.

7 Techniques in the Study of Language

Research in the study of language resembles the well-known parable of the three blind sages describing an elephant. Each sage touches a different part of the elephant and, based solely on that limited perception, each offers a unique description: one touched the trunk and said, "*an elephant is like a snake*"; another touched a leg and asserted, "*an elephant is like a tree*"; and the third touched the side and described, "*an elephant is like a wall*". While each description is partially correct, none encompasses the entirety of the animal. Similarly, research techniques in linguistics and cognitive neuroscience address specific aspects of language from different angles, offering fragments of a broader understanding. From reaction times in reading, to embodiment studies linking language with motor actions or perceptions, to neuropsychology, eye-tracking, EEG, and fMRI, each technique reveals different facets of how we process and understand language. Only by combining these diverse methodologies can we aspire to a comprehensive and holistic view of how language functions in the human brain.

7.1 Response Times

Reaction or response time in reading refers to the duration between seeing a word and recognizing its meaning. One of the premises associated with shorter reaction times is that it indexes how present a word is in memory storage. Several factors influence this time, including the frequency of word usage and the age at which words are acquired. Additionally, reaction times in reading can be influenced by various procedures, such as semantic priming, where quicker recognition of a word occurs when it is preceded by a semantically related word. The procedures are countless and aim to respond to a concrete research question. For instance, auto-paced reading of single words in a sentence, in which the reader presses a key to show solely the next word in a sentence, may be useful to capture online cognitive processes that are unfolding and also to prevent the reader from looking back to a previous word. Fortunately, besides the large number of procedures, these are usually well-documented and clearly described, ensuring that their effects on reaction time are understood and accounted for when discussed.

7.2 Behavioural Studies in Embodiment

Under the assumptions of embodied cognition, the networks associated with the processing of modal symbols may overlap with the networks associated with the very same cognitive processes the modalism asserts. As we focus on language in this chapter, we will show some of the procedures involved in the study of action language and its resonance with motor demands due to its simplicity on their premises. However, the reader should take into account that these procedures can be extensive, under the proper experimental design, to sensorial cognition. To illustrate these, we recommend the article by García and Ibáñez (2016), which presents the HANDLE model.

The model explores how hand-related language interacts with manual activities and impacts cognitive processes. It hypothesizes whether integrating hand-related expressions and manual activity results in facilitation, interference, or no effect, emphasizing the importance of timing in these interactions. Two motor measures were used to evaluate this relationship: reaction-response times and kinematic variables like hand movement speed and finger aperture. For example, Aravena et al. (2014) analysed online modulations of grip force over a sensor while subjects listened to a sentence focused on an action ("*John signs the contract*") or a mental state ("*John wants to sign the contract*"). The model considers whether this integration varies with linguistic complexity (single words versus sentences) and the level of verbal processing (lexical or semantic). Additionally, it examines motor task complexity, suggesting simple and complex tasks engage cognitive and motor systems differently, and looks at the granularity of this integration to see if it distinguishes between varied hand shapes, positions, and gestures.

7.3 Neuropsychology

Neuropsychology explores neurobehavioral disorders associated with brain dysfunction and injury. Spoken language consists of several independent components, each associated with different brain structures. The primary components are phonology, syntax, and semantics. Disruptions in phonological or semantic processing occur at the level of single words, while syntactic disruptions affect sentences. Assessing these components typically involves phonological discrimination tasks, synonym tests, and word-picture matching tasks such as the Pyramids and Palm Tree Test. For sentence comprehension, tools like the Token Test and syntactic comprehension tests are used.

Word and sentence retrieval difficulties, often seen in amnesic or nominal dysphasia, can affect specific categories like letters, colours, body parts, or proper names. These difficulties are evaluated using naming tests, such as the Boston Naming Test, which involves naming objects from verbal descriptions or pictures. Additionally, spontaneous language assessment through tasks like describing complex pictures (e.g., the cookie theft) helps identify grammatical construction impairments.

A significant contribution from neuropsychology is the double dissociation method. Tranel et al. (1995) (see Damásio, 2012) cites this method from Teuber and a study with monkeys, when discussing double dissociation: "*What is needed for conclusive proof is 'double dissociation', i.e., evidence that tactile discrimination can be disturbed by some other lesion without loss on visual tasks and to a degree comparable in severity to the supposedly visual deficit after temporal lesions*" (Teuber, 1955, p. 283).

The double dissociation method involves: (i) finding systematic correlations between brain area damage and behavioural/cognitive dysfunctions; (ii) validating findings by establishing double dissociations, where damage in area A causes dysfunction X but not Y, and damage in area B causes dysfunction Y but not X; (iii) formulating hypotheses about how a normal neural system with various subsystems performs cognitive or behavioural operations; (iv) testing these hypotheses with new cases of localized brain damage to see if the predicted effect occurs (as cited in García, 2012). Thanks to this method, several cognitive processes have been described as relatively independent, such alexia and agraphia (e.g. Greenblatt, 1973).

7.4 Eye-Tracking

Eye tracking is a method used to record eye movement and gaze location over time and during various tasks. It is often employed to study visual attention and cognitive processes such as perception, memory, language, and decision-making. According to the eye-mind principle, the fovea is directed at the stimulus we are currently processing or thinking about. Fixations and saccades are the fundamental data units for most analyses. When comparing how participants read different texts, we can analyse aggregate measures of fixation duration and location (average fixation duration, number of fixations) and saccades (average saccade amplitude).

7.5 Electroencephalography (EEG)

The procedure of Electroencephalography (EEG) involves recording the brain's electrical activity using electrodes placed on the scalp. These electrodes detect the difference in electrical potential between two sites, typically labelled as active and reference sites. During an experiment, participants are presented with a stimulus, such as a visual image, repeatedly while their brain activity is continuously recorded. We recommend the manual of Luck (2014).

Event Related Potentials. To isolate event-related potentials (ERPs), the EEG data is segmented into epochs time-locked to each stimulus presentation. By averaging these epochs, spontaneous EEG fluctuations unrelated to the stimulus -noisy- are averaged to zero, revealing the ERP wave that reflects the brain's consistent response to the stimulus. Components are classically named after the polarity against classical reference electrodes as Positive (P) and Negatives (N) and the millisecond they are usually detected.

Early components of ERPs, such as P100, N100, and P200, are linked to basic, automatic perceptual processing and are elicited as soon as a stimulus, like a word, is presented. Later components, occurring after approximately 250 milliseconds, are associated with more complex, sometimes conscious cognitive processing (Beres, 2017). The N400 component indicates rapid integration of various levels of information, from word and sentence meaning to broader discourse context. For example, in sentences like "*Dutch trains are ____ and very crowded*", the N400 amplitude to "sour" (semantically incorrect) and "white" (factually incorrect, as Dutch trains are yellow) was similar, underscoring the sensitivity of the N400 to both semantic and real-world knowledge (cited in Kutas & Federmeier, 2011).

7.6 Functional Magnetic Resonance Imaging (fMRI)

Functional magnetic resonance imaging (fMRI) is a neuroimaging technique that maps brain activity in real-time by measuring blood flow changes associated with neuronal activity. Utilizing blood-oxygen-level-dependent (BOLD) contrast, fMRI detects variations in oxygenated and deoxygenated hemoglobin, indirectly reflecting brain activity. fMRI usually is implemented on existing MRI scanners in medical facilities and most of the time research goes hand in hand with clinical procedures.

Functional neuroimaging studies have significantly advanced our understanding of the brain systems involved in language, and we are living a renewed interest in the classical neuroanatomical model proposed by Broca, Wernicke and others. Indeed, fMRI offers several advantages for language mapping. First, the small voxel size (typically two to four millimeters) provides high image quality and spatial localization capability. Also, functional data can be easily registered with high-resolution standard MRI images from the same brain location, permitting associate functional activity with specific anatomical structures (Binder, 2023).

While MRI combined with fMRI generates impressive images, they often obscure the significant complexity of the physiological processes behind them. The utility in brain research heavily depends on signal specificity, spatial and temporal resolution, and careful experimental design (Logothetis, 2008). Interpreting fMRI language studies requires a well-informed analysis of the cognitive processes activated during scanning (Binder, 2023). We recommend the review of Price (2010) which provides the methodology used to create standard reports of results and detailed description of cognitive processes associated with language, such as sentence comprehension or processing of incomprehensible sentences.

7.7 Artificial Intelligence (AI)

Computer science has progressively advanced in developing sophisticated automated methodologies for characterizing and predicting human behaviour (Bedi et al., 2015). Artificial Intelligence (AI) represents a novel approach to understanding cognition, significantly influencing the study of language (Goldstein & Papert, 1977). The authors have already indicated that AI is commonly perceived as a field focused on programming computers to appear intelligent, and that this perspective was misleading. While the use of computers to model cognitive theories is central to AI, the fundamental goal of the field is to understand intelligent processes themselves, as language, independent of their physical implementation.

For instance, in Parkinson's disease (PD), individuals exhibit not only articulatory disorders but also experience alterations in grammar, pragmatics, verbal fluency, and action-verb semantics. Traditionally, these aspects have been studied using artificial laboratory tasks. However, advances in artificial intelligence now allow for the examination of natural speech. García and Ibáñez (2016) utilized automated tools to investigate whether PD patients exhibited distinctive features at the discourse level in natural speech. They employed a process called logogenesis, which involves the accumulation of interrelated lexico-grammatical selections that form patterns beyond the word and sentence levels, offering a more ecological understanding of how this disease affects verbal communication. By analysing single-minute monologues, they successfully distinguished PD patients from controls with 77% accuracy. Furthermore, the motor system's critical role in the comprehension and production of action verbs has been demonstrated (Pulvermüller, 2005). This insight has not only enhanced the differentiation of individuals already diagnosed with Parkinson's but has also led to the development of a linguistic tool for early diagnosis. Detecting deficits in the processing of action verbs, in the absence of other cognitive deficits, may serve as a cognitive marker of early PD (Cardona et al., 2013).

Language also serves as a crucial source of clinical information in Alzheimer's disease, as its deterioration occurs in parallel with neurodegeneration. Consequently, speech and language data have been extensively examined with AI in relation to its diagnosis. De la Fuente García et al. (2020) have conducted the first systematic review on the potential application of interactive AI methods for detecting and monitoring the progression of Alzheimer's disease (AD) using natural language processing and speech technology to extract digital biomarkers for machine learning modelling.

Despite the inherent difficulties associated with the type of data used and the heterogeneity of the methods, almost all studies report relatively high performance. Compared to neuropsychological assessment methods, speech and language technology demonstrated at least equivalent discriminative power between different groups. It is reasonable to conclude that AI applied to linguistic processing holds significant promise, with the potential to gradually introduce changes into clinical practice.

8 Conclusion

> *"(...) given that human linguistic systems are rooted in the brain, they cannot be anything that the brain does not allow them to be [(...) dado que los sistemas lingüísticos de los seres humanos hallan asiento en el cerebro, estos no podrán ser nada que el cerebro no les permita ser.]"* (García, 2012, p.222)

We thank the reader for being with us in our exploration of diverse techniques and approaches to the neuroscience of language. We conclude this chapter by summarizing key insights into the intricate relationship between neuroscience and language, emphasizing that human language production and perception are deeply rooted in neuronal activity (and it cannot be otherwise). The combined methodologies of both theoretical and biological approaches to language aim to address fundamental questions about how linguistic systems are embedded within the brain's architecture. However, due to the high complexity of language, it remains only partially understood despite advancements in neuroscience.

Theoretical frameworks in cognitive neuroscience were exposed, particularly the interplay between linguistic models and neural structures. One approach used linguistic and psycholinguistic models to infer corresponding neural functioning, while another started with established neurophysiological structures to hypothesize about language functions. An integrative perspective synthesizes these methodologies, suggesting a bidirectional influence where neural structures and language functions mutually inform each other. Understanding the brain-language relationship would require both top-down and bottom-up investigative strategies.

Initially, the chapter described the evolution of neuroscientific methodologies. Historical perspectives on neurobiological models of language marked an era. The studies of Broca and Wernicke provided early evidence of localized brain functions associated with language. Broca's aphasia, characterized by speech production difficulties, and Wernicke's aphasia, marked by comprehension deficits. Even more, these models were sustained by necropsy records. Historically, that permitted left behind models like Francis Galls model. And, besides some theoretical concerns about the validity of understanding the brain from its failures, the cases laid the groundwork for more complex models that integrated extensive neural networks and pathways.

In the late 1950s, the rise of cognitive psychology introduced the 'computer metaphor,' proposing that the brain functions like a computer. Under this premise, algorithms resembling cognitive processes may be valid for psychological research. Models like the Physical Symbol System Hypothesis and various semantic memory frameworks suggested that language processing involved the manipulation of abstract symbols, and semantics could be represented mathematically as symbols associated in a high-dimensional space. These models successfully developed the basis to integrate computational models of research and accurately simulate some linguistic phenomena. However, they faced challenges such as the 'grounding problem', which questioned how abstract symbols acquired real-world meaning.

Grounded cognition theories offered an alternative, asserting that cognitive processes were rooted in sensory and motor interactions with the environment. Concepts like 'affordance' and models such as the Indexical Hypothesis suggested that language comprehension involved simulating perceptual and motor experiences related to word referents. This perspective emphasized that meaning derived from the interaction between action and environment, providing an embodied view of cognition, and thus, responding to the grounding problem.

The chapter concluded with an examination of various techniques used in language studies, including response time measurements, behavioural studies on embodied cognition, neuropsychological assessments, eye-tracking, EEG, and fMRI. The integration of these diverse approaches highlighted the multifaceted nature of language processing. The IA approach adds an improvement of our understanding of cognition, particularly language, by developing automated methods to characterize human behaviour. We have seen how in Parkinson's disease, AI analyses natural speech to detect distinctive features and early markers such as deficits in action verb processing. Similarly, in Alzheimer's disease, AI uses natural language processing to extract digital biomarkers, demonstrating high performance in diagnosis and monitoring, and shows promise for automated remote sensing and significant clinical advances.

Neuroscience has allowed us to understand the intricacies of the interface between the brain and language. Historically, cognitive neuroscience focused on four key approaches: aphasiology, symbol-based theories, embodied cognition theories and research techniques. Those methods try to understand intelligent processes themselves, such as nuanced relations between words, and nowadays it may even surpass that goal. For instance, AI can identify tacit features unnoticeable by the human eye on natural speech. This chapter does not pretend to exhaustively explore the 200-year history of the neuroscience of language. Instead, the chapter pretends to encompass how different approaches are built into each other, and how each approach makes up for the shortcomings of the others. However, a comprehensive theory of how language works is still under construction. We hope the reader has enjoyed exploring the questions we've raised about language cognition.

References

Anderson, J. M., Gilmore, R., Roper, S., Crosson, B., Bauer, R. M., Nadeau, S., et al. (1999). Conduction aphasia and the arcuate fasciculus: A reexamination of the Wernicke–Geschwind model. *Brain and Language, 70*(1), 1–12.

Aravena, P., Courson, M., Frak, V., Cheylus, A., Paulignan, Y., Deprez, V., & Nazir, T. A. (2014). Action relevance in linguistic context drives word-induced motor activity. *Frontiers in Human Neuroscience, 8*, 163.

Barsalou, L. W. (1999). *Perceptual symbol systems. The behavioral and brain sciences*. Cambridge University Press.

Barsalou, L. W. (2008). Grounded cognition. *Annual Review of Psychology, 59*(1), 617–645.

Beres, A. M. (2017). Time is of the essence: A review of electroencephalography (EEG) and event-related brain potentials (ERPs) in language research. *Applied Psychophysiology and Biofeedback, 42*, 247–255.

Bedi, G., Carrillo, F., Cecchi, G. et al. (2015) Automated analysis of free speech predicts psychosis onset in high-risk youths. *npj Schizophr 1*, 15030.

Binder, J. R. (2023). fMRI of language systems: Methods and applications. In *Functional neuroradiology: Principles and clinical applications* (pp. 565–592). Springer International Publishing.

Broca, P. (1861). Remarques sur le siège de la faculté du langage articulé, suivies d'une observation d'aphémie (perte de la parole). *Bulletin et Memoires de la Societe anatomique de Paris, 6*, 330–357.

Caplan, D. (1987). *Neurolinguistics and linguistic aphasiology: An introduction*. Cambridge University Press.

Cardona, J. F., Gershanik, O., Gelormini-Lezama, C., Houck, A. L., Cardona, S., Kargieman, L., et al. (2013). Action-verb processing in Parkinson's disease: New pathways for motor–language coupling. *Brain Structure and Function, 218*, 1355–1373.

Collins, A. M., & Loftus, E. F. (1975). A spreading-activation theory of semantic processing. *Psychological Review, 82*(6), 407–428.

Damasio, A. R. (1992). Aphasia. *New England Journal of Medicine, 326*(8), 531–539.

Damásio, A. (2012). *O erro de Descartes: emoção, razão e o cérebro humano*. Editora Companhia das Letras.

De Deyne, S., Navarro, D. J., Perfors, A., Brysbaert, M., & Storms, G. (2019). The "Small World of Words" English word association norms for over 12,000 cue words. *Behavior Research Methods, 51*, 987–1006.

De la Fuente Garcia, S., Ritchie, C. W., & Luz, S. (2020). Artificial intelligence, speech, and language processing approaches to monitoring Alzheimer's disease: a systematic review. *Journal of Alzheimer's Disease, 78*(4), 1547–1574.

De Vega, M. (1984). Introducción a la psicología cognitiva. In *Introducción a la psicología cognitiva*.

De Vega, M. (2002). Del significado simbólico al significado corpóreo. *Studies in Psychology, 23*(2), 153–174.

Domanski, C. W. (2013). Mysterious "Monsieur Leborgne": The mystery of the famous patient in the history of neuropsychology is explained. *Journal of the History of the Neurosciences, 22*(1), 47–52.

Embick, D., Marantz, A., Miyashita, Y., O'Neil, W., & Sakai, K. L. (2000). A syntactic specialization for Broca's area. *Proceedings of the National Academy of Sciences, 97*(11), 6150–6154.

Fellows, L. K., Heberlein, A. S., Morales, D. A., Shivde, G., Waller, S., & Wu, D. H. (2005). Method matters: An empirical study of impact in cognitive neuroscience. *Journal of Cognitive Neuroscience, 17*(6), 850–858.

Firth, J. R. (1957). *A synopsis of linguistic theory 1930–1955* (Studies in linguistic analysis, special volume). Blackwell.

García, A. M. (2012). La Teoría de Redes Relacionales: Correlatos neurológicos de un modelo lingüístico conexionista. *Onomázein, 26*, 221–257.

García, A. M., & Ibáñez, A. (2016). A touch with words: Dynamic synergies between manual actions and language. *Neuroscience & Biobehavioral Reviews, 68*, 59–95.

Glenberg, A. M., & Kaschak, M. P. (2002). Grounding language in action. *Psychonomic Bulletin & Review, 9*(3), 558–565.

Glenberg, A. M., & Robertson, D. A. (2000). Symbol grounding and meaning: A comparison of high-dimensional and embodied theories of meaning. *Journal of Memory and Language, 43*(3), 379–401.

Goldstein, I., & Papert, S. (1977). Artificial intelligence, language, and the study of knowledge. *Cognitive Science, 1*(1), 84–123.

Greenblatt, S. H. (1973). Alexia without agraphia or hemianopsia anatomical analysis of an autopsied case. *Brain, 96*(2), 307–316.

Greeno, J. G. (1994). Gibson's affordances. *Psychological Review, 101*, 336.

Krestel, H., Annoni, J. M., & Jagella, C. (2013). White matter in aphasia: A historical review of the Dejerines' studies. *Brain and Language, 127*(3), 526–532.

Kumar, A. A. (2021). Semantic memory: A review of methods, models, and current challenges. *Psychonomic Bulletin & Review, 28*(1), 40–80.

Kutas, M., & Federmeier, K. D. (2011). Thirty years and counting: Finding meaning in the N400 component of the event-related brain potential (ERP). *Annual Review of Psychology, 62*, 621–647.

Landauer, T. K., Laham, D., Rehder, B., & Schreiner, M. E. (1997). How well can passage meaning be derived without using word order? A comparison of latent semantic analysis and humans. In *Proceedings of the 19th annual meeting of the cognitive science society* (pp. 412–417).

Luck, S. J. (2014). *An introduction to the event-related potential technique*. MIT Press.
Logothetis, N. K. (2008). What we can do and what we cannot do with fMRI. *Nature*, 453(7197), 869–878.
Newell, A., Shaw, J. C., & Simon, H. A. (1957). Problem solving in humans and computers. *Carnegie Technical, 21*(4), 35–38.
Price, C. J. (2010). The anatomy of language: A review of 100 fMRI studies published in 2009. *Annals of the New York Academy of Sciences, 1191*(1), 62–88.
Pulvermüller, F. (1999). Words in the brain's language. *Behavioral and Brain Sciences, 22*(2), 253–279.
Pulvermüller, F. (2005). Brain mechanisms linking language and action. *Nature Reviews Neuroscience, 6*(7), 576–582.
Roelofs, A. (2024). Wernicke's functional neuroanatomy model of language turns 150: What became of its psychological reflex arcs? *Brain Structure and Function, 229*, 1–18.
Rosenbaum, R. S., Köhler, S., Schacter, D. L., Moscovitch, M., Westmacott, R., Black, S. E., et al. (2005). The case of KC: Contributions of a memory-impaired person to memory theory. *Neuropsychologia, 43*(7), 989–1021.
Simon, H. A. (1990). Invariants of human behavior. *Annual Review of Psychology, 41*(1), 1–20.
Teuber, H. L. (1955). Physiological psychology. *Annual Review of Psychology, 6*(1), 267–296.
Tranel, D., Damasio, H., & Damasio, A. R. (1995). Double dissociation between overt and covert face recognition. *Journal of Cognitive Neuroscience, 7*(4), 425–432.
Zwaan, R. A. (2004). The immersed experiencer: Toward an embodied theory of language comprehension. *Psychology of Learning and Motivation, 44*, 35–62.
Zwaan, R. A., Langston, M. C., & Graesser, A. C. (1995). The construction of situation models in narrative comprehension: An event-indexing model. *Psychological Science, 6*(5), 292–297.

Chapter 3
Linguistic Meaning and Sensory Experiences for Food

Danièle Dubois

1 Introduction: A Reformulation

This chapter aims to illustrate some fundamentals on natural language and food perception based on expertise in cognitive linguistics and psychology. It draws on nearly 40 years of research conducted within the French National Research Centre (CNRS), combining fundamental research with applied concerns in real-life contexts. The work has resulted in a collective book: "Sensory Experiences: Exploring Meaning and the Senses" (Dubois et al., 2021a), which covers research on vision, olfaction, audition, and taste.

We began in the late twentieth century, within a cognitive framework relying on the information-processing model, a model borrowed from computer science (Dubois et al. 2021c Chapter 1) We then developed a model of "perception" mainly validated in visual perception. According to this model, colour perception is a bottom-up process: some information is "captured" through sensory receptors and "up to the brain that interprets the information by comparing it to similar information previously stored in categories. These categories can be named by simple "basic names" as "veridical labels" that are assumed to map directly onto "natural" categories (Dubois, 2006; Rosch & Lloyd, 1978), as defined in natural sciences by physical parameters like wavelength in the case of color.

But when we began collecting empirical data, first on olfaction and later on taste, especially outside the laboratory and in real-world settings, this model proved inadequate. More specifically, exploring olfaction and taste in collaboration with engineers in sensory sciences proved highly productive in developing conceptual as well

D. Dubois (✉)
French National Centre for Scientific Research (CNRS), Paris, France

M. Visalli, M. V. Galmarini (eds.), *Collection and Analysis of Text Data in Sensory and Consumer Science*, https://doi.org/10.1007/978-3-032-04851-6_3

as methodological changes. Since no instrument can directly measure smell or taste as subjective experiences of food products' quality (as is the case for any sensory experience), it becomes essential to engage individuals as consumers and, when relying on their verbal productions, as speakers.

Psychology and linguistics are called for accounting and identifying how people evaluate and communicate about their subjective experience of food. Yet, when attempting to validate the classical cognitive model in the domain of taste, we found no well-established categories of "natural smells" or "natural foods" (Dubois et al., 2021a, Chap. 2). In terms of language, we found no "basic words"for smells, and only a few for flavour, many of which were not easily translatable from French to English.

This led us to propose an alternative conceptual and methodological framework (Dubois et al. 2021d; Dubois et al. 2021e -Chapters 9 and 10), shifting from information-processing to meaning constitution, drawing on philosophical and linguistic traditions to better account for the concept of sensory experiences, rather than relying on an analytic model of perception. We illustrate this paradigm shift, from language (singular) to languages[1] (plural), and from perception to experiences, using data we collected primarily on taste, though some relevant findings from other sensory modalities are also discussed. This shift also calls for a reformulation of the initial wording of our contribution. Instead of addressing "Fundamentals on natural language and food perception", we present our expertise in linguistics and psychology under the title: "Some fundamentals on languages and food experiences" (as developed in Dubois et al., 2021a, Chap. 10).

After clarifying the distinctions between language (*"langage"* in French) and languages (*"langues"* in French), and between perception and experiences, we outline the empirical consequences for studying any sense, with a focus on the specificities of taste as the experience of food. Food itself cannot be defined solely as a set of "products"; it involves multiple characteristics, identified notably through anthropological research that definitively influence eaters' judgments. These eaters themselves cannot be reduced to mere consumers.

All these considerations are reflected in language and discourse. Therefore, semantic analysis of words in context can guide the formulation of more precise questions and help identify linguistic cues as objective indicators of subjective judgments. Finally, we clarify the processes by which consensus on the meaning of taste-related terms is established, whether in ordinary language or expert discourse.

---

[1] The term "natural language" does not resolve the ambiguity. It was introduced in contrast to "artificial language", which was developed by computer (artificial) systems. There is, in fact, a certain irony in the fact that a human technological invention became the model for human language, rather than being considered as an emergence from scientific knowledge elaborated by humans from their "natural" languages.

2 Mise en Bouche, Amuse-Gueule, Starter, Antipasti[2]…

2.1 Language or Languages?

Let's start this section with the example of the word *doux* in French.

Doux is a French word (adjective) that can be used to qualify a sensation in any of the traditional five senses. For example, it can be used in the following expressions: *"une douce saveur"*, *"la douce odeur des mimosas"*, *"une chanson douce"*, *"le doux vert des prairies", or "le doux grain de sa peau".*

At this very first step, as soon as you need to translate these examples into English, you are confronted with: *"a sweet flavour", "the sweet smell of mimosas", "a sweet song"*, but *"the soft green of the meadows", "the soft grain of her skin".*

Obviously, the trivial observation is that there is no strict lexical (word-to-word) correspondence between French and English lexical forms. Depending on the sense modality, *doux* is translated into English as *sweet*, when describing taste, smell and audition, (and sometimes vision, as in *a sweet face*), and as *soft* when describing vision, and especially touch.

Here are some more examples from English to French: a "soft drink" is translated in dictionaries as *"une boisson gazeuse, limonade, ou soda"*, but never as "*boisson molle*"; likewise, a sweet face is not translated as "*un visage sucré*". Many more examples could be given, and eventually, this could become a fun game!

For our current concern, the question is whether the differences in using one or two words (such as *doux* versus *sweet* and *soft*) induce differences in meaning across languages and ultimately refer to similar or different realities. How can we determine whether *doux* as *sweet* is similar to *doux* as *soft*, or, conversely, whether *soft* is as *doux* as *sweet*? From an empirical standpoint, the challenge is to discern the actual reality each of these words denotes. In any case, it is difficult to consider that a single word directly corresponds to a specific (objective) property of the world, here, a sensory characteristic of a product, that would exist *per se*, independently of its wording in one language.

These examples reveal that, when processing verbal data, we must decide whether the relevant empirical analysis concerns the specific semantic properties of one language (named in French as *langue*—for example, French, English, or any other language among the diversity of languages) or some universal property of language (named in French as "*langage*").[3] *"Langage"* refers to language as a universal faculty (for humans) to use one language as one langue, which is an observable, specific linguistic system enabling communication and the elaboration of knowledge about the world. In other words, it cannot be stated that a word has some

[2]The metaphoric phrasing of certain titles and subtitles points to the diversity of expressions of what taste is about, within the diversity of languages and cultures.

[3]The distinction between *langue* and *langage* in French versus the single term language in English is another example of the non-correspondence in wording between the two languages, which carries significant conceptual and empirical consequences.

meaning that directly refers to an objective reality. Rather, meaning is given to a word through its usage by a speaker who interprets and communicates about his/her experience of the world in his/her language.

When considering that the two French words may refer to two different concepts, what are the empirical consequences for exploring the sensory properties of food as a product? In the absence of a direct one-to-one mapping between a word and a perceived property of the product, verbal data should be processed as reflecting sensory properties inferred from the semantics of a specific language and identified in an observable realization in text or discourse from one speaker's judgment. The identification of a product's properties must take into account the mediation of psychological processes involved in tasting, as well as the linguistic abilities of the individual to master the resources of his/her language ("langue") when producing discourse or responding to questions. However, it remains to be clarified what reality is actually targeted by such questioning: perception or subjective experience as psychological processes, or through such processes the identification of some objective property of the food or product?

2.2 *Taste: Perception or Experience?*

Similarly to the rationale developed for the word "language", the word "perception" refers to the concept of perception as a universal process that relates an organism to the external world, by means of processing information coded by different receptors. Such psychological model is physiological based on the nature of the receptors in defining the (five) senses as well as the bottom-up information processing along neural trails. The point is not to dismiss the intrinsic value of such a model or its usefulness for specific purposes (such as explaining certain diseases or physiological impairments), but to question its relevance and adequacy when it comes to capturing both the linguistic semantics of the words (such as *doux*) in linguistics as well as of the subjective (psychological) reality of experiencing something as *doux*.

Furthermore, remaining within linguistics, the semantics of *doux* includes reference to additional characteristics of the qualia,[4] as the subjective experience. *Doux* carries an hedonic value, expressing a positive judgment on the objects described as such. The consensual dictionary definition of *doux* is: "*qui fait sur les sens une impression agréable*" (in English, "*which makes a pleasant impression on the*

[4] The term "quale" (plural "qualia") refers to the qualitative properties of experience insofar as they elicit in the subject the experience of a distinctive impression. "*This blue that I perceive*", "*this pain felt*", "*this coffee fragrance*", these are all qualia. Since the mid-twentieth century, several words have competed as designations for these properties: they have been called "subjective qualities", "sensuous qualities", "phenomenal properties," "phenomenological properties", or even "immediate impressions ("raw feels"; see Feigl, 1967). The Latin word prevailed in English-language philosophy and has since been adopted in German and French translations, probably because of its symmetry with the quantum/quanta pair, one representing a qualitative differential, the other a quantitative differential. (From Cassin et al., 2014; French edition: Cassin, 2004)

senses"). It also conveys an evaluation of low intensity of stimulation across senses, for example: *"saveur ni forte ni piquante", "agréable à l'oreille en raison de sa faible intensité", "odeur délicate, légère et agréable", "agréable au toucher, léger, moelleux"* (in English: *"neither strong nor pungent flavor", "pleasant to the ear due to its low intensity", "delicate, light and pleasant smell" "pleasant to the touch, light, soft"*).

In short, the semantics of *doux* does not refer solely to an analytic perception of some property of the product, nor to the coding of information from a specific sense defined on a physiological basis, even if such properties can be described along some physical dimensions measurable by natural sciences and instrumentation. Rather, *doux* refers to a judgment of a sensory experience made by an individual, a human being, and also a speaker of a particular language, who attempts to articulate his/her subjective experience. The scientific analysis of such an experience thus relies primarily on psychological and linguistic investigations, rather than exclusively on physiological or neurological knowledge.

2.3 The "Chocolaty" Tone of a Violin?

We confirm this positioning by reporting another unexpected but meaningful example of the cross-modal referential value of sensory words: the history of our analysis of the word (adjective) *chocolaty*, used to qualify the sound of a violin. While working in an acoustic laboratory on the evaluation of the quality of musical instruments, our expertise in semantics was challenged by the editorial board of the *American Lutherie Journal*. Their question was the following: "*Instrument reviewers often use colourful descriptions of tone that make me crazy. I recently read that one reviewer used the word 'chocolaty' to qualify the tone of a Sitka/Indian rosewood violin. What the heck is that? Why can't we come up with standardized descriptions of tone*?"

This question reflects a recurring tension between the semantic richness of languages (*"langues"*) in their ordinary use (of metaphors or metonymy, among other devices) and the pursuit of standardization in technical or scientific evaluation—particularly when sensory words are used across modalities, such as taste and hearing. From a strictly linguistic point of view, *chocolaty* is an adjectival form derived from the noun *chocolate*. According to dictionary definitions, it is generally used to describe "*anything that looks, feels, or tastes like chocolate*". Notably, no mention is made of its use to qualify a sound. Looking to understand why *chocolaty* might have been used to describe the sound of a violin, we conducted a quick survey (see Dubois & Fritz, 2015). We asked twenty-one French native speakers, most of them acousticians, the following question: "*Qu'est-ce que, pour vous, un son chocolaté ?*" (*"What is a chocolaty sound for you?"*). Even if the question may at first seem surprising, the answers revealed a consensus around certain semantic features associated with the word *chocolaty* when used to describe a musical tone. Below are sample answers describing the "chocolaty" tone of a violin, with the number of occurrences indicated in brackets.

A strong consensus is observed on the use of *doux* (7), other semantically related words such as *rond* (2) (*round*), *sucré* (1) (*sweet/sugary*), *lisse* (1) (*smooth/sleek*), *riche* (2) (*rich*), *onctueux* (2) (*smooth/creamy*), *savoureux* (1) (*tasty*), *moelleux* (1) (*mellow*), *chaud* (1) (*warm*).

Some other adjectives from other semantic fields were cited, such as *épais* (1) (*thick*), *lourd* (1) (*heavy*), *profond* (1) (*profound/deep*), *sourd* (1) (*dull/muted*), *sombre* (1) (*dim/dark*).

Finally, adjectives referring to a global hedonic judgment were also mentioned, such as *intime* (3) (*intimate*), *agréable* (1) (*pleasant/nice*), *réconfortant* (1) (*comforting*), *joyeux* (1) (*joyful*), *plaisant* (1) (*pleasant*). To summarize, when asked "*What is a chocolaty tone?*", people largely agree, making it possible to say that a tonal quality of a (violin) sound can indeed be described as *chocolaty*, that is, a "*soft, round, low-frequency tone, characteristic of an instrument producing a pleasant feeling of intimacy*".

Relating these lexical answers to psycholinguistic processes, we can consider the use of a diversity of sensory words crossing over different sensory modalities as a regular (psycho)linguistic phenomenon in everyday communication. This supports the idea that the semantics of lexical meaning mirrors the multimodal processing of food, as extensively documented by Spence (see Spence, 2021,[5] among many others). This observation further reinforces the view that exploring lexical meaning can serve as a valuable tool for accessing the multimodal characteristics of the psychological experience of food.

Remaining at the psycholinguistic level, the communication of such a sensory experience can be successful if the following conditions are met: (i) there is shared knowledge of the word (here, *chocolate*) and a shared sensory experience of the product (chocolate) between speaker and listener; (ii) it is this shared knowledge of the word and shared experience of the world that allow the transfer of certain semantic features attributed to the word (*chocolate*) onto the relevant selected features of the previously experienced product. Our psycholinguistic interpretation is thus as follows: in the absence of "ready to use" and consensual simple words to express the sensory experience involved in perceiving and evaluating a particular musical tone, the use of *chocolaty* can be seen as a "do-it-yourself" attempt to communicate an individual's unobservable subjective experience. This is done by referring to a somewhat similar and commonly shared sensory experience of a familiar object (chocolate), using a widely known word (*chocolate*).

This example is not an exception but rather typical of most common-sense adjectives used even in descriptive sensory analysis. Such a syncretic qualification of sensory experience—which includes hedonic or evaluative judgments, as seen with *doux*—is actually more frequent for smell and taste, but it is not limited to these

[5] This converges with Spence's observations, who, however, focuses directly on psychological processes without taking into account the linguistic constraints of the responses.

senses. Rather than inappropriately applying models developed for colour perception to smell and taste, it is suggested to revisit the conceptualization and naming of colours based on the more generic model derived from smell and taste (see Dubois & Grinevald, 2000; Dubois & Cance, 2012, and Sect. 3.3 below on colour words for marketing purposes).

From an epistemological point of view, these results invite us to consider colour naming as an exception rather than the generic model for any sensory experience. Instead of relying exclusively on an "official" terminology, the investigation of the linguistic resources available in any given language (*langue*) is encouraged, across the diversity of situations (and languages) in which the qualification of objects takes place. More fundamentally, these examples, illustrating the complexity of linking the semantics of sensory linguistic expressions with the subjective experiences of the senses, question the conception of the sensorium based on the analytic physiological distinction of the five senses (Classen, 1993; Howes, 2003, 2011, 2019; Dubois et al., 2021a, b—Chaps. 2 and 10). This challenge is even more pronounced in the case of taste, for which there is no full consensus, as reflected in the following definitions. According to Spence et al. (2014), for example, taste should be distinguished from flavour: "*Taste characterizes one of the (five) senses as a perceptual experience, and flavour characterizes the taste of food, as the property of a product, comprised of combinations of taste and odour.*" Auvray and Spence (2008) proposed that the multisensory perception of flavour may indicate that the current taxonomy used to define our senses is simply not appropriate. According to them, "*the act of eating allows different qualities of foodstuffs to combine into unified percepts; flavour can be used to describe the combination of tastes, smells, trigeminal and tactile sensations, as well as the visual and auditory cues, that we perceive when tasting food*" (p. 1016).

To go further, and remaining within linguistics, we attempted to clarify the distinction between taste and flavour by exploring the linguistic expressions available for enumerating *tastes* versus *flavours*, based on the (risky) *a priori* hypothesis of a correspondence with the French words *goûts* and *saveurs*.

3 Words for Taste?

What are the linguistic resources available in any language to talk about taste or flavour? These are two words precisely defined as terms by experts in sensory sciences, but also widely known and used in their common-sense meanings. Here, we report on some work we carried out for French (see Dubois & Cance, 2021, p. 296; Dubois & Giboreau, 2006).

3.1 Linguistic Resources for Taste (or Flavour?)

When teaching psychology and linguistics in relation to sensory analysis, we systematically began each lesson by asking French-speaking students to produce lists of words they associate with either *goûts* (*tastes*) or *saveurs* (*flavours*). We regularly collected the following responses, with words associated with *goût* including: *chocolat* (*chocolate*), *orange*, *artichaut* (*artichoke*), *noisette* (*hazelnut*), *raisin* (*grape*), *banane* (*banana*), *sauce soja* (*soy sauce*), *vinaigre* (*vinegar*), *fruité* (*fruity*), *floral*, *épicé* (*spicy*), *lacté* (*milky*), *herbacé* (*herbaceous*), *boisé* (*woody*), *iodé* (*iodic*), *minéral* (*mineral*), *empyreumatique* (*empyreumatic*).

Linguistically speaking, the words for *goût* fall into two main categories:

- nouns referring to tasty substances or objects, such as fruits or vegetables, or edible ingredients used in culinary preparations (e.g., vinegar, soy sauce);
- adjectives mainly constructed from nouns referring, as above, to tasty objects or substances, such as *fruity*, *floral*, or *spicy*. These are generally common words, with the exception of a few technical terms like *empyreumatic*, which come from participants with expert knowledge.

For *saveurs*, participants regularly listed the four canonical flavours, typically expressed as adjectives: *sucré* (*sweet*), *salé* (*salty*), *amer* (*bitter*), *acide* (*sour*), and occasionally *umami*. In contrast to the words for *goût*, which show a high diversity of answers, including many hapax (answers occurring only once), *saveur* (*flavour*) elicits significant consensus, with participants consistently producing the four traditional terms *learned* as tastes (*salty*, *sweet*, *sour*, *bitter*), along with the more recent inclusion of *umami*,[6] and only a few other terms occurring at low frequency.

Linguistically speaking, *sucré* and *salé* are also adjectival forms derived from nouns or substances, respectively *sucre* (*sugar*) and *sel* (*salt*) (this observation applies in English only to the adjective *salty* but not too *sweet*). It can also be noted that *amer* and *acide* are simple adjectives or "free" words (as defined by Kristol, 1978, for basic colour terms), meaning that their lexical forms are not morphologically (in the language/*langue*) marked to be referring to any specific substance or product.

Technically speaking, *saveur* (*flavour*) does not refer to any specific product but to a small set of taste specific properties (of food). Similar to how seven colours of the rainbow or eleven basic colour terms in cognitive science are enumerated, *flavour* is commonly described using only the four "traditional" terms. This attests to an "official" and consensual terminology based on acquired knowledge, rather than denoting some individual sensory experience. However, these four or five words do not exhaust the discriminative power of taste experiences, which therefore must be

[6]Umami was recognized as a basic taste only in 1985, during the first Umami symposium in Hawaii, where it was introduced as a technical (scientific) term referring to the taste of certain glutamates and nucleotides.

communicated by means of other linguistic resources and psycholinguistic processes.

On cognitive grounds, the distinction between "free" adjectives and constructed words correlated to different conceptualizations:[7] the regular linguistic process of using adjectival forms morphologically constructed from nouns, the name of a source, actually drives an interpretation including some properties of this source and prevents from the abstraction of some analytical property of the product.

Such ambivalence of words for taste can also be observed in the English word *taste* itself, which can be either a noun or a verb. As a noun, *taste* can refer either to the taste of something or to the subjective experience of "it". As a verb, the polysemy is even more clearly marked in its syntactic properties: in the phrase "*it tastes of banana*", the banana is understood to *have* some taste as a property of the object, whereas in "*I taste banana*", the focus is on the psychological, subjective experience. This ambiguity in referring must be clarified. Is *taste* referring to something in the world external to the subject (objective in some sense), such as sensory stimulation described in terms of physics or chemistry as within an experimental setting? Or to the sensory psychological experience of "it", which belongs to the domain of psychology?

When processing text data, and being aware of these syntactic variations in the use of linguistic forms, it can be important to use these linguistic marks to identify in verbal reports some related conceptual différences, before any automatic processing.

3.2 *Taste for Food, Food to Taste*

Even within the canonical conception of the senses, the case of taste is trickier than that of vision or audition, and even more complex than smell, due to the great variability of food, even when defined at the level of the organic molecules responsible for gustatory stimulation. The difficulty in defining taste as a simple perceptual psychological process (one of the five senses) may stem from its very close linguistic and cognitive relation to what is tasted, and what is tasted is food. It is actually the concept of food that unifies the complexity of taste, rather than any physiological coding. It therefore seems necessary to define more precisely what food is, as one further reason for sensory sciences to focus on characterizing the properties of food as a product, before investigating consumers' behaviour and representations of food in their everyday lives.

Food can be defined as "*any nutritious substance that people or animals eat or drink or that plants absorb in order to maintain life and growth*" (definition adapted from the Oxford Dictionary). Even as a substance, food is defined in relation to

[7] Even free adjectives can refer to different conceptualizations once their semantics are analyzed in discourse (see Dubois & Cance, 2012, for color words).

humans (or animals), albeit restrictively considered from a biological perspective "in order to maintain life and growth". The limits of this definition quickly become apparent when considering humans, *a fortiori* when considering them as consumers, who face an immense diversity of edible products. This diversity is infinite, as are the many ways in which these products can be prepared: raw, grilled, broiled, stewed, not to mention the wide variety of seasonings, spices, and processing methods. Additionally, food undergoes further transformation in the mouth during chewing and before swallowing, after sniffing, and so forth (Köster et al., 2014; Mainland & Sobel, 2005).

This is even more relevant when considering some extreme contrasting examples from anthropological research. Many people from the Andean altiplano consume certain types of clay, known as *choco* in Quechua or *pasa* in Aymara. Browman and Gundersen (1993) regarded this practice of eating edible earths as a 2500-year-old custom rather than the result of biological determinism. Shifting from "geophagy" to anthropophagy (i.e., cannibalism; see, for example, Bonnassie, 1989; Montanari, 2018), human flesh presents a clear counterexample: although it is edible, it cannot be considered as food, at least by "us". In between these extremes, anthropological and linguistic evidence can help us better define what counts as food. For instance, Ricaud Oneto (2024) explored cultural differences in the perception of sweetness and bitterness to understand food acceptance among two indigenous groups in the Peruvian Amazon. She showed that while the vocabulary and cultural representations for daily drinks, such as manioc beer, were shared by both communities, they differed in their acceptance of sweet foods for children. The Napuruna, for example, viewed sweetness as a sign of *laziness* (far from being a sensory descriptor!) Closer to us, what about insects, dogs, horses, rabbits, snails, or even pigs and pork?[8] In short, food cannot be considered solely as nutrients. Rather, it is a word pointing to concept to be defined as alimentation within the human and social sciences. This is explored in history (Ariès, 2016; Braudel, 1961), anthropology (Howes, 2011; Lévi-Strauss, 1964), and sociology (Fischler, 1990; Poulain, 2002; among many others). As such, the properties of food cannot be reduced to (universal) biological characteristics but must take into account cultural variations and eating habits.

To explore more closely the properties involved in defining food from a psychological perspective, let us take a detour through the food classifications proposed by Ross and Murphy (1999). The authors highlighted the limitations of categorizing food within a single hierarchy, a single function, or as isolated knowledge. Below is a short excerpt from their extensive paper:

> *"The current work examines the representation and use of a real-world concept, food. (…) Our knowledge of food is extensive and is accessed many times per. It is not some isolated body of knowledge but part of many aspects of our physical and social life. We know which foods to eat for energy, and which may upset our stomachs. We know the foods that are likely to be served at various holidays and social events. We know which foods we can afford to buy and how long it will take to prepare them. Our knowledge of food is connected*

[8] Beyond the religious taboo about pork meat, it is interesting to note that the English language lexically distinguishes between the living animal (pig) from its edible meat (pork).

> *to much of our other knowledge. Besides eating food, many of us plan meals, cook, and shop for food. Newspapers and magazines are filled with articles on cooking and on the health implications of different foods. The representation of foods is bound to be affected by the large number of interactions we have with them, as well as the wide diversity of these interactions."* (Ross & Murphy, 1999, p. 297)

Based on our observations concerning the linguistic resources used to talk about food, we can add two considerations.

There are no longer just animals, but as meat, which cannot be defined simply as (animal) proteins. Consider the earlier example of pork, but also ask: what about insects, cockroaches, slugs, worms, larvae, shrimps and snails? What is considered "meat" depends on cultural context and is eventually marked in the language itself.

The categorization of fruits and vegetables, clearly defined in biology, becomes problematic when considered as food (products). If, in biology, a tomato is a fruit, in culinary and cultural contexts, it becomes a vegetable, grouped with zucchini and eggplant on supermarket shelves and in the pot when preparing *ratatouille*. What about avocado, or melon served with prosciutto. These are as examples of categories of dishes, served as starters rather than fruits as desserts, according to different scripts for meals.

I cannot fail to mention the story of the carotte, as reported by Jean-Marie Pelt (1994, p. 45). What is a carrot? "*La carotte, qui est une racine, a été promue administrativement au rang de fruits par la communauté européenne à compter du 1er janvier 1991 afin de permettre aux Portugais d'exporter de la confiture de carotte en se conformant à la définition qui veut que toute confiture soit faite à base de fruits.*" ("*The carrot, which is a root, has been administratively promoted to the rank of fruits by the European Community since January the 1st 1991, in order to allow Portuguese to export carrot jam, according to the definition of jam that requires to be made out of fruits*"). The carrot can be classified in various ways: as a root in biological science, as food, as a vegetable in sensory sciences, and even as a fruit for official economic reasons. But then, what is the taste of carrots? It may be sweeter (sucrée) in Portugal than when served grated with oil and vinegar as a starter or when cooked in a veau Marengo (a veal stew with onions and carrots).

Furthermore, food, at least in our Western cultures and within the field of food sciences, tends to be seen as a highly transformed product. From its production through various forms of agriculture, to its processing and distribution by food industries and markets, food undergoes multiple stages before reaching our kitchens. In this process, human eaters are themselves transformed into "consumers" (at least in our cultures and as participants or subjects in sensory sciences; see Dubois et al., 2021a, Chap. 12). Once in the kitchen, food is further altered through a wide range of culinary practices, cooking, frying, steaming, broiling, and more, before being served on our plates and finally eaten, within an equally diverse range of eating contexts: alone, at home, with family, with friends, and so on.

As a direct consequence, empirical research on taste should consider the full diversity of processes involving food, as well as the variety of criteria used to classify it. These go far beyond biological criteria such as nutrient content. Instead, food must be understood within the broader context and diversity of human cultural and culinary practices, ranging from its production, transformation, distribution, and purchase, to its preparation, consumption, and, ultimately, its explicit tasting, evaluation, and appreciation (or rejection). When evaluating the sensory qualities of food, it is essential to consider the sociocultural symbolic meanings associated with it. Food is undeniably a cultural product, and this necessitates close attention to anthropological research before examining how such cultural constraints influence psychological processing, particularly when investigating individual judgments about food.

4 Main Course: Collecting and Analysing Verbal Data

4.1 From "Fundamentals" to Empirical Research

We have shown that the holistic nature of sensory experience does not align with the analytic conception of the senses grounded in a physiological model. Instead, it is rather reflected in the semantics of (natural) languages (*les langues*), which better fit with the syncretic and holistic character of subjective experience. The question remains to draw the empirical consequences of this observation and to make explicit how people's sensory experiences, specifically in relation to food, can be explored. We will report on research we conducted that proposes guidelines for designing open-ended questions and analysing the verbal (text) data thus collected. These are not methodological recipes, since we demonstrated that there is no direct *a priori* agreed-upon link between common-sense words and any specific sensory (subjective) experience of food. The description of food's sensory properties is inevitably mediated by the psychological experience of any individual.

Even the analytical judgments of trained experts - who develop their analytic knowledge and skills through specialized procedures such as panel training and thus rely on precise, consensually defined terminology—cannot serve as instruments to "measure" the holistic sensory experience and the common-sense vocabulary of consumers. To access such subjective experience, the analysis of semantics of natural languages can be an option, especially when combined with attention to the psychological and psycholinguistic processes involved in experiencing and naming.

Once you acknowledge that there is no sensory property of food *per se* that can be directly accessed independently of a (human) experiencer and that access to it necessarily involves the mediation of subjective experience and its verbal expression, it becomes essential to clearly define what is being investigated. This includes specifying the limits of generalization: to other populations, to other types of products (whether similar or different), and to linguistic expressions in other languages. The main question, therefore, is to precisely articulate these empirical constraints before attempting any form of generalization.

As far as linguistic expression is concerned, a basic requirement is to control and test the *a priori* formulation of the questions asked, to properly evaluate the meaning of the answers. In the absence of a validated, absolute reference to objective sensory properties, the methodological approach relies on analysing differential answers, across different or similar products, contrasting populations or cultural contexts, and, of course, different languages. These variables must be defined in accordance with the specific aims of the research.

The main question, then, is one of segmentation within each of these domains, rather than relying solely on *a priori* formulations, derived from a restrictive, expert-centered perspective. In food science, unlike in the physical sciences, we cannot assume the availability of ready-made instruments like thermometers, sonographs, or colorimeters, that can be directly applied to an objective physical world. Even in the case of colour, one must ask: is the meaning of what we call *blue* reducible to its wavelength value? (see below).

It is not desperate though. Based on what we have presented above, a first suggestion is to adapt investigative methods to the specific nature of what is being studied and to the available "tools". In the present case, this means considering linguistic expressions as the result of psycholinguistic processes that reflect sensory experiences, shaped by the resources of a given language *(langue)*, the identity of the speaker, and the communicative purpose (e.g., identifying perceived properties or judgements on a product, pleasing the consumer, or promoting a product, etc.) In other words, since involving human participants is unavoidable, why not begin by paying close attention to who is being questioned? Their answers become the "measure" through which the product is better known, revealing their representations, eating habits as consumers, and language use as speakers. This kind of investigation is indeed time-consuming, but it may ultimately improve the relevance and accuracy of a later automatic language processing applied to verbal data.

4.2 *Eating Well? A Sociological Classification of Eating Behaviours*

As a productive methodological example, we would like to mention the seminal psychosociological work of Lahlou (1998). His analysis is based on the verbal answers of 2000 French eaters to the open-ended question: "*Si je vous dis 'bien manger', à quoi pensez-vous*? " (*"If I ask you what 'eating well' means to you, what comes to your mind?"*). An automated textual analysis conducted using the Alceste software (Reinert, 1983) enabled the identification of eight thematic classes, which were summarized using the following labels: 1: "*entrée-plat principal-fromage-dessert*" ("starter-main course-cheese-dessert"); 2: "*manger à sa faim*" ("eat to prevent hunger"); 3: "*manger ce qu'on* aime" ("eat what you like"); 4: "*pas trop de graisse et de sucre" ("not too much fat and sugar");* 5: "*équilibré" ("balanced");* 6: "*petits plats" ("small dishes");* 7: "*convivial*"; 8: "*restaurant*".

Based on this classification of eating situations, Lahlou was able to infer a typology of eating behaviours, which in turn allowed for a classification of *eaters*. He identified seven population groups, each clearly characterized by distinct strategies in purchasing, storing, preparing, and eating food. These were organized into two broad strategic types: family and non-family. The family category includes multi-person households with children and encompasses profiles such as rural domestic, well-settled, and family-oriented. The non-family category includes smaller households, such as childless couples (traditional elderly and modern urban) and single individuals (single camper or isolated).

When focusing on products rather than on behavioural or sociological variables, the characteristics defining these categories can be summarized as follows:

- Single campers: their meals are purely utilitarian, and they're mainly interested in what's on the plate. They are nibblers.
- Modern urbanites: They're quite concerned about dieting, but don't follow many diets. They are adept at service products and snacking.
- Rural domestics: The products they buy more than the average are essentially "basic" products.
- Family-oriented: Their diet is strongly influenced by their children's tastes, or those attributed to them by their parents.
- Well-established: They over-consume mostly transformed and festive products.
- Traditional aged: The products they significantly over-consume are coffee/chicory mixes, rusks, sliced ham and fresh bulk potatoes.
- Isolated: Their diet is characterized above all by monotony: fresh bread, pasta, rice, potatoes and fresh bulk vegetables, meat by the slice and fresh fruit make up the bulk of their diet.

What is interesting for sensory sciences is that this classification showed that behavioural strategies have a predictive effect on product consumption. This is supported by the strong statistical interaction between the behavioural categories and the consumption frequencies of 150 commonly consumed products, as reported in the data. This finding validates the idea that psychosociological investigations into consumers' behaviour (considering factors such as meal duration, places of purchase, commensality, eating alone or with family, kitchen equipment, cooking practices, etc.) can ultimately lead to identifying specific products likely to be purchased, going well beyond their sensory properties. It is no surprise, then, that this classification, which was developed with co-financing from major agri-food companies, gained wide circulation within marketing and among consultants. Such a classification can be productively used to segment consumers, tailoring questionnaires to elicit more relevant verbal answers to be segmented and automatically processed, aiding both product development and marketing strategies.

Returning more specifically to linguistic issues, we will provide below some examples[9] illustrating how the wording of questions can influence responses, whether the questions are explicitly focused on the sensory experience itself or on describing the product through evaluation, by both experts and "ordinary" (yet precisely defined) consumers.

4.3 *Ingredients: Wording the Product*

The first step is to identify the available linguistic resources (primarily lexical ones) that can account for the sensory experience, given that there is no consensus on how to name the (unobservable) subjective experience in any sensory modality. Although scientific knowledge and terminology (and instruments) exist to describe the physical reality of stimulation, such as acoustic waves for sound, there is no direct instrumental measurement for food. Much food is not a "natural" reality easily accounted for by natural sciences, but rather a cultural one, similar to music, perfumes, or paintings. Therefore, it is necessary to rely on linguistic expressions to convey subjective experience. As a first step (as previously mentioned), it is essential to inventory the lexical resources among which word (lexical units) fundamental to refer to, being either to an "objective" (i.e., physical) or "subjective" (i.e., psychological) reality.

We report here some investigations we conducted on colour naming for different purposes and practices as an illustrative detour.[10] We collected colour naming data in French from different contemporary colour charts, which are designed and commonly used primarily for the sale of various products with distinct purposes: (i) for decorative home use or artistic paints (watercolour or oil paints), (ii) for cosmetics, such as lipsticks, nail polish (from different brands), or hair dye, intended either for customers in supermarkets or for professional use (see Dubois & Cance, 2012; Dubois & Grinevald, 2000).

The colour chips found in drugstores or art supply stores are typically arranged along a spectrum that loosely mirrors the colours of the rainbow, with a name (in French) printed below each chip. However, they differ in at least two meaningful ways for our purposes. Their primary aim is not to reproduce the full range of visible colours exhaustively, but rather to present a repertoire of coloured or colouring products available for purchase within specific practices. Artistic paint charts offer basic colours intended for mixing to create other colours. Decorative charts, such as those for cosmetics or car paints, provide ready-made colours designed for application on surfaces (e.g., walls, furniture, nails, lips, hair, cars). The nature of the

[9] More examples are given in the methodological part II of the book "Exploring sensory experience" (Dubois et al., 2021a) in Chapter 12 ("From stimulation to stimuli"), Chapter 13 ("Procedures and outcomes"), Chapter 14 ("Making sense of the outcomes").

[10] To our knowledge, there is no such an inventory that could be done for different practices on food related to different professionals.

Table 3.1 Naming of "red" (in French) across different colour charts

Common name	Colour (paints) for artists	Colour chart for decoration
Rouge	Rouge vermillon français	Rouge
Rouge violet	Laque Garance rose	Rouge d'orient/grenadine
Rouge brique	Rouge de cadmium orange véritable	Poterie
Rouge foncé	Rouge de Venise	Rouge basque/paprika
Marron	Laque carminée	Brun van Dijk

materials or coloured objects is different (e.g., homes, body parts, cars). As a result, the concept of colour in these contexts goes beyond the three standard characteristics of Munsell colour chart (hue, intensity, saturation), incorporating other characteristics such as the matter of the colour itself (e.g., oil vs. watercolour), or shiny vs. satiny gloss paint for the decorative paint or metallic for car paints.

These differences underscore the fact that colour charts are situated into cultural practices (here commercial and/or artistic ones), contrasting with the preconception of universality of the Munsell chart, which reflects a particular form of scientific knowledge rooted in a physical definition of color.

Table 3.1 shows that beyond lexical and syntactic forms, the semantics of artistic charts differ notably from those of decorative ones. In artistic colour charts, the names do not refer directly to specific colour names, but rather to pigment names (e.g., cadmium, carmine, indigo), which are *technically* associated to qualifiers referring to a colour name and to the quality of the pigment (*rouge cadmium clair véritable, bleu cobalt imitation*). These terminological denominations refer to some *knowledge* of the pigments rather than to common word of colour as hue as well as its perception.

In decorative colour charts, monolexemic forms do not refer to basic colour names but to concrete object names whose colour is considered typical. This naming strategy is based on metonymy, where the name of a coloured object stands in for the colour itself: for example, *prune* (*plum*) for a shade of brown, *jonquille* (*daffodil*) for a yellow, *sable* (*sand*) for beige, and *liège* (*cork*) for a light brown. Polylexical units are also built on the name of sources typically associated with the colour term, such as *bleu canard* (*duck blue*). These source-based overall represent 50% of the processes of designation for the decorative colour charts. This mechanism closely mirrors what we observed in the domains of taste and smell—where source-based or object-based naming frequently replaces abstract sensory descriptors.

It is the practice of colour use that introduces meaningful distinctions. For artistic paints, what matters is the materiality of the pigment itself in the manipulation of the colour to produce the palette of colours and their fixation. Colour names refer to technical denominations as terminology which integrates some knowledge of the pigments and their properties as colour-as-matter, this wording being addressed to painters as colour experts using and sharing a consensual technical terminology. For decorative paints, the material origin of the pigment is irrelevant to the user. Here, colour is some appearance spread on the objects which are labelled by common words.

For lipstick (in French *rouge à lèvres*, literally "red for lips") or nail polish, the naming strategy is primarily driven by marketing objectives aimed at appealing to consumers, rather than by the referential value of the product's colour.

The names include nouns referring to objects: *tulipe* (*tulip*), *cœur* (*heart*), *brugnon* (*nectarine*); concepts: *flirt*, *amour* (*love*); proper nouns referring to people or places: *Venise*, *Marilyn*; adjectives referring to qualities: *ardent* (*intense*), *incandescent*, *ethnique* (*ethnic*); even verbs: *j'adore* (*I love it*). These names draw from a wide range of domains and may even appear arbitrary, as they are not directly related to the colour itself, but rather to culturally associated concepts, such as fire, heat, or love. Actually, each brand creates its own semantic space, crafting a distinct commercial image for what is essentially the same type of product. The aim is not to describe the specific shade of red, but to evoke emotional values such as love and passion, exotic places, all traditionally related in the shared cultural background of consumers in French supermarkets. Thus, colour naming here is less about referential description of "red" and more about customer seduction, a marketing strategy designed to enhance appeal and drive sales (cf. Dubois & Grinevald, 2003; Dubois, 2009, among others).

The analysis of colour chip naming in French, along with the various uses and types of targeted consumers, reveals a wide range of colour naming practices and significant variability influenced by multiple factors, such as the coloured or colouring matter, its properties, and the intended marketing of the product. Naming, in this context, serves more as a means of communicating semantic effects than as providing a precise or "veridical" label that directly refers to the colour of the product itself. The concept of colour and the categorization of colours become meaningful only when considered in relation to the specific practices involving coloured materials, and the linguistic forms used in verbal interactions motivated by those practices.[11]

4.4 The Formulation of the Question and the Analysis of the Answers

In this section, we report some results from our investigations in other sensory modalities (e.g., sound), but we believe these insights can be extended to research on food experiences.

[11] See also Cance, 2021; for colors in car interiors.

4.4.1 The Noise of Doors Abroad Trains: Is It "*désagréable*" or "*gênant*"?

An investigation on acoustic comfort aboard trains (Cance et al., 2021) questioned passengers about "*what is for you unpleasant abroad this train*" and were tested through two French question formulations that appeared similar[12] and can both be translated as "unpleasant" but differ subtly in wording:

- Q1: *Qu'est-ce qui pour vous est désagréable?*
- Q2: *Qu'est-ce qui pour vous est gênant?*

Q1, which asked about what people found "*désagréable*" regarding comfort aboard the train, elicited short answers such as "*le bruit des portes*" ("the noise of the doors"). These responses pointed to specific, objective acoustic phenomena caused by physical objects.

Q2, which inquired about what they found "*gênant*", also mentioned "*le bruit des portes*" ("the door noise"). However, the answers were more complex, including contextual details such as "*le bruit des portes au passage d'un voyageur*" ("the door noise when a passenger is passing by") witnessing an event actually including an acoustic stimulation but resulting in a subjective perceptual effect evaluated as negative. The noise of the door is not negative as such (as a noise) but when it is interpreted as a cue revealing that someone (another passenger) has opened the door and will pass with his/her luggage or with a cup of coffee coming from the bar.[13]

Three comments arising from this brief report may be relevant to the question at hand. First it reveals the great sensitivity of lexical wording of the question, even when these are oriented toward hedonic evaluation. Second, the answers suggest that the analysis of linguistic data should account for larger segments of discourses than simple words, emphasizing the strong relationship between open questioning and analysis verbal data. Third, it also reveals that the same acoustic stimulation may lead to different interpretations: one focusing on the signal itself (the noise), and another on its effect as a semantic cue signalling the occurrence of an event. This distinction aligns conceptually with Schaeffer's differentiation between "causal listening" or "ordinary listening" (Chion, 1983, 1994) and "reduced listening" or "musical listening" (Castellengo, 2015). The former entails identifying the sound

[12] Once again, we must adapt our approach to the specifics of each language. For example, the French terms *agréable* and *plaisant* are both commonly translated as pleasant in English, losing the subtle distinction between them. This is the case even though English morphology similarly forms negatives through prefixation (e.g., *disagreeable/displeasing* as counterparts to *désagréable/déplaisant*).

[13] *Gênant,* as the present participle of the verb *gêner* (*to bother*), accounts for the actual nature of "being bothered", while the suffix *-able* in the adjective *désagréable* (*unpleasant*) indicates a property of the sound as the *potential* effect it can produce as a stimulation. This interpretation has been corroborated in other corpora contrasting the use of present or past participle forms versus forms constructed with the suffix *-able* (e.g., *siège incliné* vs. *siège inclinable* — *inclined* vs. *inclinable seat*).

source within an everyday context, differing from the reduced listening focusing on the sound quality itself, abstracted from its context (as for music listening).[14]

4.4.2 Would You Taste some Jam? Coupling Instructions and Experimental Situations

Here we provide a final illustration of how different ways of questioning people influence their sensory experience, through variations in instructions and tasks, as well as a framework for interpreting the answers. This example comes from an inquiry conducted with a group of engineers[15] involved in designing sensors for applied industrial purposes. These engineers sought to acquire knowledge about human senses as inspiration for incorporating "cognitivity" into the design of artificial "cognitive" sensors, going beyond their expertise in signal and information processing. A group of 23 engineers from different nationalities and native speakers of different languages had to identify and characterize their perception of taste through verbal descriptions (in English) of 12 samples of preparations (from French supermarkets) and named as *confitures* (jams) or *gelées* (jellies). The products were tested by three different groups of participants under three conditions of tasting.

- Group A had to taste each "real" product in its original jar, with the visual label visible (in French), providing information on "what it is" and how it is named.
- Group B had to taste the product in an anonymous transparent glass container, allowing visual perception before tasting.
- Group C proceeded to a blind test (tasting without seeing the product): the experimenter gave a spoonful of each product into the hand of the subjects who introduced it in their mouth and tasted it, then they reopened their eyes to write down their answer to the same questions as Groups A and B.

Different formulations of the questions were tested, designed to be either subject-oriented (1) or object-oriented (2), with the aim of eliciting either a subjective, personal feeling and judgment or an objective characterization of the product: Q1—*Describe the taste you experience* and Q2—*Characterize the sample in detail in as many aspects as you can.*

Examples of naming jams and jellies in three conditions includes for Question 1, Gr.A (full product): *"strange taste somewhat familiar but no idea strange*

---

[14] In other words, "reduced listening" is oriented toward the qualification of the sound itself, as in music, while "causal listening", the ordinary mode of processing, aims at identifying "what's going on" in the world, such as on a TGV train or in soundscape studies (Castellengo, 2015; Gaver, 1993; Guastavino, 2021; Paté & Gaillard, 2021). Processing sounds in one way or the other does not rely solely on the sound itself but on how it is interpreted by humans (see, for example, the reduced processing of Henry's musical piece *Variations pour une porte et un soupir* (1963)).

[15] This study was performed as a teaching course in September 2013 when the first author was a research fellow at INCAS3 (Assen, Netherlands), a research institute that contributes to fundamental research in the field of sensors and sensor systems.

structure", "reminds me of the self-made jelly of my grandmother which we made in the summer of the fruit in her garden", "I am not familiar with this taste but I know it is olive"; Gr. B (vision + taste): *"sweet fruity summer kind of feeling", "dry", "sandy", "weird taste", "sweet but salty", "peaches but probably another"*; Gr. C (blind, taste only): *"with seeds, sour, familiar but don't remember what it is", "jelly/jam source, bad feeling did not like that taste, unknown little scaring", "familiar, salty, sour, bit of sweet Italian taste, pieces"; and f*or Question 2, Gr. A (full product): *"sweet, bit sour, jelly chunky, sticky, green, tasty", "smooth, brown, honey like texture"*; Gr. B (vision + taste): *"brown fruity a mash", "ginger, lemon, sugar, bright yellow, bits of lemon", "sour and sweet, orange colour"*; Gr C (blind, taste only): *"creamy", "very compact homogeneous", "sugar, jelly, whole pieces of berry"*.

A rapid linguistic analysis reveals that the three groups used different wordings in their answers to Q1. In Group A, answers mainly consisted of long sentences featuring personal marks such as nominal pronouns or possessive adjectives (e.g., *my grandmother*). In Group B, answers were lists of adjectives with few syntactic connections (e.g., *dry, sandy, sweet but salty*). In Group C, answers tended to be shorter, using more generic terms (e.g., *familiar*, *bad feeling*, *strange*) and often included expressions of uncertainty or incongruity, reflecting less familiarity. For instance, important variations for chestnut purée or green tomato jam were observed among this group of young people living in the Netherlands, but a strong consensus emerged around peanut butter, perhaps reflecting its status as a universal (?) food category.

In all conditions, the verbal answers revealed references to subjects' personal memory of previous sensory experiences, particularly in terms of familiarity or weirdness, with some notable differences: the full product condition (A) evoked vivid remembrance of past experiences (including many mentions of "grandmothers"), whereas condition B elicited more "objective" descriptions. The blind condition (C) produced more global answers, reflecting the participants' difficulty in identifying the product, which triggers uncertainty and even emotional reactions such as curiosity or even fear. In short, given the same question, the different presentations resulted in varied answers, expressed through distinct linguistic forms that draw on different combinations of knowledge types, whether individual experiential or learned knowledge.

In response to Q2, Group A provided descriptions of the products mainly using nouns referring to objects (such as sources or ingredients) or adjectives, often derived from nouns (e.g., *chunky*), that described properties related to various sensory modalities, reflecting the multisensorial nature of tasting (e.g., *sandy, sharp, hot, green, yellow*, etc., relating to texture or colour). Group B gave shorter responses, primarily adjectives, referring to analytical and abstract properties, most frequently visual ones (mainly colours). Group C also used adjectives, but these mainly referred to smell and texture, and, unsurprisingly, did not include any colour terms.

More generally, from a psychological perspective, this highlights the importance of vision,[16] allowing source recognition, and naming in shaping the interpretation of sensory stimuli by incorporating prior knowledge. This is especially important for taste, when the stimulation is not visible or unknown and can lead to uncertainty and even anxiety.

A major lesson from this experiment is that by combining different product experience conditions with varied question types, we can elicit distinct kinds of knowledge involved in taste. These range from subjective, sometimes emotional accounts of sensory experience (reflecting the psychological effects of stimulation, especially in blind tests when questions explicitly orient the question to personal answers from individual memories) to consensual "objective" descriptions of the product, particularly when the stimuli are visually presented, and the questions focus on identifying "what it is".

Regarding the influence of question formulation, an "object-centered" question (as in Q2) elicits more consensual answers. Participants often try to provide the experimenter with the expected answer, or, if possible, the "true" answer, also named as the "veridical label" (see Dubois & Rouby, 2002). However, this approach does not reveal much about the subject's actual subjective experience of the product.

Consequently, the extreme sensitivity of human processing to different question orientations and experimental conditions requires careful attention to the ecological validity of the experimental settings. This is essential both in designing the task, particularly when defining the instructions, and in analysing the resulting verbal data.

4.4.3 You Said "Clear"? How to Get at the Polysemy of a Word

The last example addresses the analysis of verbal data, which prevents a purely quantitative (or statistical) approach to word frequencies that fails to account for the textual context of occurrences. In the absence of a consensus on naming subjective experiences, how can we capture the variations in meaning as expressed by different individuals? We suggested a method for identifying the diversity of word meanings in discourse, initiated by the seminal work of Cheminée (2009); see also Cheminée et al., 2005), who developed a framework for analysing the semantics of *clair* (light, pale, clear, bright—illustrating the polysemy of the word in French) when used in evaluating piano sounds.

The analysis was conducted on transcriptions of oral comments made by 18 pianists as they evaluated the quality of nine pianos during performance. First, a quantitative analysis of lexical units (primarily adjectives) was carried out, and their frequency of occurrences was calculated. Next, the most frequently cited words were selected, with *clair* emerging as the most commonly used. Finally, by examining equivalences, similarities, contrasts, and oppositions in the distribution of these

[16] Well known by experts in wine who include it as a first step in the protocol of wine degustation.

words across different textual contexts, it was possible to identify both regularities and variations in meaning for the same lexical forms within their respective contexts. Below are some examples of answers (S: subject; P: piano).

(S18P3) *"C'est un piano qui a une définition très bonne à la fois très claire (...) enfin, quand je dis claire, on entend bien, on a une audition facile des notes, par contre c'est pas un son clair dans le sens vert, le côté clinquant, brillant du son"* (*"It's a piano that has a very good definition at the same time very clear (…) well, when I say clear, we hear, we easily hear the notes, for cons it's not a clear sound in the sense of green, the flashy side, brilliant sound*").

(S7P1) *"Il est, je crois, le plus clair, il a vraiment une certaine transparence"* (*"It is, I think, the clearest, it really has some transparency"*).

(S1P9) "... *dans l'aigu, par contre, c'est très clair, c'est très sec*" ("... *in the treble, on the other hand, it's very clear, it's very dry*").

This process, focusing on the reformulations and oppositions of *clair* within its sentential contexts of occurrences, led to the identification of three distinct meanings of *clair*. These were associated (and paraphrased) with closely related words used by the pianists themselves, as well as with their (near) antonyms, as reported in Table 3.2.

The method developed here relies more strictly on linguistic marks of equivalence and opposition, approximations, and rephrasings found within the discourse. It also draws on meta-comments made by the speakers themselves, such as "*comme on dit dans le métier*" ("*as we say in the profession*") or "*dans le jargon des pianistes*" ("*in the jargon of pianists*"), which indicate semantic variations in their

Table 3.2 the three semantisms of "*clair*" for the evaluation of piano sounds by pianists

Adjective	Clair = défini Clear = defined	Clair = lumineux clear = bright	Clair = sec Clear = dry
Associated adjectives	*Net* (neat)	*Brillant* (bright)	*Brillant* (brilliant)
	Précis (precise)	*Riche* (rich)	*Percussif* (percussive)
	Détachés (detached)	*Profond* (deep)	*Percutant* (impactful)
	Définies (defined)	*Incisif* incisive)	*Métallique* (metallic)
		Cuivré (copper)	*Sec* (dry)
		Chaud (warm)	*Vert* (green)
		Rond (round)	*Clinquant* (clinking)
		Ouvert (open minded)	*Perçant* (piercing)
			Dur (hard)
			Aigu (high-pitched)
			Criard (shrill)
			Agressif (aggressive)
Antonyms	*Imprécis* (imprecise) *Brouillon* (muddle) *Sourd* (deaf) *Pâteux* (mushy) *Confus* (confused)	*Sombre* (dark) *Profond* (deep)	*Profond* (deep) *Riche* (rich) *Chaud* (warm) *Rond* (round) *Doux* (soft)

word usage. These variations may reflect either a singular individual's subjectivity or a standard consensual form used as a member of the community.

The main conclusions that can be drawn from this analysis and which are of relevance to sensory analysis are the following: it confirms the difficulty to *a priori* differentiate between adjectives that carry an axiological (evaluative) value from those with a strictly denotative (descriptive) meaning, as illustrated earlier with the example of *doux*. The method allows for the identification of not only the "common-sense" meanings of words, but also their more specific uses: as specific terminology functioning within a particular community (in this case, pianists), as well as their technical meanings in domains such as acoustics.

The linguistic analysis of words in discourse makes it possible to identify linguistic marks that account for the specificity of discourse used to express sensory experiences, such as personal marks (David, 1997), or other enunciative procedures that distinguish between established knowledge and purely subjective individual statements (see Vion, 2009). This distinction is crucial in sensory analysis, as it enables differentiation between objective discourse (referring to the product) and subjective discourse (reflecting the consumer's personal sensory experience). This opposition is correlated with (and observable through) the use of personal pronouns: third-person pronouns (*il/elle—he/she* or *it* in English) when referring to entities in the world, first- and second-person pronouns (*je/nous—I/we*; *tu/vous—you*; *moi/toi/nous/vous—me/you/us/you*) indicating the subject's involvement in the discourse.

The use of personal marks may also correlate with a second opposition, between individual statements (*I, me, my, you, your*) and collective ones (*we, us, you (pl.), our, your (pl.)*). This distinction reflects different types of knowledge: personal (individual judgment), or shared knowledge within a group. The latter is often used to assert group identity or the authority of expertise, for example, "*we, as musicians*" or "*we, as perfumers*".

These observations encourage the use of personal marks as reliable heuristics for accessing the diversity of conceptualizations of sensory experiences—ranging from individual subjective perceptions, which may be more or less shared within a community of practice and language, to their constitution as objective knowledge about the external world.[17] This approach also enables tracking the construction of objectivity emerging from discourse of subjective experience, both as processed by expert panels and as expressed in everyday life, as illustrated in the following example.

[17] Such processes, from singular expressions of personal observations to objective (scientific) statements parallels the historical process of establishing scientific evidence, as described by Licoppe (1996). This process moves from an individual researcher's observation to the collection of experimental data validated by a community or institution, such as an academy of sciences or through peer evaluations. Licoppe also traced this progression through the gradual elision of personal marks in scientific reports and official publications, where there is no longer anybody (no observer as no enunciator) but "the results show that…".

5 Meaning Constitution and the Birth of a Term as a Descriptor

To conclude this chapter, we would like to illustrate one way of tuning the semantics of a word so that it can effectively capture a sensory experience. How can we build a relationship between a word and a sensory experience that could get a "true" referential value? How does the meaning of a word become consensual, eventually transforming it into a term that can be used as an "objective" descriptor of a sensory property of some food? The processes observed here among ordinary consumers, aimed at establishing objectivity in product description, are in fact commonly employed by panellists in standardized sensory science practices, particularly within the ritual of degustation, as seen in wine expertise, for example.

The case of dark bread and meaning constitution, borrowed from Bieler (see Bieler & Dubois, 2011), explicitly illustrates the ordinary communicative processes used to convey individual subjective taste experiences from a psycholinguistic perspective. This analysis highlights the interaction between linguistic and experiential processes in establishing a commonly shared meaning for descriptive words (in German) that account for the sensory experience of tasting dark bread (Bieler, 2011; Bieler & Dubois, 2011).[18]

We present in Table 3.3 transcriptions from a focus group where three friends discuss their feelings while tasting dark bread. This serves as an illustration of meaning constitution through the complex processes of co-constructing an intersubjective experience and reaching an agreement on naming, that is, assigning a name to this shared experience (for more details, see Dubois, 2017).

Eight steps can be identified in the process of elaborating the meaning of a term from ordinary words, illustrated here with the example of tasting dark bread (the participants' original wording is in italics, in German). The column on the right corresponds to the eight psycholinguistic processing steps involved in reaching a consensus on the meaning of a term based on the other proposed words.

1. The participants first introduce a consensus on the reference to be tasted and named: the taste of dark bread.
2. They exchange and communicate a subjective (memorized) previous experience that the present tasting reminds them of; for example, while tasting the bread, one participant made reference to a similar sensation and said "*it is then as what occurred to me*".
3. This personal memory is established to be shared with the other participants: "*it is like when you go into the forest*". The speaker shifts from first to second person, creating a reference to a situation that other interlocutors can fill with their

[18] From the Swiss project, a corpus consisting of 34 focus group discussions was audio-visually recorded and transcribed according to GAT9. These focus groups collected discussions centered on the tasting of various foods and beverages, such as apples, bread, coffee, yogurt, Coca-Cola, beer, and chocolate.

Table 3.3 Excerpt from a focus group discussion on tasting of dark bread: co-construction of meaning and naming of a shared sensory experience

01	Ida:	*alsoichhabsjetztnochmalsproBIERT-*	well, I have now once again tried	1
02		*undichhabdieGANzezeitvorherversucht-*	and earlier I tried the whole time	1
03		*nach irgendwA:s zu SUchen,*	to look for something.	1
04		*aber wie das (.) um das BROT-*	but how to do it… the bread	1
05		*irgendwie den geSCHMACK zu beschreiben;*	somehow to describe the taste	1
06		*und dann sind mir so (-)-*	and then what occurred to me	2, 4, 5, 7
07		*eben so BAUMpilze in den sinn gekommen;*	was tree fungi	4,
08		*also wenn du in den WALD gehst-*	like when you go into the forest	2, 3, 4, 5
09	Ria:	*[((hustet))]*	[coughs]	
10	Ida:	*[und dann-]*	[and then]	
11		*«mitdenHändenPilzeanzeigend>überall-*	«demonstrating with her hands> everywhere	
12		*die PILze an den BäUmen hängen>-*	the fungi hanging on the trees.	4,
13		*so riecht (.) also so wie dass RIECHT-*	smells like this I mean, it smells like this	4, 5, 6
14		*[und so SCHMECKT auch das brot.]*	[and that's how the bread tastes.]	4,
15	Liv:	*[((lacht))]*	[laughs]	
16	Ida:	*also (–) vielleicht auch ein bisschen so-*	well maybe also a bit like	
17		*wie TRüffel oder so (-);*	like truffles or something.	7
18		*aber irgendwas das aus dem BOden kommt;*	but something that comes out of the ground	7
19		*und aus dem wald-*	and out of the forest.	7
20	Joy:	*ERdig;*	earthy	8
21	Ida:	*[mhm (.) also ähm GANZ genau;]*	[mhm (.) well, umh, precisely]	8
22	Ada:	*[ERdig ist gut;]*	[earthy is good.]	8
23	Liv:	*ja-*	yes	8

own experiences, thereby constructing an "interworld" of shared subjective experiences (here, the generic experience of someone walking in the forest).

4. The discussion then moves toward generalization and shift to an objective situation: "*there was tree fungi*" followed by "*it smells like…*", moving from an individual past experience to a generic type, creating common ground as an objectivity situated in the outside world.
5. Objectivization is further achieved by the deletion of personal marks, moving from "*it occurs to me*" to "*it smells like this*".
6. From this agreement emerges the statement, "*and that's how the bread tastes*", establishing this "common ground" as an objective description of the bread's taste.
7. The group then attempts to refine this common ground through categorization and naming processes (Rosch & Lloyd, 1978; Dubois, 2000, 2006). Each participant proposes additional wordings, such as "*also, well maybe a bit like…*", "*like*

truffles or something", "*but something that comes out of the ground*", or "*and out of the forest*", eventually proposing a generic category: *fungi*.

8. Finally, they converge on Joy's proposal: *earthy*. The two other participants confirm this with Ida saying, "*mhm... well, umh, precisely*", and Ada explicitly making an evaluative statement regarding the adjective proposed by Joy, "*earthy is good*", which Liv further endorses by saying "*yes*", thus confirming consensus.

To summarize, in the absence of an immediately available consensual lexical form, subjective experiences are named indirectly through a detour and the mediation of a common shared experience in an ordinary situation (such as walking in a forest) that everyone may have lived and can relate to. Given the condition of shared experience (of tasting), any lexical form denoting such sensory experience can be adopted if the group converges on it, thereby constructing objectivity through psycholinguistic processes of categorization and naming. This is, in fact, the process that regularly operates within expert panels, although with a task oriented more toward the product than the sensation itself, within codified procedures of degustation. Similar cases can be made for any product; for instance, Vion provides a comparable and detailed linguistic analysis of wine tasting (Vion, 2004; Vion, 2009).

6 Conclusion

When exploring verbal comments from open-ended questions targeting sensory experiences, it is important to recognize that a word does not have a single meaning as some true value referring to an objective reality in the external physical world (and considering "subjective" experience a mere representation of "it"). People use words available as resources within their own language, to express their individual subjective experiences, which may differ from those of others. Therefore, it is important to create a more precise inventory of the linguistic resources available to speakers in each language. In the absence of ready-made lexical forms, which is often the case for senses like smell and taste, speakers productively use what is available and what they consider relevant. For example, they may refer to the name of the source ("*it tastes like banana*"), or they may negotiate and reach consensus on any word within a small group during a degustation session, as experts commonly do in sensory panels, or more generally within a community.

Such statements, however, call for a reconsideration of some traditional oppositions, such as the one between subjectivity and objectivity. We have shown that objectivity can be achieved by negotiating and establishing consensus on a lexical form among a group of people. Such a lexical form, whether arbitrary or motivated,[19] can be a commonly used word becoming a term, or even a neologism (for example, *chocolaty*). Similarly, the negotiation of *earthy* to qualify the taste of dark bread

[19] Physicists named the elementary units of the matter as atoms meaning they could not be breakable, such naming becoming inappropriate by the end of the XIX century.

illustrates this process (see also Cance & Dubois, 2017, for a more theoretical development). For technical terms or *jargon*, the validity of the consensus relies on the authority of the people questioned. In sciences, as in expert panels, this consensus is given as the objective meaning, as a truth value (of a term) under the authority of the scientists, or the expertise of panellists. This raises an important question: why question consumers, if not to acknowledge the validity of their verbal comments on their subjective experience, and to track the complex psychological and linguistic processes that relate their wording to their experience of the product considering a direct link between the semantics of word and what it is expected to refer to?

A second opposition concerns the validity of qualitative versus quantitative data. Quantitative research typically reduces phenomena to numerical values to enable statistical analysis (Smith, 2015). This approach is common when verbal data, such as responses to questionnaires or semantic scales,[20] are converted into numbers for statistical processing.

In contrast, qualitative research involves collecting data in the form of more naturalistic verbal reports, such as interview transcripts or written reports, and analysing these texts. However, there is often a temptation to apply automatic processing to large volumes of text data, driven by the assumption that the bigger, the more precise and reliable. But this can be misleading. We have shown that different individuals, having different experiences, assign different meanings to the "same" words, and conflating these meanings by mixing all data together is ultimately meaningless.

Despite the prevailing trend of applying quantitative methods, often borrowed from the natural sciences, to the cognitive sciences, new perspectives are emerging that recognize the limitations of reducing individual and collective human phenomena to mere "numbers", "information processing", or "rational thinking". This shift is evident in fields such as economics, linguistics, communication, marketing, sensory sciences, and even psychology. We can mention some of these, applied to exploring sensory experience, in the work of Spinelli (2018).

However, the opposition between quantitative and qualitative data cannot be reduced to a simple contrast between numerical data and symbolic (primarily verbal) data. Rather, it reflects deeper differences in assumptions, worldviews, and motivations for conducting research, differences that ultimately lead to distinct methodologies.

As pointed out by Avenier and Thomas (2011) and Yanow and Schwartz-Shea (2015), efforts to bridge the divide between quantitative and qualitative approaches expose deeper epistemological issues. The "quanti-quali" division demarcates a distinction between epistemological claims rooted in positivist philosophical presuppositions and those influenced by schools of thought that put humans as agents producing meaning at the center of their concerns. While this qualitative–quantitative opposition may appear clearly defined at first, it becomes increasingly blurred

[20] Raimbault (2006) showed that participants give highly different meaning to the labels of scales designed to explore soundscapes, along with the numerical value of the scales (the middle value of the scale being ambiguous and even meaningless when statistically processed).

when investigating sensory experiences. The methods used to acquire knowledge in this domain vary widely depending on prior (scientific) knowledge drawn from different disciplines, as well as on different sensory modalities, populations, cultures and practices. This raises a critical question: is *a priori* knowledge, whether borrowed from scientific fields other than psychology, or elaborated from different cultures or practices, directly relevant to the interpretation of sensory data, regardless of the research question or the people involved? Our research adopts a situated cognitive approach, which leads us to consider data as constructs, as responses are constrained by the different procedures through which they are elicited which have to be also questioned (Dubois et al., 2021b). They cannot stand as directly interpretable "raw data" but as observable mediations from which meaning can be negotiated and eventually reach at a consensus among observers and participants.

As an empirical conclusion, these results suggest that before engaging in any automatic processing of verbal data, several key considerations must be addressed. It is essential to clearly define who is targeted establishing the consensus, what kind of food is concerned, under what conditions are participants being questioned? For what purpose is the data being collected (descriptive analysis vs marketing for example)? Then, it is important to establish a segmentation on the variables that are relevant, especially regarding who is questioned and for what purpose? This segmentation should draw on insights from the human sciences, rather than relying solely on traditional distinctions between male and female, young or old, etc. The focus should shift toward the cultural diversity of food-related behaviours: purchasing, cooking, and eating habits for food. This is why we have emphasized defining food as a cultural product rather than merely in terms of its nutritional properties and have cited the sociolinguistic work of Lahlou (1998) for designing the relevant characteristics in defining the population. We must also remember to consider each respondent as a speaker of a specific language; with all the linguistic and cultural variability this entails. Third, linguistic analysis should be grounded in the specific properties of one language and validated as semantic cues pointing to the subjective reality rather that directly pointing to the physical description of the product. Therefore, the analysis should be differential, that is, it should contrast the different meanings that emerge from different corpora across the segmented populations and purposes and ways of questioning. It should not be assumed that verbal data are directly referential to the product or anchored to expert-defined categories.

In other words, such an analysis could be processed on a relatively small number of participants, identifying their own frames of reference or world views, relying on a precise linguistic analysis of their verbal reports, rather than directly processing a large sample of data along a preconceived hypothesis and key words, or selected on their frequencies, but remaining undefined given their highly polysemy or their metaphoric acception and critically, given their non correspondence from one langue (*langue*) to another one. Then, the automatic processing of larger corpora may be relevant and meaningful.

References

Ariès, P. (2016). *Une histoire politique de l'alimentation: du paléolithique à nos jours*. Max Milo.

Auvray, M., & Spence, C. (2008). The multisensory perception of flavor. *Consciousness and Cognition, 17*(3), 1016–1031.

Avenier, M. J., & Thomas, C. (2011). Mixer quali et quanti pour quoi faire? Méthodologie sans épistémologie n'est que ruine de réflexion. In *Journée de l'atelier méthodologie de recherche de l'AIMS. Caen, France*.

Bieler, L. M. (2011). On understanding taste. Cooperative strategies of meaning constitution in conversations about taste. In *Proceedings of Talking about taste – Sensory semantics and the language of taste. Zurich, Switzerland*.

Bieler, L. M., & Dubois, D. (2011). Communicating taste: A dialogical perspective on meaning constitution in conversation about taste. In *Proceedings Les descripteurs du vin, Vino lingua* (pp. 1–19).

Bonnassie, P. (1989). Consommation d'aliments immondes et cannibalisme de survie dans l'Occident du Haut Moyen âge. In *Annales. Histoire, Sciences Sociales* (Vol. 44, No. 5, pp. 1035–1056). Cambridge University Press.

Braudel, F. (1961). Alimentation et catégories de l'histoire. *Annales. économies, Sociétés, Civilisations, 16*(4), 723–728.

Browman, D. L., & Gundersen, J. N. (1993). Altiplano comestible earths: Prehistoric and historic geophagy of highland Peru and Bolivia. *Geoarchaeology, 8*(5), 413–425.

Cance, C. (2021). Chapter 3: Experiencing and talking about colors. In D. Dubois, C. Cance, M. Coler, A. Paté, & C. Guastavino (Eds.), *Sensory experiences: Exploring meaning and the senses*. Benjamins ed.

Cance, C., & Dubois, D. (2017). *Diversité des ressources linguistiques et des discours dans différentes modalités sensorielles: de l'expression de l'expérience sensible à la construction de l'objectivité* (manuscript).

Cance, C., Dubois, D., & Coler, M. (2021). Chapter 14: Making sense of the outcomes: Tools for analysis. In D. Dubois, C. Cance, M. Coler, A. Paté, & C. Guastavino (Eds.), *Sensory experiences: Exploring meaning and the senses*. Benjamins ed.

Cassin, B. (2004). *Vocabulaire européen des philosophies: Dictionnaire des intraduisibles*. Éditions Seuil, Dictionnaires Le Robert.

Cassin, B., Apter, E., Lezra, J., & Wood, M. (2014). *Dictionary of untranslatables: A philosophical lexicon*. Princeton University Press.

Castellengo, M. (2015). *Écoute musicale et acoustique*. Eyrolles.

Cheminée, P. (2009). Est-ce bien "clair"? Stabilité, instabilité et polysémie d'une forme lexicale en contexte. In D. Dubois (Ed.), *Le Sentir et le Dire* (pp. 309–338). L'Harmattan.

Cheminée, P., Gherghinoiu, C., & Besnainou, C. (2005). Analyses des verbalisations libres sur le son du piano versus analyses acoustiques. In *Colloque interdisciplinaire de musicologie (CIM05). Montreal, QC, Canada*.

Chion, M. (1983). *Guide des objets sonores – Pierre Schaeffer et la recherche musicale*. Institut national de l'audiovisuel & Buchet-Chastel.

Chion, M. (1994). *Audiovision – Sound on screen*. Columbia University Press.

Classen, C. (1993). *Worlds of senses: Exploring the senses in history and across cultures*. Routledge.

David, S. (1997). Représentations sensorielles et marques de la personne: contrastes entre olfaction et audition. In D. Dubois (Ed.), *Catégorisation et Cognition: De la perception au discours* (pp. 211–242). Kimé.

Dubois, D., Grinevald, C. (2003). En voir de toutes les couleurs : processus de dénomination des couleurs et constructions cognitives. In Vandeloise C., éd. : Traité des sciences cognitives. Paris, Hermès, 80–114.

Dubois, D. (2006). From psychophysics to semiophysics. Categories as acts of meaning. A case study from olfaction and audition, back to colors. In M. Plümacher & P. Holz (Eds.), *Speaking of colors and odors. An interdisciplinary approach to cognitive and linguistic categorization of color vision and olfaction* (pp. 167–184). John Benjamins Publishing.

Dubois, D. (2000). Categories as acts of meaning: The case of categories in olfaction and audition. Cognitive Science Quarterly, 1, 35–68.

Dubois, D. (2009). *Le Sentir et le Dire – Concepts et méthodes en psychologie et linguistique cognitives*. L'Harmattan.

Dubois, D. (2017). How wor for sensory experiences become terms. *Terminology. International Journal of Theoretical and Applied Issues in Specialized Communication, 23*(1), 9–37.

Dubois, D., & Cance, C. (2012). Vers une sémiotique du sensible: des couleurs en discours et en pratiques. *Histoire Épistémologie Langage, tome, 34*(fascicule 1), 63–95. http://www.persee.fr/doc/hel_0750-8069_2012_num_34_1_3237

Dubois, D., & Cance, C. (2021). Chapter 8: Taste as a holisensory experience. In D. Dubois, C. Cance, M. Coler, A. Paté, & C. Guastavino (Eds.), *Sensory experiences: Exploring meaning and the senses*. Benjamins ed.

Dubois, D., & Fritz, C. (2015). You said "chocolaty"? What do you mean? *American Lutherie, 122*, 70–71.

Dubois, D., & Giboreau, A. (2006). Descriptors: Attributes? Labels? Terms? Names? *Food Quality and Preference, 17*, 671–672.

Dubois, D., & Grinevald, C. (2000). Denomination of colors in practice. In A. Melby & A. Lowels (Eds.), *Proceedings of LACUS forum XXVI: The lexicon, Vancouver, Canada* (pp. 237–246).

Dubois, D., & Rouby, C. (2002). Odor and their names, the "veridical label" revisited. In C. Rouby, B. Schaal, A. Holley, D. Dubois, & R. Gervais (Eds.), *Olfaction, taste, and cognition* (pp. 47–66). Cambridge University Press.

Dubois, D., Cance, C., & Coler, M. (2021a). Chapter 8: Subjects or participants? In D. Dubois, C. Cance, M. Coler, A. Paté, & C. Guastavino (Eds.), *Sensory experiences: Exploring meaning and the senses*. Benjamins ed.

Dubois, D., Cance, C., Coler, M., Paté, A., & Guastavino, C. (2021b). *Sensory experiences: Exploring meaning and the senses*. Benjamins ed.

Dubois, D., Cance, C., Coler, M., & Paté, A. (2021c). Chapter 1: The five senses and the cognitivist approach to perception. In D. Dubois, C. Cance, M. Coler, A. Paté, & C. Guastavino (Eds.), *Sensory experiences: Exploring meaning and the senses*. Benjamins ed.

Dubois, D., Cance, C., Coler, M., Paté, A., & Guastavino, C. (2021d). Chapter 10: Questioning sensory experience. In D. Dubois, C. Cance, M. Coler, A. Paté, & C. Guastavino (Eds.), *Sensory experiences: Exploring meaning and the senses*. Benjamins ed.

Dubois, D., Cance, C., Coler, M., & Paté, A. (2021e). Chapter 9: From perception to sensory experiences: A paradigm shift. In D. Dubois, C. Cance, M. Coler, A. Paté, & C. Guastavino (Eds.), *Sensory experiences: Exploring meaning and the senses*. Benjamins ed.

Fischler, C. (1990). *L'homnivore: le goût, la cuisine et le corps*. Odile Jacob.

Gaver, W. W. (1993). What in the world do we hear? An ecological approach to auditory event perception. *Ecological Psychology, 5*(1), 1–29.

Guastavino, C. (2021). Chapter 4: Experiencing soundscapes. In D. Dubois, C. Cance, M. Coler, A. Paté, & C. Guastavino (Eds.), *Sensory experiences: Exploring meaning and the senses*. Benjamins ed.

Howes, D. (2003). *Sensual relations: Engaging the senses in culture and social theory*. University of Michigan Press.

Howes, D. (2011). Sensorial anthropology. In D. Howes (Ed.), *The varieties of sensory experience: A source book in the anthropology of the senses* (pp. 167–191). Toronto University Press.

Howes, D. (2019). Multisensory anthropology. *Annual Review of Anthropology, 48*(1), 17-28.

Köster, E., Moller, P., & Mojet, J. (2014). A "misfit" theory of spontaneous conscious odor perception (mitscop): Reflections on the role and function of odor memory in everyday life. *Frontiers in Psychology, Cognitive Science, 5*(Special issue "Applied olfactory cognition"), 29–43.

Kristol, A. M. (1978). *Color: les langues romanes devant le phénomène de la couleur, Thèse université de Zurich, et Romanica Helvetica, 88*. Francke.

Lahlou, S. (1998). *Penser manger. Alimentation et représentations sociales*. P.U.F. 239p. ISBN-13:978-2130491361.
Lévi-Strauss, C. (1964). *Le Cru et le Cuit (Mythologiques 1)*. Plon.
Licoppe, C. (1996). *La formation de la pratique scientifique*. La découverte.
Mainland, J., & Sobel, N. (2005). The sniff is part of the olfactory percept. *Chemical Senses, 31*(2), 181–196.
Montanari, A. (2018). *Cannibales: histoire de l'anthropophagie en Occident*. Arkhê.
Paté, A., & Gaillard, P. (2021). Chapter 6: Exploring and talking about music. In D. Dubois, C. Cance, M. Coler, A. Paté, & C. Guastavino (Eds.), *Sensory experiences: Exploring meaning and the senses*. Benjamins ed.
Pelt, J.-M. (1994). *Des fruits*. Fayard.
Poulain, J. P. (2002). *Sociologies de l'alimentation*. P.U.F.
Raimbault, M. (2006). Qualitative judgments of urban soundscapes: Questionning questionnaires and semantic scales. *Acta Acustica united with Acustica, 92*(6), 929–937.
Reinert, M. (1983). Une méthode de classification descendante hiérarchique: application à l'analyse lexicale par contexte. *Les cahiers de l'analyse des données, VIII*(2, Dunod), 187–198.
Ricaud Oneto, E. (2024). Exploring cultural differences of sweetness and bitterness to understand food acceptance among two indigenous people in Peruvian Amazon. *Science Talk, 10*, 100325.
Rosch, E., & Lloyd, B. (1978). Principles of categorisation. In E. Rosch & B. Lloyd (Eds.), *Cognition and categorization* (pp. 27–48). Lawrence Erlbaum.
Ross, B. H., & Murphy, G. L. (1999). Food for thought: Cross-classification and category organization in a complex real-world domain. *Cognitive Psychology, 38*(4), 495–553. https://doi.org/10.1006/cogp.1998.0712
Smith, J. A. (2015). *Qualitative psychology: A practical guide to research methods*. Sage.
Spence, C. (2021). Sonic seasoning and other multisensory influences on the coffee drinking experience. *Frontiers in Computer Science, 3*. Human-Media Interaction. https://doi.org/10.3389/fcomp.2021.644054
Spence, C., Piqueras-Fiszman, B., Michel, C., & Deroy, O. (2014). Plating manifesto (II): The art and science of plating. *Flavour, 3*(4).
Spinelli, S. (2018). Semiotics and sensory sciences: Meaning between texts and numbers. In *Quantitative semiotic analysis* (pp. 75–100). Springer.
Vion, R. (2004). Modalités, modalisations et discours représentés. *Langages, 156*(4), 96–110.
Vion, R. (2009). De la subjectivité en discours aux discours sur la subjectivité. In D. Dubois (Ed.), *Le sentir et le dire* (pp. 367–391). L'Harmattan.
Yanow, D., & Schwartz-Shea, P. (2015). *Interpretation and method: Empirical research methods and the interpretive turn*. Routledge.

Part III
Handling Text Data

Collecting text data often appears to be the easiest part of the research process since it requires no specialized software or technical expertise. It is because of this simplicity including open-ended questions and free comment fields in consumer survey becomes especially appealing. In addition, vast corpora of consumer-generated text, such as online reviews and internet comments, are readily available without the need to solicit any direct responses. However, analysing this type of data is far more complex and without the proper tools, researchers may find this stage frustrating and difficult to navigate.

This section addresses these challenges. Chapter 4 focuses on the organization of vocabulary, introducing lexicons and ontologies used in sensory and consumer science. Chapter 5 presents a comprehensive overview of techniques drawn from content analysis and natural language processing (NLP), providing the foundational methods necessary to approach the chapters that follow.

Chapter 4
Vocabulary Organization: Lexicons and Ontologies in Sensory and Consumer Science

Magalie Weber

1 Introduction

Communicating is a universal process that we all engage in without even thinking about it, like breathing! Every communication scheme is based on simple principles (Shannon & Weaver, 1949), which form the basis of any exchange between what we commonly call a sender (or emitter) and a receiver. The sender has an intention that he will transmit to the receiver, who then will interpret it. Language is a tool that links words or expressions to the meaning they are intended to convey. In fact, language serves as the code used to communicate the message, even though much of what we transmit is also unconscious and non-verbal. The set of words or expressions in a language constitutes a vocabulary.

Knowledge, in the context of information science, encompasses the understanding, interpretation, and application of the information conveyed by data (Stock & Stock, 2013). Knowledge is information that has been processed by the human mind through learning, experience, or instruction. When processed effectively, knowledge is largely cumulative. Once acquired, it can be used to gain a broader and deeper range of additional knowledge, as well as to correct the mistakes we make as we go along. In this way, knowledge contributes to its own development and refinement. However, this is only possible if what we know is recorded in a way that allows it to be quickly and easily retrieved and understood by those who need it. The relationship between information and knowledge is dynamic and essential: information acts as the foundation upon which knowledge is built. To fully grasp information, it is necessary to understand not only lexical meanings, but also semantics.

M. Weber (✉)
BIA, National Research Institute for Agriculture, Food and the Environment (INRAE), Nantes, France
e-mail: magalie.weber@inrae.fr

M. Visalli, M. V. Galmarini (eds.), *Collection and Analysis of Text Data in Sensory and Consumer Science*, https://doi.org/10.1007/978-3-032-04851-6_4

Indeed, in a formalized language, as opposed to syntax, which defines the structure and symbols (the alphabet), semantics refers to the rules for constructing well-formed expressions (the grammar), as well as the deduction rules operating from axioms. Semantics deals with structures that satisfy formulas or logical axioms, and, in this context, is then called model theory.

Automated information systems (AIS) offer unique opportunities for representing knowledge, as they have the capacity to handle enormous quantities of it. AIS refers to computer-based methods and tool that performs functions such as collecting, processing, storing, transmitting, and displaying information to support decision-making (Khosrow-Pour, 2024). As a result, AIS must store knowledge in formats that can be processed by non-human agents. AIS must also be capable of sharing information. To achieve this, each system must represent information in a standardized manner. For an AIS to serve as a repository for information collected by researchers, it must be pre-programmed to accommodate that information. This means that, for every type of input an AIS may host, it must include a corresponding category. Consequently, an AIS must have a ready-to-use categorical structure that allow it to classify each piece of programmed information under the appropriate heading. Ideally, this structure should align with those of other information systems to facilitate seamless information exchange. For this interoperability to be possible, a shared categorical structure must exist across all information systems.

Over the last decade, the rise of Big Data and the social web (Web 2.0) has prompted the community of knowledge scholars and practitioners to rethink the standpoint from which knowledge organization systems are designed (Ibekwe-SanJuan & Bowker, 2017). In essence, Big Data does not eliminate the need for human-built knowledge organization systems, but it does require new methods for integrating massive amounts of information through machine learning. Today, the field of knowledge organization must shift from establishing apodictic classifications (those considered universally and eternally valid) to embracing the evolving nature of findings in the new world in both social and natural sciences. This can be achieved by creating maximally flexible schemes, such as faceted classifications.

The Semantic Web, also known as the "Web of Data" or Web 3.0, is a concept that aims to make the vast amount of information on the web machine-readable at scale (Yu, 2014). Linked Data is a set of design principles for sharing machine-readable interlinked data on the Web. In other words, it establishes links between datasets that are understandable not only to humans, but also to machines. To this end, a comprehensive suite of standards, technologies, and related tools has been developed and specified by the World Wide Web Consortium (W3C), forming the foundation for numerous new applications (Semantic Web Standards). Compared to the traditional early web, Web 3.0, which is the version we use today, is a more "intelligent" web, especially evident in search engines that are increasingly capable of understanding natural human language. Another major change is the growing mobility of internet use. Currently, a large proportion of internet activity occurs on mobile devices, and tools such as "the cloud" are expanding the ways users interact with online resources. For this reason, Web 3.0 could also be referred to as the "mobile web" or "flexible web". The next evolution, Web 4.0, is expected to bring radical change. Once fully developed, it will eliminate many of the steps required to use Web 3.0, making it more direct and user-friendly. While traditional search engines may not disappear, they will be integrated into virtual assistants. These virtual assistants will be able to understand natural language much like, allowing users to interact with them conversationally and submit queries in the same way they would speak to a person.

The focus of this chapter is the different types of resources that are used to structure information and knowledge. In the first section, we present the various types of vocabulary, ranging from the most basic lexical level to the most formal organizational system: ontologies. We then examine the various dictionaries and taxonomies employed in sensory and consumer sciences. Next, we describe the uses of knowledge organization systems, such as taxonomies, thesauri, and ontologies. Finally, we demonstrate the use of a knowledge graph coupled with an ontology to access data from various sources in the field of sensory and consumer sciences. Finally, this chapter will conclude with a look at how we can exploit textual data through the coupling of ontologies and automatic language processing.

2 What Is "Vocabulary"

2.1 Overview

In the context of data, vocabulary refers to a set of terms used within a specific domain or subject area. These terms often come with definitions that standardize their usage, ensuring that everyone interpreting the data understands them in the same way. Vocabularies are also known as linguistic or semantic resources, semantic artifacts, or termino-ontological resources. Glossaries, lexicons, terminologies, taxonomies, thesauri and ontologies are all different types of knowledge processing resources. However, they differ in terms of their structure, function and level of formalization. A controlled vocabulary can be defined as an explicitly enumerated list of terms, each with unambiguous and non-redundant definitions (National Information Standards Organization, 2005). Its main function is to support the indexing of content and/or its retrieval through browsing or searching. Controlled vocabulary can take a variety of forms, with different levels of formalization as illustrated in Fig. 4.1. This typology is more akin to a continuum than a set of clearly defined categories, but it does present the different functions that resources can perform according to their level of semantics.

2.2 Low Formalization Level Lexical Resources

An authority file or record is a set of headings with cross-references to and from each heading. Types of headings include proper name headings (which can be called identifiers), subject headings, and a preferred name or term. They can include a

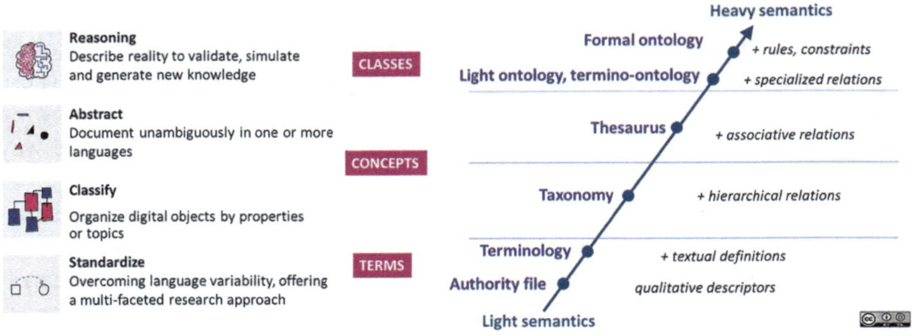

Fig. 4.1 Scale of semantics artifacts. (Adapted from: INRAE, 2024, Les types de vocabulaires et leurs usages, Vocabulaires Ouverts@INRAE, https://vocabulaires-ouverts.inrae.fr/types-de-vocabulaires-et-usages)

qualifier, generally enclosed in parentheses, to distinguish homographs (National Information Standards Organization, 2005).

A terminology is, first and foremost, a set of specialized terms or nomenclature relating to a particular field of activity or specific to a subject, science, or art. It is typically presented as an alphabetical list of terms within a domain of knowledge, accompanied with definitions. Lexicons, glossaries and dictionaries are related resources. These lexical resources are mainly used for linguistic tasks such as syntactic analysis, translation, or basic semantic analysis. They are also useful for applications such as spellcheckers and word processors.

A glossary is an alphabetical list of terms within a particular domain of knowledge, each with accompanying definitions. Traditionally, glossary appear at the end of a book and includes terms within that book that are either newly introduced, uncommon, or specialized in the context of that text.

A lexicon is a collection of words or multi-word expressions in a language, accompanied by their definitions, grammatical properties (such as categories or genders), and sometimes examples of usage. Lexicons aim to comprehensively describe all terms within a specific field.

Dictionaries are intended to be exhaustive, referential and precise in describing the words of one or more languages. They focus on definitions and grammatical information, but do not usually represent complex logical relationships between concepts.

2.3 *Knowledge Organization Systems*

On the other hand, taxonomies, thesauri and ontologies provide a consistent vocabulary that is key for effectively communicating, analysing, and processing data across different systems and contexts.

A taxonomy is a collection of controlled vocabulary terms organized onto a hierarchical structure (National Information Standards Organization, 2005). In biology, taxonomy refers to the science of describing, identifying, naming, and classifying organisms. Within a taxonomy, each term is related to one or more terms through parent/child (broader/narrower) relationships.

A thesaurus (plural: thesauruses, thesauri) is a controlled vocabulary arranged in a known order and structured so that the various relationships among terms are displayed clearly and identified by standardized relationship indicators (National Information Standards Organization, 2005). It serves as a structured directory of terms (subject headings or keywords) also known as a simple knowledge organization system in information science. It provides standardized relationships for a given language: hierarchical relationships between terms that depict broader (generic) to narrower (specific) or whole-part relationships, associative relationships between terms (related, but not hierarchically), synonyms or variants (non-preferred terms), and scope notes, or other metadata on terms (Dextre., 2012).

An ontology is a formal, explicit specification of a shared conceptualization (Studer et al., 1998). It is representation of knowledge, or formalized knowledge, in a particular domain. The notion of "shared conceptualization" refers to the collective understanding of a domain by a group of experts. The "formal and explicit specification" refers to the use of a formal language to represent that understanding in a way that is machine-readable. Ideally, the proposed conceptualization should be reviewed and accepted by the relevant expert community.

Ontologies first emerged in the Artificial Intelligence (AI) community, borrowing the term "ontology" from philosophy, where ontology is concerned with the study of being or existence. The term was adopted by the AI community in the 1980s for computational models that can enable automated reasoning (Stuart, 2016).

3 Use of Lexicons and Taxonomies in Sensory and Consumer Science

3.1 Sensory Lexicons

Sensory perception involves breaking down experiences into the five main senses: sight (visual perception), hearing (auditory perception), touch (tactile perception), taste (gustatory perception), and smell (olfactory perception). Each sense relies on different types of stimuli and receptors. Some sensory perceptions involve multiple senses; for example, the perception of texture engages visual, acoustic, and tactile sensation mechanisms. Consistency in sensory evaluation is essential to ensure that results are comparable between different panellists and to minimize personal or subjective biases. However, sensory analysis is inherently more subjective than instrumental measurements, as it depends on human perception, which can vary across individuals and cultures. For this reason, training is crucial—especially when it comes to describing sensory qualities. Through training, a shared and standardized terminology can be established. Without such training, different assessors are likely to describe the same sensation in inconsistent or conflicting ways.

Sensory lexicons are lists or repositories of descriptive terms (mainly adjectives) that refer to the attributes or characteristics of objects (generally foods or beverages) perceived by the human senses.

The International Organization for Standardization (ISO) has developed several international standards that define sensory evaluation methods and associated descriptors. In addition to ISO standards, many national organizations such as AFNOR (Association Française de Normalisation) and ASTM (American Society for Testing and Materials) have also created standards for sensory testing, including the standardization of sensory descriptors. These standards are widely adopted in the food industry and other sectors to ensure reliable and comparable assessments. ASTM DS72 (A Sensory Vocabulary | ASTM) summarizes lexicon development into five key steps: (i) establishing a "frame of reference" using a wide array of products in the category; (ii) developing and generating descriptive terms for the products; (iii) clarifying terms by using references and definitions; (iv) providing examples to ensure that panellists fully understand the terms; and (v) creating the final list of agree-upon descriptors for inclusion in the lexicon. Additionally, Civille and Lawless (1986) emphasize that descriptive terms should be as orthogonal as possible (that is, uncorrelated with one another) and should, where possible, be grounded in the product's underlying structure (Civille & Lawless, 1986).

An important technique often used during lexicon and wheel development is the consensualization method. This approach draws on the collective expertise of trained panellists, researchers, or industry professionals—depending on the product being analysed. Participants suggest sensory descriptors they believe best represent the product's characteristics. These descriptors are then collected, grouped, and analysed to identify commonly occurring terms. Frequently mentioned descriptors are retained and subsequently validated by tasters to ensure they are well understood and sufficiently comprehensive. This may include conducting consistency tests to compare the evaluation across different panellists. Once all samples have been evaluated and the descriptive terms have been generated, discussed and refined, the final list is compiled. Each term is then clearly defined to ensure accuracy and consistency in future evaluations (Drake & Civille, 2003; Lawless & Civille, 2013).

3.2 Sensory Wheels or Taxonomies

Sensory lexicons are generally organized by sensory categories, with specific subcategories, forming taxonomies that enable a fine-grained classification of sensory perceptions. These taxonomies are often represented as "wheels", namely circular diagrams representing different categories of a food's sensory properties. Each sector of the wheel is divided into more specific sub-categories, enabling a more in-depth description. By offering a shared frame of reference, sensory wheels help standardize participants' responses and reduce subjective variation. They serve as visual guides that assist evaluators (panellists) in recognizing and articulating their sensory perceptions more clearly. This process is simplified by associating specific, agreed-upon terms with particular sensory experiences. For example, in a wine tasting session, a panellist can use a sensory wheel to identify fruity, woody, or spicy notes and link them to validated descriptors previously established by experts in the field. In the wine industry, these wheels are typically memorized and internalized by trained panellists, rather than simply suggested as optional tools.

In sensory and consumer science, many wheels have been developed for describing aromas/odours and flavours/taste of specific products or beverages. Most wheels are product-dependent, as for example the Ann Noble's iconic wine aroma wheel (Noble et al., 1984, 1987), or other beverages such as coffee (Spencer et al., 2016; World Coffee Research I Sensory Lexicon), whisky or bourbon (Bourbon Tasting Wheel; Lee et al., 2001), herbal infusions or tea (Koch et al., 2012; Theron et al., 2014), beer (Parker, 2012), and so on. Other popular food products have their own wheels, such as cheese (Berodier et al., 1997; Heisserer & Iv, 1993), chocolate (De Pelsmaeker et al., 2019), olive oil (Mojet & de Jong, 1994), spices (Lawless et al., 2012), pulses and pulse—derived ingredients (Chigwedere et al., 2022), as well as a variety of plant cultivars and animal-derived food such as beef and pork (Rajic et al., 2022), fishes (Warm et al., 2000), or even caviar (Baker et al., 2014). Additionally, texture descriptors are used to assess or predict mouthfeel, which encompasses a number of inter-related tactile sensations, especially for liquids such as beer, wine, dairy- or fruit-based beverages, and for creamy products such as mayonnaises, sauces, custard desserts and yogurts (Ditschun et al., 2025; Gawel et al., 2018). Other lexicons, taxonomies or wheels exist for hedonic perceptions of food (Panagiotou & Gkatzionis, 2022) and emotional response to wine consumption (Ferrarini et al., 2010).

Emotional lexicons are used in areas like marketing, UX/UI design, and psychology, where sensory experiences are tied to emotional reactions. However, emotion-related sensory descriptors or hedonic evaluations do not yet have strict global standards. Existing resources include Robert Plutchik's Wheel of Emotions (Plutchik, 1980), which organizes basic emotions into eight main or primary categories, often represented in the form of a flower. The web is home to a variety of emotional word wheels, also known as "feeling wheels." One thing to note is that there is not a single wheel that can be considered as universal. However, the Arousal-Valence model of emotions (Mneimne et al., 2010) provides a visual aid that organizes emotions based on their level of arousal (how calm or agitated you feel) and valence (whether positive or negative). Valence refers to the pleasant–unpleasant quality of a stimulus and ranges from negative to positive, whereas arousal refers to the intensity of a stimulus and ranges from dull to arousing (Heilman, 1997).

3.3 Focus on Texture Analysis

Although lexicons and wheels are generally product-specific, there have been attempts to standardize the terms used in sensory description in a generic way. Food texture analysis is a scientific field that originated about a century ago, but it began to gain significant momentum in the early 1960s. Texture analysis involves evaluating a food product based on the sensations it elicits in the mouth before ingestion. Traditionally, this concept has been extended to include upstream perceptions—such as how the product looks and how it feels when held in the hand or positioned on a fork. In practice, all five senses are engaged: touch when handling the product and placing it in the mouth, sight when selecting it, and hearing during chewing. While taste and smell are theoretically excluded from texture analysis, psychorheological studies have shown that their influence is difficult to eliminate in sensory panel testing contexts (Bagot, 1998). A widely accepted definition of texture was proposed by A. Szczesniak: "*Texture is the sensory and functional manifestation of the structural and mechanical properties of foods, detected through the senses of vision, hearing, touch, and kinesthetics.*" (Szczesniak, 1988). This definition is notable for bridging the gap between the two main approaches to texture evaluation: sensory analysis and instrumental measurement. Finally, texture perception in the mouth has been formally defined in the ISO standard for sensory analysis (ISO 5492:2008) as: "*All of the mechanical, geometrical, surface, and body attributes of a product perceptible by means of kinaesthetic and somaesthetic receptors and, where appropriate, visual and auditory receptors, from the first bite to final swallowing.*" From the outset, the field faced challenges related to terminology, particularly in defining the concept of texture within the food industry. One of the first objectives of texture analysis was to name and describe sensory experiences as objectively as possible, that is, to interpret the sensory field through language (Daniel & Roudot, 2007). Each culture has developed its own terminology for the products it commonly uses. While this terminology often suffers from geographical or sensory overlaps, it is usually sufficient in everyday contexts, where meaning is inferred from the situation. However, the terminological problem had to be addressed if food texture is to be treated as a clearly defined and manageable characteristic. This involves distinguishing between concepts that are physically different, and understanding the links between food structure and texture. It also requires deconstructing complex notions into simpler, unitary phenomena, each of which must be clearly named. For example, the term "crispy" is associated with a compound sensation that includes physical elements such as fracture, sound, structure, aspect. Each one of these elements must be precisely described and named in order to reconstruct the textural profile of a food sample clearly and without risk of misunderstanding. It also means creating new culturally-neutral terms that avoid connotations likely to lead to inaccurate sensory description (Roudot, 2004).

Despite these needs, consensus on standardized terminology is still lacking. Some researchers argue that, given the strong influence of the individual's culture on food perception, loosely defined glossary is preferable to allow panellists to express their sensations in their own terms. However, from an analytic perspective, this position is untenable for two main reasons. First, scientific communication requires precision, which is only possible through a shared vocabulary. Second, while texture analysis is rooted in sensory evaluation, it increasingly involves instrumental techniques, particularly given the role of texture in food processing technologies. The sensory (often considered subjective) and instrumental (considered objective) approaches must be aligned, as they are evaluating the same properties using different methods (Szczesniak, 1963). For this reason, a precise and standardized glossary is essential to bridge the two approaches and ensure coherent interpretation and communication of results.

This idea led to the development of a generic wheel and lexicon of food textures, created in collaboration with a dedicated group of experts. The project was initially grounded in the definitions provided by the ISO standard for sensory terminology and further enriched through research drawn from scientific literature and relevant databases. These sources were used to compile a comprehensive list of texture-related attributes (Bondu et al., 2022). A first level of texture descriptor categories was defined, corresponding to body attributes, geometrical attributes, mechanical attributes, and a category for gathering multidimensional texture attributes. Surface attributes listed in the ISO standard were intentionally excluded, as they were considered more relevant to visual perception. New, precisely defined subcategories were then added to capture specific modalities or degrees of perception. For instance, the descriptor "sticky" was refined into more granular terms such as "sticky to the fingers", "sticky in the mouth", "sticky to the lips", "sticky to the palate", "sticky to the spoon", and "sticky to the teeth". Similarly, "hard" was contextualized as "hard in the mouth", "hard to the fingers", or "hard with a utensil", while "firm" included variants such as "firm at first bite", "firm in the mouth" and "firm to the fingers". Figure 4.2 shows the hierarchy of texture-related sensory attributes.

4 Use of Knowledge Organization Systems in Sensory and Consumer Science

4.1 Taxonomies and Thesauri

As previously mentioned, controlled vocabularies and knowledge organization systems exist along a continuum, and the appropriate choice depends on the specific context and use case. At one end of the spectrum, terminologies provide simple lists of terms with minimal semantic structure. They are primarily used for basic labelling and have low semantic richness.

Taxonomies go a step further by categorizing or clustering data into a hierarchical structure of classes and subclasses. Thesauri manage multilingualism and synonymy, making them valuable tools for content analysis and document classification. By incorporating associative relationships and a comprehensive inclusion of

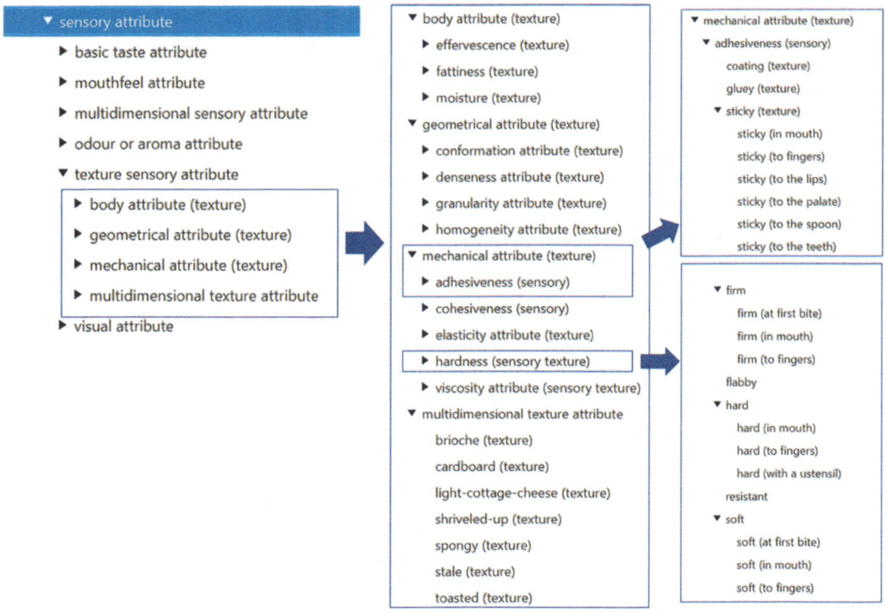

Fig. 4.2 Hierarchy of texture sensory attributes. The first two levels are shown, along with details of the declination of the adhesiveness and hardness attributes (Bondu et al., 2022)

synonyms or non-preferred, thesauri support more effective manual or automated tagging by offering suggestions to indexers or providing contextual cues to classification tools. A relevant standard for creating a thesaurus is the ISO 25964 standard (ISO 25964-1:2011). Part 1 of the standard, published in 2011, covers all aspects of developing a thesaurus, monolingual or multilingual. It essentially advises on the selection and fitting together of concepts, terms and relationships to make a good thesaurus. To encourage networking interoperability, it includes a data model and recommended format for data exchange. Part 2 of the standard, published in 2013, recommends the sort of mappings that can be established between one knowledge organization system and another. SKOS, the Simple Knowledge Organization System, is a W3C recommendation published in 2009 that present a way of expressing thesauri, classifications or other types of controlled vocabularies when published on the Web (SKOS Simple Knowledge Organization System, n.d.).

4.2 *Ontologies and Knowledge Graphs*

At the most formal end of the spectrum are ontologies. They are mainly used in artificial intelligence, to model complex knowledge, support automatic reasoning and facilitate interoperability between systems. In short, ontologies are providing a structured way to organize information, facilitating semantic understanding, while other hierarchically arranged vocabularies such as thesauri and taxonomies are focusing on the standardization of words and expressions, with less formalization and few complex relationships between terms. The current version of OWL, also referred to as OWL2 (OWL Web Ontology Language) is a standard developed by the world wide web consortium (W3C), along with the Resource Description Framework (RDF), the standard model for data interchange on the Semantic Web (RDF Schema 1.1).

Ontologies vary in their level of formalism, ranging from termino-ontologies or lightweight ontologies to formal and foundational ontologies. Importantly, there are semantic artifacts that blur the boundaries between these categories, and it might be difficult to decide if an ontology is lightweight or formal. For instance, a well-structured thesaurus with logically defined relationships may function as an ontology—hence the frequent confusion between the two. Similarly, if an ontology provides a hierarchy of classes but lacks defined relationships among them, it may be more accurately described as a taxonomy. Resources such as WordNet (Fellbaum, 2005) lie on the borderline between a lexicon and an ontology. Termino-ontologies combine both conceptual and lexical components: the ontology represents the domain's concepts, while the terminology links these concepts to terms. This structure enables semantic annotation, document indexing, and supports multilingualism and synonym handling. At the opposite end of the formalism spectrum are folksonomies—a portmanteau of folk and taxonomy. These systems allow users to collaboratively tag digital resources with keywords ("tags") to facilitate retrieval. Also known as collaborative tagging, social classification, or social indexing, folksonomies evolve over time into informal classification systems based on user behaviour, rather than predefined structures. There is ongoing debate among web and information science professionals about the advantages and limitations of folksonomies compared to controlled vocabularies curated by experts. Each approach offers distinct benefits depending on the context—folksonomies encourage user participation and adaptability, while controlled vocabularies ensure consistency, accuracy, and interoperability.

An ontology-based dataset is called a Knowledge Graph (KG). It is a graph-structured data model used to represent, integrate and store interlinked descriptions of entities as a set of subject-predicate-object statements, typically expressed in Resource Description Framework (RDF) format (Hogan et al., 2021). It encodes not only the data but also the semantics of the vocabulary or terminology used, enabling advanced understanding and reasoning. KGs are stored in a central knowledge base and are exploited for a variety of applications, such as integrated data access, semantic search, browsing, querying, information extraction, and insight generation. Ontologies are playing a crucial role in improving semantic understanding.

Since ontologies are used as reference in knowledge bases, they represent a critical resource in the information system (Munn & Smith, 2008). They consist of (i) concepts (or object classes): abstract categories representing types of entities; (ii) semantic relations and custom-defined relationships that connect the concepts; (iii) attributes or properties: characteristics associated with concepts or relationships; (iv) logical rules or axioms: formal expressions that define constraints and logical dependencies. Abstract concepts, also known as universals, define categories or classes of entities and are designated by labels (names or terms) that best express the ideas they represent. Instances (or individuals) are the concrete realization of the abstract concepts. These represent actual entities in the domain that populate the ontology. Ontologies also allow for expressing highly specific relationships between entities. These include taxonomic relationships (i.e., hierarchical relation between classes) and other specific relations that link universals and individuals. Higher-level concepts (superclasses) encompass more general ideas, while more refined concepts (subclasses) inherit all the attributes and constraints of their superclasses. Axioms are formal logical statements that define rules governing classes and properties. Examples include domain and range constraints, class disjointness, value restrictions, and cardinality rules. This formal representation enables data to be organized according to abstract concepts and relationships in a way that is interpretable and actionable by computer systems.

Ontology engineering (also called ontology building) is a set of tasks related to the development of ontologies for a particular domain. It refers to the process of developing ontologies, which involves the use of methodologies, tools, and languages to build ontologies. It includes various development methodologies, such as building ontologies from scratch, reusing existing ontologies, and the distributed construction of ontologies (Ashraf et al., 2015). The Linked Open Terms (LOT) methodology is an overall and lightweight methodology for building ontologies based on existing methodologies and oriented to semantic web developments and technologies in the Linked Data and Agile software development context (Poveda-Villalón et al., 2022).

4.3 Ontologies in Sensory and Consumer Sciences

A large number of ontologies are available in the food domain. In a recent survey published by the European Food Safety Agency (EFSA), 40 ontologies and seven other knowledge organization systems have been collected (Tzitzikas et al., 2024). To date, however, only a few ontologies that can describe all the characteristics of foodstuffs in their entirety are available, and many resources are not maintained.

Among the ontologies that are actively maintained, FoodON is one of the most comprehensive and exhaustive resource, enabling food description from farm to fork (Dooley et al., 2018). FoodOn is part of the open-source OBO Foundry consortium, which consists of interoperable life science-oriented ontologies (Dooley et al., 2024). At its launch in 2018, FoodOn was largely based on LanguaL, a 14-faceted descriptive food indexing system whose origins can be traced back to the U.S. FDA's 1975 CFSAN Factored Food Vocabulary (Ireland & Møller, 2010). Each facet represents a branch of vocabulary devoted to describing some aspects of a food, using standard, controlled terms that describe nutritional and/or hygienic qualities of food, the organism it came from or the part of the organism it mainly consists of, and for multi-component foods, ingredient composition, or the type of process involved in its creation. FoodON relies on the OBO Foundry community of practice (Smith et al., 2007) and is built upon the Basic Formal Ontology (BFO) foundational model (Basic Formal Ontology (BFO), n.d.). Each facet of FoodOn corresponds to a BFO ontology class, generally representing a material entity, quality, information about a food product, or process leading to creation of food. On the other hand, the European Food Safety Agency (EFSA) also developed FoodEx2, a standardized food classification and description system based on European legislations called FoodEx2 (EFSA, 2015). FoodEx2 consists of descriptions of a large number of individual food items aggregated into food groups and broader categories, based on the identification of the most relevant treatments or process to create new natures of products from raw materials. In addition to hierarchical parent-child relationships, FoodEx2 also provides additional descriptors organized into facets, which have been mapped to LanguaL's facets. FoodEx2 is an authoritative reference and is used by national agencies in European countries for database annotation and data exchange (D'Addezio et al., 2024; Jean-Cédric et al., 2013).

To solve the data harmonization problem, a termino-ontological resource called TransformON has recently been developed (Weber et al., 2023). It is based on the FoodEx2 classification and W3C standards, while also aligning with FoodON and the OBO Foundry principles. TransformON consists of two complementary parts: (i) concept hierarchies that form the branches of the ontology described in OWL; and (ii) thesauri available for each branch in SKOS format. The SKOS format supports multilingualism and groups different labels as alternative terms (synonyms) linked to a preferred label, which is designated as an OWL class in the ontology. The ontological part serves as the conceptual model for representing knowledge and is built upon a generic core model called the Process and Observation Ontology (PO2). This model was designed to (i) describe any biomass transformation process and (ii) capture observations made on the process and its input or output products—whether food, non-food, by-products, or residues produced within the food chain. The observation component includes descriptors (or attributes), which represent characteristics or properties observable, measurable, or calculable from experimental measurements or evaluations. PO2 reuses the lightweight Sensor, Observation, Sample, and Actuator (SOSA) ontology (Janowicz et al., 2019), along with other W3C ontologies and standard data schemas such as the Time Ontology in OWL (https://www.w3.org/TR/owltime/), PROV-O (https://www.w3.org/TR/owltime/), QUDT (https://qudt.org/). Other schemata such as schema.org, DCAT (https://www.w3.org/TR/vocab-dcat-3/), and UCUM (https://ucum.org/ucum) are also used to provide additional metadata. Together, these provide a unified architecture to represent provenance, quantities, quantity kinds, units, dimensions, and data types (Fig. 4.3).

Much of the vocabulary is drawn from the "Food" and "Feed" branches of the FoodEx2 classification system. The current version of the ontology (V17.0) gathers 8900 classes, 20 individuals and 94 properties. Figure 4.4 shows the different hierarchies which are specializing each core concept of the PO2 core model.

In addition, terms and concepts essential for knowledge representation in sensory science have also been integrated in PO2/TransformON. Notably, a

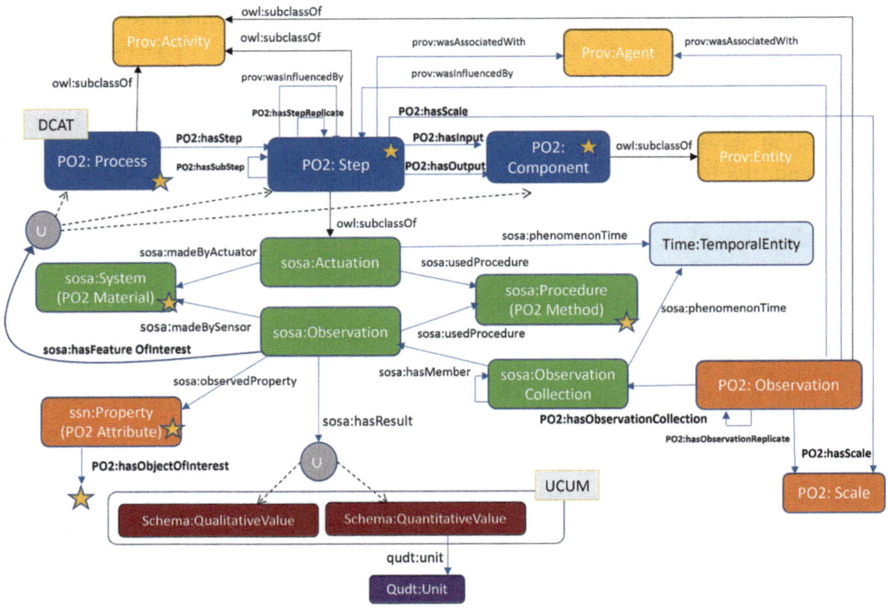

Fig. 4.3 Core model of the process and observation ontology (V2.4)

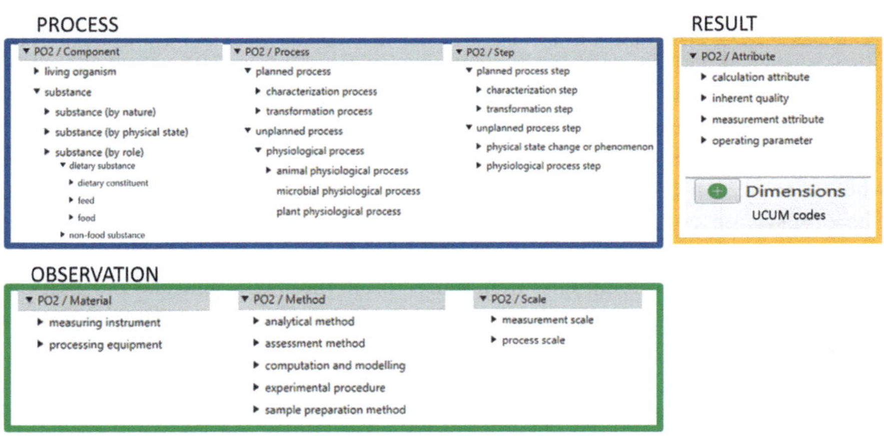

Fig. 4.4 Extract of the PO²/TransformON hierarchies (only the first two levels are shown, except for the Substance hierarchy that displays two additional sub-levels)

data-centric typology of sensory evaluation measures has been included, enabling the classification of sensory evaluation methods according to this typology, along with minimum information standards for reporting sensory measures (Visalli et al., 2023). The generic wheel of textures described in Sect. 3.3 has been incorporated as a branch under "sensory texture attributes", and similar work has been undertaken for flavour attributes. To develop this, terms related to odour and aroma were added to the ontology through a collaborative effort involving a panel of experts and researchers in the field. A generic typology was created based on existing sensory wheels collected from the literature. The process involved analysing the various categories of descriptors to identify commonalities among the aroma wheels. For instance, most wheels include categories such as green or vegetative, fruity, floral, and earthy, although sub-categories and their positions in the wheel may vary. When a category appeared in the majority of the wheels, it was included in the generic version. The same principle was applied to sub-categories—for example, the "Fruity" category includes sub-categories such as citrus, berries, and exotic fruits. To validate the relevance of these categories, experts were asked to classify each descriptor into a main category and its associated sub-category. If a descriptor did not fit the proposed structure, they were invited to suggest a new category. Each expert completed this exercise individually. At the end of the consensualization process, more than 450 descriptors were included into a hierarchy comprising a total of 19 categories and 34 sub-categories. Figure 4.5 shows the hierarchy of concepts which corresponds to a refinement of the current sensory terminology. In this case, we are dealing with a type of taxonomy, since the vagueness of the sensory terminology will be preserved for ease of tasting analysis. Then, when annotating the dataset, the ontology makes it possible to associate the product with the characteristics that describe it, thanks to the relationships defined in the PO2 core model. These relationships also make it possible to associate the material and method used

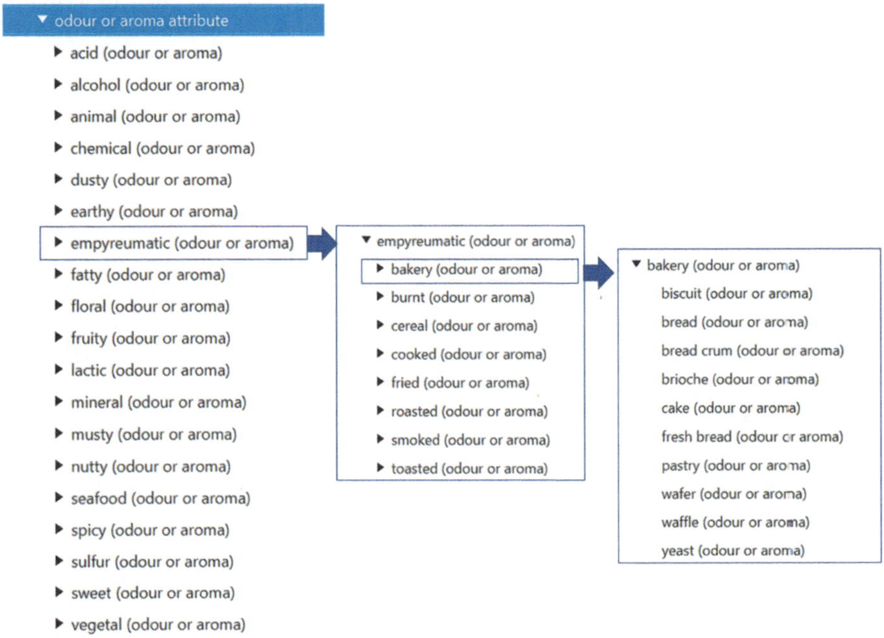

Fig. 4.5 Hierarchy of odour/aroma attributes. The first two levels are shown, along with details of the declination of the empyreumatic category, and two additional sub-levels

for characterization (sensory or instrumental). The ontology thus makes it possible to contextualize the data, and to produce clear, knowledge-based information. Subsequently, we could associate the mechanisms linked to the phenomena observed, generally only attainable by fine instrumental methods. The firmness of a melon, for example, will be expressed instrumentally as "firmness due to high skin elasticity", while the firmness of a cheese will become "firmness due to high viscosity".

These hierarchies constitute the terminological part of the ontology that enables the standardization of terms used in sensory analysis across all food categories. These hierarchies can be exported in SKOS, which allows for flexibility in the vocabulary management. The integration of multiple dimensions relevant to food acceptability—including sensory, nutritional, environmental, economic, practical, social, ethical, and emotional aspects—is also possible as new needs arise.

At the instance level, when describing actual data, the relationships between different concept classes enable linking descriptors to specific food products and associating them with their composition and production methods (see Sect. 4.4).

Although PO2/TransformON includes descriptors for many food characteristics, the current coverage of sensory attributes is largely confined to those derived from food-centric research, where evaluations are treated as analytical measurements. To provide a more comprehensive perspective, a logical framework is currently being developed to formalize knowledge within the broader context of sensory and consumer science. This framework will be implemented in a dedicated ontology module called ConsomON (Weber et al., 2025). ConsomON is designed to structure knowledge and model the biological, cognitive, and behavioural processes involved in consumers' food choices. These processes span the entire decision-making pathway, from initial exposure to food stimuli through to the final act of consumption.

4.4 Ontologies, Data fairification and Food Design

Ontologies are particularly helpful for ensuring data are Findable, Accessible, Interoperable and Reusable, as defined by the FAIR principles (Wilkinson et al., 2016). Ontologies enable improved search and question answering. By using the standardized vocabulary provided by the ontology, data can be better indexed and discovered, enhancing findability. Moreover, the ontology provides clear and machine-readable descriptions of data, making it easier for systems to retrieve and interpret the data across different platforms, thus improving accessibility. Ontologies help resolve ambiguities in language by providing context and distinguishing between different meanings of the same word. By using well-defined terms and structures, ontologies ensure that data are described in a way that others can understand and reuse without ambiguity, thereby enhancing reusability. This disambiguation also improves the reproducibility and long-term value of datasets. When different systems use the same ontology, they can seamlessly share and process data, enabling cross-disciplinary research and analysis, thereby enhancing interoperability.

Facing growing sustainability challenges such as global warming, increased obesity and an ageing population, the food industry must develop new strategies to formulate well-balanced products that meet nutritional requirements (e.g. reduced fat, sugar and salt content), ensure a sustainable supply of animal and vegetable protein and appeal to consumers in terms of their sensory properties, all while using eco-friendly transformation processes. Combining data and knowledge from different scientific domains (e.g., nutrition, sensory perception, eco-design, microbiology, biochemistry, process engineering) with environmental analysis data and knowledge is key to tackling this challenge. To provide answers to sustainability challenges, a database called BaGaTel, guided by the PO2 ontology, has been built to integrate data to reformulate dairy products considering nutritional and sensory properties together with environmental concerns (Pénicaud et al., 2019). The PO2 ontology's common vocabulary and structure allowed data from different projects to be combined and integrated into the knowledge base (KB) providing relevant answers to various questions and proving its suitability as a support tool for the multi-criteria assessment of food systems. The KB made possible to estimate missing data in a dataset supposing that samples have common characteristics, thus enabling the formulation of hard cheese. In terms of environmental concerns, it provided information on relevant data that needed to be acquired and enabled the estimation of missing data. Another work (Guichard et al., 2021a, b) aimed to gather data obtained in different projects that used similar methodologies, and to reach a critical number of samples from similar food products with different compositions. Thanks to the harmonized data structure and vocabulary, the KB enabled data to be gathered from different projects along with the associated metadata (e.g. materials, methods, and experimental conditions).

5 Conclusion and Perspectives

Ontologies belong to the area of Knowledge Representation and Reasoning (KRR), one of the classical areas of Artificial Intelligence (AI). As knowledge bases, they are very useful resources for many artificial intelligence applications (Khadir et al., 2021). However, behind the term AI we used to distinguish Symbolic AI (ontologies, knowledge graphs, and reasoning belong to this category), and Sub-Symbolic AI that includes Machine Learning (ML) and Deep Learning (DL). Modern AI (neural network-based machine learning) has showcased impressive results in NLP (Natural Language Processing), machine translation, image understanding, and other fields. Nevertheless, there are methods that require the combination of pure ML and knowledge-based reasoning, e.g., for the task of entity recognition and linking over texts and images. NLP systems that use ontologies can better understand the meaning of words in context, because they can link terms to their relationships and meaning in a structured domain. In short, ontologies and KGs can be leveraged for (i) improving the quality of training data (since the same data can have different representations, while duplicates and conflicts could be difficult to detect), and (ii) for improving the quality of predictions, e.g., in a scenario for entity and relation disambiguation they can provide links to entities/relationships in a KG (Mountantonakis & Tzitzikas, 2017).

Ontologies are by nature flexible and adaptable as new classes or relationships can be added as scientific disciplines develop, and categories may have different ontological properties in related disciplines. They are at the core of Linked data, as alignment between resources can be achieved thank to mappings between classes or concepts. However, one challenge that remains lies in the fact that changes must be able to propagate swiftly across multiple interrelated classifications. Ontologies serve as a backbone for knowledge graphs, enabling to infer relationships between concepts (knowledge representation and reasoning). When an application understands a given ontology, it can parse the ontology and generate a list of axioms based on it, with all facts expressed as RDF statements. When an application makes inferences, it can add new RDF statements into the existing collection, expanding the knowledge base. The ontology provides a standardized and disambiguated vocabulary. When used to annotate data, the ontology and its associated graph together form a comprehensive knowledge base that can be used for modelling purposes. The ontology is also a powerful tool for gathering detailed information from sensory or instrumental experiments.

Ontologies offer reasoning capabilities based on class inferences. When combined with natural language processing and other machine learning techniques, they facilitate the integration and generation of new knowledge (Mountantonakis & Tzitzikas, 2017; Nhat-Vinh et al., 2020; Perrot et al., 2023; Saini & Ramanathan, 2022; Zushi et al., 2023). This also opens up new possibilities for text mining (Zhao et al., 2025; Zhong & Lotfi, 2020) and natural language querying (Apidianaki et al., 2024; Ashraf et al., 2015).

References

Apidianaki, M., Fourtassi, A., & Padó, S. (2024). Language learning, representation, and processing in humans and machines: Introduction to the special issue. *Computational Linguistics, 50*(4), 1201–1210. https://doi.org/10.1162/coli_e_00539

Ashraf, J., Chang, E., Hussain, O. K., & Hussain, F. K. (2015). Ontology usage analysis in the ontology lifecycle: A state-of-the-art review. *Knowledge-Based Systems, 80*(34), 47. https://doi.org/10.1016/j.knosys.2015.02.026

ASTM. A Sensory Vocabulary | ASTM. https://www.astm.org.

Baker, A. K., Vixie, B., Rasco, B. A., Ovissipour, M., & Ross, C. F. (2014). Development of a lexicon for caviar and its usefulness for determining consumer preference. *Journal of Food Science, 79*(12), S2533 S2541. https://doi.org/10.1111/1750-3841.12703

Basic Formal Ontology (BFO). https://basic-formal-ontology.org/.

Bagot, J. (1998). Information, sensation et perception. FeniXX. ISBN/EAN: 9791037636195

Berodier, F., Lavanchy, P., Zannoni, M., Casals, J., Herrero, L., & Adamo, C. (1997). A guide to the sensory evaluation of smell, aroma and taste of hard and semi-hard cheeses. *Food Science and Technology-Lebensmittel-Wissenschaft & Technologie, 30*(7), 653–664.

Bondu, C., Salles, C., Weber, M., Guichard, E., & Visalli, M. (2022). Construction of a generic and Evolutive Wheel and lexicon of food textures. *Food, 11*(19), 3097. https://doi.org/10.3390/foods11193097

Bourbon Tasting Wheel. The Council of Whiskey Masters: Scotch and Bourbon Certification & Education Program, Home of the Whiskey Sommelier. https://www.whiskeymasters.org/bourbon-tasting-flavor-wheel.

Chigwedere, C. M., Wanasundara, J. P. D., & Shand, P. J. (2022). Sensory descriptors for pulses and pulse-derived ingredients: Toward a standardized lexicon and sensory wheel. *Comprehensive Reviews in Food Science and Food Safety, 21*(2), 999 1023. https://doi.org/10.1111/1541-4337.12893

Civille, G. V., & Lawless, H. T. (1986). The importance of language in describing perceptions. *Journal of Sensory Studies, 1*(3 4), 203 215. https://doi.org/10.1111/j.1745-459X.1986.tb00174.x

Daniel, C., & Roudot, A.-C. (2007). La terminologie de la texture des aliments. *Meta: journal des traducteurs /Meta: Translators' Journal, 52*(2), 342–351. https://doi.org/10.7202/016075ar

D'Addezio, L., Sette, S., Piccinelli, R., Donne, C. L., Turrini, A. (2024). FoodEx2 Harmonization of the Food Consumption Database from the Italian IV SCAI Children's Survey. Nutrients, 16(7), 1065. https://doi.org/10.3390/nu16071065

De Pelsmaeker, S., De Clercq, G., Gellynck, X., & Schouteten, J. J. (2019). Development of a sensory wheel and lexicon for chocolate. *Food Research International, 116*(1183), 1191. https://doi.org/10.1016/j.foodres.2018.09.063

Dextre. (2012). From ISO 2788 to ISO 25964: The evolution of thesaurus standards towards interoperability and data modeling. *Information Standards Quarterly, 24*(1), 20–26. https://doi.org/10.3789/isqv24n1.2012.04

Ditschun, T. L., Riddell, E., Qin, W., Graves, K., Jegede, O., Sharafbafi, N., Pendergast, T., Chidichimo, D., & Wilson, S. F. (2025). Overview of mouthfeel from the perspective of sensory scientists in industry. *Comprehensive Reviews in Food Science and Food Safety, 24*(2), e70126. https://doi.org/10.1111/1541-4337.70126

Dooley, D. M., Griffiths, E. J., Gosal, G. S., Buttigieg, P. L., Hoehndorf, R., Lange, M. C., Schriml, L. M., Brinkman, F. S. L., & Hsiao, W. W. L. (2018). FoodOn: A harmonized food ontology to increase global food traceability, quality control and data integration. *npj Science of Food, 2*(1), 23. https://doi.org/10.1038/s41538-018-0032-6

Dooley, D., Andrés-Hernández, L., Bordea, G., Carmody, L., Cavalieri, D., Chan, L., Castellano-Escuder, P., Lachat, C., Mougin, F., Vitali, F., Yang, C., Weber, M., McGinty, H. K., Lange, M. (2024). OBO Foundry food ontology interconnectivity. Semantic Web, 15(4), 1239–1258. https://doi.org/10.3233/sw-233458

Drake, M. A., & Civille, G. V. (2003). Flavor Lexicons. *Comprehensive Reviews in Food Science and Food Safety, 2*(1), 33–40. https://doi.org/10.1111/j.1541-4337.2003.tb00013.x

European Food Safety Authority (EFSA). (2015). The food classification and description system FoodEx 2 (revision 2). *EFSA Supporting Publications, 12*(5). https://doi.org/10.2903/sp.efsa.2015.EN-804

Fellbaum, C. (2005). Wordnet and Wordnets. In K. Brown (Ed.), *Encyclopedia of language and linguistics* (pp. 2–665). Elsevier.

Ferrarini, R., Carbognin, C., Casarotti, E. M., Nicolis, E., Nencini, A., & Meneghini, A. M. (2010). The emotional response to wine consumption. *Food Quality and Preference, 21*(7), 720–725. 8th Pangborn Sensory Science Symposium. https://doi.org/10.1016/j.foodqual.2010.06.004

Flavor Wheel for Bread. Michael Kalanty. https://www.michaelkalanty.com/flavor-wheel-for-bread

Gawel, R., Smith, P. A., Cicerale, S., & Keast, R. (2018). The mouthfeel of white wine. *Critical Reviews in Food Science and Nutrition, 58*(17), 2939–2956. https://www.tandfonline.com/doi/full/10.1080/10408398.2017.1346584

Guichard, E., Thomas-Danguin, T., Buchin, S., Perret, B., Guillemin, H., Pénicaud, C., & Salles, C. (2021a). Compilation of data on model cheeses composition, rheological and sensory properties, from six research projects exported from the BaGaTel database. *Data in Brief, 36*, 106971. https://doi.org/10.1016/j.dib.2021.106971

Guichard, E., Thomas-Danguin, T., Buchin, S., Perret, B., Guillemin, H., Pénicaud, C., & Salles, C. (2021b). Relationships between cheese composition, rheological and sensory properties highlighted using the BaGaTel database. *International Dairy Journal, 118*, 105039. https://doi.org/10.1016/j.idairyj.2021.105039

Heilman, K. M. (1997). The neurobiology of emotional experience. *The Journal of Neuropsychiatry and Clinical Neurosciences, 9*(3), 439–448. https://doi.org/10.1176/jnp.9.3.439

Heisserer, D. M., & Iv, E. C. (1993). Determination of the sensory flavor attributes of aged natural cheese. *Journal of Sensory Studies, 8*(2), 121–132. https://doi.org/10.1111/j.1745-459X.1993.tb00207.x

Hogan, A., Blomqvist, E., Cochez, M., D'amato, C., Melo, G. D., Gutierrez, C., Kirrane, S., Gayo, J. E. L., Navigli, R., Neumaier, S., Ngomo, A.-C. N., Polleres, A., Rashid, S. M., Rula, A., Schmelzeisen, L., Sequeda, J., Staab, S., & Zimmermann, A. (2021). Knowledge graphs. *ACM Computing Surveys, 54*(4), 71:1–71:37. https://doi.org/10.1145/3447772

INRAE (2024), Les types de vocabulaires et leurs usages. Accessed 5 July 2025, https://vocabulairesouverts.inrae.fr/types-de-vocabulaires-et-usages/

Ibekwe-SanJuan, F., & Bowker, G. C. (2017). Implications of big data for knowledge organization. *Knowledge Organization, 44*, 187–198. Special issue on new trends for Knowledge Organization, Renato Rocha Souza (guest editor). https://hal.science/hal-01489030v1

Ireland, J. D., & Møller, A. (2010). LanguaL food description: A learning process. *European Journal of Clinical Nutrition, 64*(S3), S44–S48. https://doi.org/10.1038/ejcn.2010.209

ISO 25964-1:2011. Information et documentation—Thésaurus et interopérabilité avec d'autres vocabulaires—Partie 1: Thésaurus pour la recherche documentaire. https://www.iso.org/fr/standard/53657.html

ISO 5492:2008. *Analyse sensorielle—Vocabulaire*. https://www.iso.org/fr/standard/38051.html

Janowicz, K., Haller, A., Cox, S. J. D., Le Phuoc, D., & Lefrançois, M. (2019). SOSA: A lightweight ontology for sensors, observations, samples, and actuators. *Journal of Web Semantics, 56*(1), 10. https://doi.org/10.1016/j.websem.2018.06.003

Khadir, A. C., Aliane, H., & Guessoum, A. (2021). Ontology learning: Grand tour and challenges. *Computer Science Review, 39*, 100339. https://doi.org/10.1016/j.cosrev.2020.100339

Khosrow-Pour, D.B.A. M. (Éd.). (2024). Encyclopedia of information science and technology, 6th: IGI Global. https://doi.org/10.4018/978-1-6684-7366-5.

Koch, I. S., Muller, M., Joubert, E., Van Der Rijst, M., & Næs, T. (2012). Sensory characterization of rooibos tea and the development of a rooibos sensory wheel and lexicon. *Food Research International, 46*(1), 217–228. https://doi.org/10.1016/j.foodres.2011.11.028

Lawless, L. J. R., & Civille, G. V. (2013). Developing lexicons: A review. *Journal of Sensory Studies, 28*(4), 270–281. https://doi.org/10.1111/joss.12050

Lawless, L. J. R., Hottenstein, A., & Ellingsworth, J. (2012). The Mccormick spice Wheel: A systematic and visual approach to sensory lexicon development. *Journal of Sensory Studies, 27*(1), 37–47. https://doi.org/10.1111/j.1745-459X.2011.00365.x

Lee, K. Y. M., Paterson, A., Piggott, J. R., & Richardson, G. D. (2001). Origins of flavour in whiskies and a revised flavour wheel: A review. *Journal of the Institute of Brewing, 107*(5), 287–313. https://doi.org/10.1002/j.2050-0416.2001.tb00099.x

Mneimne, M., Powers, A. S., Walton, K. E., Kosson, D. S., Fonda, S., & Simonetti, J. (2010). Emotional valence and arousal effects on memory and hemispheric asymmetries. *Brain and Cognition, 74*(1), 10–17. https://doi.org/10.1016/j.bandc.2010.05.011

Mojet, J., & de Jong, S. (1994). The sensory wheel of virgin olive oil. *Grasas y Aceites, 45*(1–2), 42. https://doi.org/10.3989/gya.1994.v45.i1-2.967

Mountantonakis, M., & Tzitzikas, Y. (2017). How linked data can aid machine learning-based tasks. In I. J. Kamps, G. Tsakonas, Y. Manolopoulos, L. Iliadis, & I. Karydis (Eds.), *Research and advanced Technology for Digital Libraries* (pp. 155–168). Springer International Publishing. https://doi.org/10.1007/978-3-319-67008-9_13

Munn, K., & Smith, B. (2008). *Applied ontology: An introduction*. De Gruyter.. https://doi.org/10.1515/9783110324860

National Information Standards Organization. (2005). *ANSI/NISO Z39.19-2005 (R2010), Guidelines for the Construction, Format, and Management of Monolingual Controlled Vocabularies*. NISO. https://doi.org/10.3789/ansi.niso.z39.19-2005R2010

Nhat-Vinh, L., Tansuchat, R., Yuizono, T., & Van-Nam, H. (2020). Incorporating active learning into machine learning techniques for sensory evaluation of food. *International Journal of Computational Intelligence Systems, 13*(1), 655–662. https://doi.org/10.2991/ijcis.d.200525.001

Noble, A. C., Arnold, R. A., Masuda, B. M., Pecore, S. D., Schmidt, J. O., & Stern, P. M. (1984). Progress towards a standardized system of wine aroma terminology. *American Journal of Enology and Viticulture, 35*(2), 107–109. https://doi.org/10.5344/ajev.1984.35.2.107

Noble, A. C., Arnold, R. A., Buechsenstein, J., Leach, E. J., Schmidt, J. O., & Stern, P. M. (1987). Modification of a standardized system of wine aroma terminology. *American Journal of Enology and Viticulture, 38*(2), 143–146. https://doi.org/10.5344/ajev.1987.38.2.143

OBO Foundry. http://obofoundry.org/

OWL Web Ontology Language. https://www.w3.org/TR/owl-guide/

Panagiotou, M., & Gkatzionis, K. (2022). Lexicon development to measure emotions evoked by foods: A review. *Measurement: Food, 7*, 100054. https://doi.org/10.1016/j.meafoo.2022.100054

Parker, D. K. (2012). Chaoter 6: Beer: Production, sensory characteristics and sensory analysis. In J. Piggott (Ed.), *Alcoholic beverages* (pp. 133–158). Woodhead Publishing. https://doi.org/10.1533/9780857095176.2.133

Pénicaud, C., Ibanescu, L., Allard, T., Fonseca, F., Dervaux, S., Perret, B., Guillemin, H., Buchin, S., Salles, C., Dibie, J., & Guichard, E. (2019). Relating transformation process, eco-design, composition and sensory quality in cheeses using PO2 ontology. *International Dairy Journal, 92*(1), 10. https://doi.org/10.1016/j.idairyj.2019.01.003

Pennington, J. A. T., & Butrum, R. R. (1991). Food descriptions using taxonomy and the 'Langual' system. *Trends in Food Science & Technology, 2*(285), 288. https://doi.org/10.1016/0924-2244(91)90723-V

Perrot, N. M., Roche, A., Tonda, A., Lutton, E., & Thomas-Danguin, T. (2023). Predicting odor profile of food from its chemical composition: Towards an approach based on artificial intelligence and flavorists expertise. *Mathematical Biosciences and Engineering: MBE, 20*(12), 20528–20552. https://doi.org/10.3934/mbe.2023908

Plutchik, R. (1980). A general psychoevolutionary theory of emotion (p. 3–33). https://doi.org/10.1016/b978-0-12-558701-3.50007-7

Poveda-Villalón, M., Fernández-Izquierdo, A., Fernández-López, M., & García-Castro, R. (2022). LOT: An industrial oriented ontology engineering framework. *Engineering Applications of Artificial Intelligence, 111*, 104755. https://doi.org/10.1016/j.engappai.2022.104755

Rajic, S., Simunovic, S., Djordjevic, V., Raseta, M., Tomasevic, I., & Djekic, I. (2022). Quality multiverse of beef and pork meat in a single score. *Foods, 11*(8), Article 8. https://doi.org/10.3390/foods11081154

Reninger J.C., Bruschini, J., Naudin, S., Denys, S. (2013). On the implementation of FoodEx2 as part of the Standard Sample Description for the electronic transmission of harmonised chemical occurrence data to EFSA. EFSA Supporting Publications, 10(5).https://doi.org/10.2903/sp.efsa.2013.en-431

RDF Schema 1.1. https://www.w3.org/TR/rdf-schema/

Roudot, A.-C. (2004). Terminology in food quality management. In *Proceedings of the 5th international conference on quality, reliability, and maintenance, QRM*.

Saini, K., & Ramanathan, V. (2022). Predicting odor from molecular structure: A multi-label classification approach. *Scientific Reports, 12*(1), 13863. https://doi.org/10.1038/s41598-022-18086-y

Semantic Web Standards. https://www.w3.org/2001/sw/wiki/Main_Page

Shannon, C. E., & Weaver, W. (1949). *The mathematical theory of communication*. The University of Illinois Press.

SKOS Simple Knowledge Organization System. (n.d.). https://www.w3.org/2004/02/skos/

Spencer, M., Sage, E., Velez, M., & Guinard, J.-X. (2016). Using single free sorting and multivariate exploratory methods to design a new coffee taster's flavor Wheel. *Journal of Food Science, 81*(12), S2997–S3005. https://doi.org/10.1111/1750-3841.13555

Smith, B., Ashburner, M., Rosse, C., Bard, J., Bug, W., Ceusters, W., Goldberg, L. J., Eilbeck, K., Ireland, A., Mungall, C. J., Leontis, N., Rocca-Serra, P., Ruttenberg, A., Sansone, S., Scheuermann, R. H., Shah, N., Whetzel, P. L., Lewis, S. (2007). The OBO Foundry : coordinated evolution of ontologies to support biomedical data integration. Nature Biotechnology, 25(11):1251–1255.https://doi.org/10.1038/nbt1346

Stock, W. G., & Stock, M. (2013). Handbook of information science: De Gruyter Saur. https://doi.org/10.1515/9783110235005.

Stuart, D. (2016). *Practical ontologies for information professionals*. Facet Publishing. https://dokumen.pub/practical-ontologies-for-information-professionals-9781783301522-9781783300624.html

Studer, R., Benjamins, V. R., & Fensel, D. (1998). Knowledge engineering: Principles and methods. *Data & Knowledge Engineering, 25*(1 2), 161 197. https://doi.org/10.1016/S0169-023X(97)00056-6

Szczesniak, A. S. (1963). Classification of Textural Characteristicsa. Journal Of Food Science, 28(4), 385–389. https://doi.org/10.1111/j.1365-2621.1963.tb00215.x

Szczesniak, A. S. (1988). The meaning of textural characteristics - crispness. Journal Of Texture Studies, 19(1): 51-59. https://doi.org/10.1111/j.1745-4603.1988.tb00924.x

Theron, K. A., Muller, M., van der Rijst, M., Cronje, J. C., le Roux, M., & Joubert, E. (2014). Sensory profiling of honeybush tea (Cyclopia species) and the development of a honeybush sensory wheel. *Food Research International, 66*(12), 22. https://doi.org/10.1016/j.foodres.2014.08.032

Tzitzikas, Y., Marketakis, Y., Fafalios, P., Kritsotaki, A., Axaridou, A., Mountantonakis, M., Theodoridou, M., & Bekiari, C. (2024). Ontologies and case studies. *EFSA Supporting Publications, 21*(12). https://doi.org/10.2903/sp.efsa.2024.EN-9120

Visalli, M., Schlich, P., Mahieu, B., Thomas, A., Weber, M., & Guichard, E. (2023). First steps towards FAIRization of product-focused sensory data. *Food Quality and Preference, 104*, 104765. https://doi.org/10.1016/j.foodqual.2022.104765

Warm, K., Nelsen, J., Hyldig, G., & Martens, M. (2000). Sensory quality criteria for five fish species. *Journal of Food Quality, 23*(6), 583–601. https://doi.org/10.1111/j.1745-4557.2000.tb00583.x

Weber, M., Buche, P., Ibanescu, L., Dervaux, S., Guillemin, H., Cufi, J., Visalli, M., Guichard, E., & Pénicaud, C. (2023). PO2/TransformON, an ontology for data integration on food, feed, bioproducts and biowaste engineering. *npj Science of Food, 7*(1), 47. https://doi.org/10.1038/s41538-023-00221-2

Weber, M., Duclos, F., Guillemin, H., Dervaux, S., Cufi, J., & Visalli, M. (2025). *ConsomON, an ontology for structuring knowledge and modeling consumer food behavior*. Integrated Food Ontogology Workshop (IFOW), JOWO 2025; FOIS 2025, Sep 2025, Catane, Italy. ⟨hal-05146634⟩.

Wilkinson, M. D., Dumontier, M., Aalbersberg, I. J., Appleton, G., Axton, M., Baak, A., Blomberg, N., Boiten, J., Da Silva Santos, L. B., Bourne, P. E., Bouwman, J., Brookes, A. J., Clark, T., Crosas, M., Dillo, I., Dumon, O., Edmunds, S., Evelo, C. T., Finkers, R.,. . . Mons, B. (2016). The FAIR Guiding Principles for scientific data management and stewardship. Scientific Data, 3(1).https://doi.org/10.1038/sdata.2016.18

World Coffee Research | Ensuring the future of coffee. https://worldcoffeeresearch.org/

Yu, L. (2014). *A developer's guide to the semantic web*. Springer Berlin Heidelberg. https://doi.org/10.1007/978-3-662-43796-4

Zhao, Q., Xia, Y., Long, Y., Xu, G., & Wang, J. (2025). Leveraging sensory knowledge into text-to-text transfer transformer for enhanced emotion analysis. *Information Processing & Management, 62*(1), 103876. https://doi.org/10.1016/j.ipm.2024.103876

Zhong, J. & Lotfi, A. (2020). Sensor2Vec: An embedding learning for heterogeneous sensors for activity classification. 2020 international symposium on community-centric systems (CCS), New York. https://doi.org/10.1109/ccs49175.2020.9231478.

Zushi, N., Takeuchi, G., Ogawa, M., Gotow, N., Kakeya, H., Koabayakawa, T., & Ayabe-Kanamura, S. (2023). A new systematic collection and classification of odour words by using a product review dataset. *PLoS One, 18*(8), e0289368. https://doi.org/10.1371/journal.pone.0289368

Chapter 5
Transforming Raw Text into Structured Data: An Overview of Content Analysis and Natural Language Processing Techniques for Sensory and Consumer Science

Michel Visalli

1 Introduction

In sensory and consumer science, understanding consumer perceptions, preferences, and behaviours is essential for developing successful products and improving customer satisfaction. Traditionally, such insights have been obtained through verbal-based methods (Valentin et al., 2012), which typically involve closed-ended questions targeting specific aspects of the consumer experience.

In recent years, verbal methods have been enriched with new approaches that enable consumers to express their feedback in their own words (Bécue-Bertaut, 2014). Indeed, valuable information can be found in text data derived from consumer feedback collected experimentally or expressed spontaneously, ranging from survey responses to online product reviews or social media comments and forum discussions. These data sources offer rich, direct insights into consumer experiences and can reveal subtle nuances that closed-ended questionnaires may miss.

Unlike structured data such as numerical values or categorical responses predefined by the experimenter, text data must be pre-processed before they can be analysed statistically, as the relevant features (e.g., sensory or affective attributes, consumer attitudes, etc.) need to be identified and extracted from the raw text. Traditionally, manual content analysis was the go-to method for exploiting text data, relying on human coders to identify, categorize, and interpret relevant content. While effective in certain contexts, this method is labour-intensive, prone to human error, and therefore not well-suited for handling large datasets. With technological

M. Visalli (✉)
Centre for Taste and Feeding Behavior (CSGA), National Research Institute for Agriculture, Food and the Environment (INRAE), Dijon, France
e-mail: michel.visalli@inrae.fr

M. Visalli, M. V. Galmarini (eds.), *Collection and Analysis of Text Data in Sensory and Consumer Science*, https://doi.org/10.1007/978-3-032-04851-6_5

advances, Natural Language Processing (NLP) has emerged as a powerful alternative, offering automated solutions for processing text. NLP techniques use algorithms and statistical models to clean, structure, and extract meaningful information from raw text, transforming it into a format appropriate for quantitative analysis. Thus, these tools enable more efficient, scalable and systematic exploration of text data. However, given the wide range of available NLP approaches, the choice of technique can significantly influence the outcomes an analysis, as the results of preprocessing largely determine its semantics (Balbi & Triunfo, 2013).

This chapter explores key techniques used in sensory and consumer science to convert raw text data into structured formats. It covers manual, semi-automated, and automated methods, highlighting how these approaches can be used in combination to enhance data preparation. While the focus remains on preprocessing rather than advanced statistical analysis, the chapter offers a practical foundation for working with text data in sensory and consumer science. Readers will gain a clear understanding of how to apply both content analysis and NLP methods within the field. The chapter concludes by examining future directions in text preprocessing, especially in the context of emerging artificial intelligence (AI) technologies.

2 Data Collection: From Natural Language to Text Data

2.1 What Is Natural Language?

According to Boyd and Schwartz (2021), "*Human (natural) language is truly fascinating. When we speak, our words provide structure to an otherwise intangible relationship between the self and the world. When we write, we give shape to the very substance of our innermost, subjective experiences. Language is quintessential human potential: it has relatively fixed rules and forms, yet we can string together a combination of words that are completely unique, formulating ideas that have never been shared before—not ever—in the entire history of humankind. […] Language is at once obvious and perplexing. We can write epic novels, sing songs, and communicate intimate thoughts and feelings through language, all without explicit awareness of the psycholinguistic machinery at play.*"

In the context of sensory and consumer science, natural language refers to the way consumers express their thoughts, perceptions, experiences, and attitudes using the everyday language they naturally use in speech or writing. When responding to open-ended questions in natural language, consumers often provide more authentic answers, unbounded by the limitations of predefined numerical or categorical scales. It encourages deeper reflection, making the task more engaging for respondents. By allowing such open-ended responses, natural language helps reduce several common biases often associated with structured response formats (see Choi & Pak, 2004, for a list of the most common biases associated to questionnaires). It avoids priming bias, which occurs when participants are influenced by the provided answer options and may select choices they might not have otherwise considered. It also eliminates acquiescence bias, where respondents tend to agree with statements regardless of their true beliefs, and dumping bias, where respondents choose the closest available option when their exact answer is not included among the options. Furthermore, natural language removes the impact of order effect, where the order of options influences responses, and limits straightlining, characterized by respondents selecting the same answer, as well as central tendency bias, where extreme points on a scale are avoided. Overall, it may help mitigate social desirability bias, where respondents provide answers they believe are more socially acceptable.

Natural language allows for great flexibility in how consumers interpret and respond to questions, making it particularly valuable for exploring complex or multidimensional constructs, especially in contexts where consumer perspectives vary widely. Respondents can frame their answers in ways that are personally meaningful, offering a more personalized and nuanced view of their experiences. They can also express emotions, sentiments or attitudes that go beyond liking. More generally, consumers may report aspects of a product or experience that experimenters may not have anticipated, revealing insights that might otherwise go unnoticed. Moreover, natural language responses often come with additional context, giving researchers a deeper understanding of the "why" behind consumer opinions. Finally, text data can be curated and aggregated from different datasets related to the same objects (as often done with online reviews), potentially leading to the uncovering of new insights through data reuse or data-driven research (Visalli et al., 2023b).

2.2 How Are Natural Language Responses Transcribed into Text Data?

Natural language responses to open-ended questions are generally recorded through computerized programs in typed text, which is already in a usable format and can be directly processed. Written text, such as responses on paper forms or surveys, needs to be scanned and converted into digital format for processing. This can involve optical character recognition technology to transcribe handwritten text into machine-readable text. When responses are initially recorded in audio format, speech-to-text technology (Mahieu et al., 2019) can be used to transcribe the spoken language into text data. However, both handwritten text and transcription from audio may introduce inaccuracies, such as misinterpretations of handwriting or speech.

In addition to direct data collection, natural language responses can also be obtained through web scraping, a technique used to automatically extract text data from websites such as online reviews, blogs, forums, tweets, or social media platforms (Lotfi et al., 2021). This approach allows researchers to gather large volumes of unsolicited, real-world consumer feedback (see Part V of this book for case studies). Text data collected from web scrapping typically requires additional specific preprocessing, including the conversion of HTML into plain text and the handling of platform-specific formatting. Furthermore, legal and ethical considerations (e.g., respecting terms of service, privacy policies, and data ownership) must also be carefully addressed.

3 Text Data Preprocessing: From Raw to Structured Data

3.1 What Is Text Preprocessing?

Unlike responses from close-ended questions, text data do not follow a standardized format that can be directly exported into a structured, tabular form suitable for immediate analysis. The goal of preprocessing is to transform noisy, ambiguous, inconsistent, and unstructured raw text into structured, ready-to-analyse data. In this process, the response variables (i.e., the features deemed relevant for inclusion in the quantitative statistical analysis), are identified *a posteriori*, that is, during the preprocessing stage, rather than being predefined by the researcher. For example, in a free comment descriptive sensory evaluation task (ten Kleij & Musters, 2003), one observation consists of text data feedback provided by a single panellist evaluating a specific product. This observation can be represented as a vector in the form [panellist, product, attribute 1, attribute 2, …, attribute j], where each attribute is encoded as 1 if it was identified in the text data and 0 otherwise. The complete set of observations can then be aggregated into a data table, with each row representing an individual observation and each column corresponding to either an independent variable (panellist and product) or a binary response variable (attribute). This structure, presented in Table 5.1, is comparable to the data format obtained using methods such as Check-All-That-Apply (CATA, Adams et al., 2007).

Table 5.1 Example of text data to after preprocessing of free comment data

Panellist	Product	Attribute 1	Attribute 2	…	Attribute J
Panellist 1	Product 1	1	0	…	1
Panellist 1	Product 2	1	1	…	1
Panellist 1	…	…	…	…	…
Panellist 1	Product P	1	0	…	1
…	…	…	…	…	…
Panellist I	Product 1	0	0	…	1
Panellist I	Product 2	1	0	…	1
Panellist I	…	…	…	…	…
Panellist I	Product P	1	1	…	0

3.2 What Type of Features Can Be Extracted from Text Data?

Features extracted from text data refer to words, word groups, or broader concepts related to product or consumer characteristics that are identified within the text and considered valuable for understanding the consumer experience. Simple features typically consist of attributes that directly capture consumer experiences toward a product, such as sensory descriptors (e.g., "sweet", "fat", "creamy", "crispy", or "bitter"), hedonic terms (e.g., "liked", "unpleasant"), emotional expressions (e.g., "boring", or "surprising"), or other type of comments (e.g., "expensive", "convenient", "filling"). Simple features can also consist of compound words or expressions (e.g., "easy to open", "long-lasting", "red fruit").

Complex features may involve combinations of attributes and contextual words that convey the intensity of a perception (e.g., "very sweet", "not crispy"), the level of appreciation (e.g., "too bitter", "not creamy enough"), specify the sensory modality associated to the attribute (e.g., "fat taste", "fat texture"), or include multiple elements together (e.g., "very fat texture"). Complex features can also result from grouping several individual features into broader conceptual features not explicitly stated in the text data, for instance, coding "raspberry" and "strawberry" into "red fruit", or "too sweet" and "very sweet" into "sweet+".

The degree of granularity in feature extraction refers to how detailed the list of features is. It determines the preprocessing operations to be performed and can influence the selection of specific techniques. Therefore, it should be chosen according to the nature of the text data and the research objectives. Fine-grained feature extraction offers a more detailed interpretation of the text, capturing contextual subtleties, but requires considerable preprocessing effort and may dilute information, potentially reducing the statistical power of subsequent analyses. In contrast, a coarse-grained approach simplifies preprocessing and provides a broader view of the text, but may overlook important details (Visalli et al., 2025).

3.3 Why Is Text Preprocessing Challenging?

As presented, text data can originate from two main sources: experimental data and online data. Experimental data is purposefully collected under controlled conditions, allowing participants to receive specific instructions during data collection that may help simplify subsequent preprocessing, such as entering only words or adjectives into separate text boxes rather than writing full sentences in a single field. In contrast, online data is collected from public platforms and can vary widely in format and quality.

Text data is most often noisy. It frequently contains inconsistent punctuation and typographical errors that must be corrected without altering the general meaning of the sentences. Additionally, it typically includes irrelevant or redundant content that needs to be filtered out to avoid interfering with subsequent statistical analyses.

Text data is inherently unstandardized. Consumers may use slang, informal language, abbreviations, regional dialects, emojis, sarcasm, irony, metaphors, or context-specific meanings that pose challenges for most automated preprocessing techniques. Different individuals may describe similar experiences using synonymous words or expressions, complicating efforts to consolidate responses into unified categories. Natural language responses are qualitative by nature, making it difficult to develop a quantitative understanding of subjective experiences.

Text data is also ambiguous. Polysemy occurs when a single word has multiple meanings, while homographs are words spelled identically but with different meanings. For example, in the sentence "*The sharp taste of the lemon zest contrasted well with the sweetness of the cake*", the word "sharp" describes flavour intensity; whereas in "*The knife is sharp enough to easily slice the lemon*" it refers to the quality of the blade. Meanings can also vary by cultural background or personal experience: for some consumers, terms like "spicy", "hot", and "burning" may be interchangeable, while for others, "spicy" specifically refers to seasoning. Similarly, "sweet" might refer to taste, texture, or preference, and "light" could describe texture, sensory intensity, or brightness.

Lastly, the existence of numerous natural languages presents challenges for cross-cultural research, with several languages still underrepresented or completely unsupported by current NLP tools (Wang, 2023).

4 Using Content Analysis for Manual Feature Extraction

4.1 What Is Content Analysis?

Before the emergence of NLP, only humans were capable of reliably extracting features from text data. In manual coding, all decisions regarding data transformations and feature extraction are manually made by human analysts. In sensory consumer science, it is estimated that 50–90% of the overall duration of the project is spent on manual preprocessing the data (Jaeger & Rasmussen, 2021).

Content analysis is a widely used technique to identify patterns and extract features from raw text. It is often considered the gold standard when the text corpus is manageable for human analysis (Antons et al., 2020). A wide range of books on qualitative research methods include comprehensive overviews of content analysis procedures which can be applied to preprocess text data in sensory and consumer science (Bengtsson, 2016; Erlingsson & Brysiewicz, 2017; Krippendorff, 2019). However, no universal step-by-step protocol exists.

Deductive coding begins by identifying relevant features based on existing theories or prior research (Fife & Gossner, 2024). This approach is typically employed when the researcher has a clear theoretical framework or set of predefined categories and wants to test those concepts in the data. However, in practice, inductive, bottom-up approaches are more commonly employed (Thomas, 2006), where relevant features naturally emerge from the text through a reflective, flexible, and iterative process involving multiple rounds of coding to refine and achieve satisfactory results. This approach, consistent with grounded theory (Glaser & Strauss, 2017), provides a flexible way to extract relevant features without predefined categories, which explains why it has been widely applied in sensory and consumer science (Aguirre et al., 2019; Arellano-Covarrubias et al., 2022; Ares et al., 2010a, b, c, 2008, 2011, 2015, 2016; de Andrade et al., 2016; Olson, 2017; Vidal et al., 2015).

4.2 *What Are the Main Steps Involved in Manual Feature Extraction?*

4.2.1 Initial Human Impregnation (Familiarization with Context)

Before diving into formal coding, it is crucial for annotators to immerse themselves in the context of the text data and familiarize themselves with the nature and structure of the descriptions provided by consumers. To start, annotators should select and read a representative subset of descriptions. While reading, they should take preliminary notes on the types of information present in the text. This approach enables them to gain a comprehensive understanding of the data's quality, complexity, and granularity. It also ensures that subsequent decisions on standardization rules are based on a solid understanding of the data, providing a foundation for more systematic and analytical steps in data preprocessing.

4.2.2 Definition of Guidelines for Coding Text Data

Manual processing is prone to human error, as factors like fatigue or lapses in judgment can cause inconsistencies within the same annotator over time. To minimize these issues, annotators should be provided with clear and detailed guidelines that they can refer to at any time. Since content analysis demands domain expertise, the initial coding framework ought to be developed by team members with relevant skills and experience in data preprocessing. These guidelines should explicitly outline the types of information to retain or not, and provide instructions on how to handle ambiguous cases. Including examples of complex descriptions can better prepare annotators for various situations.

4.2.3 Training and Calibration of Annotators

In addition to establishing clear guidelines, building a strong foundation for annotators is essential to ensure the reliability of the manual coding process. Proper training is crucial to help annotators fully understand the task. Training sessions should use a sample subset of the text data (typically around 10% of the full corpus of text data) allowing annotators to become familiar with both the guidelines and the coding procedures. These sessions also create a space for annotators to discuss challenges, refine decision-making criteria, update the guidelines as needed, and ensure consistent alignment across the team.

Calibration exercises, in which multiple annotators independently code the same subset of data, play a crucial role in ensuring consistency by aligning interpretations and minimizing inter-annotator variability. After coding, the results should be compared, and any discrepancies discussed to guarantee a shared understanding among annotators. A common method to assess this consistency is by calculating inter-coder agreement using the Kappa coefficient (McHugh, 2012).

To improve inter-coder agreement and enhance the reliability of annotations, Bayerl and Paul (2011) recommend using a limited number of features, selecting annotators with similar domain expertise, providing intensive training, involving large groups of annotators (five or more for highly critical tasks, three or four for less critical ones), and resolving disagreement based on consensus rather than pairwise or majority methods.

4.2.4 Manual Feature Extraction

Bengtsson (2016) outlines the content analysis process in four stages: decontextualization, recontextualization, categorization and compilation.

Decontextualization involves the researcher carefully reading through the data and selecting meaning units, which are the smallest fragments of text that can convey information. A meaning unit can be a word, a group of words, or, more rarely, a sentence. Once the meaning units are identified, they are often annotated or tagged with initial codes. In cross-cultural research, this step also includes translation and back translation to ensure consistency across different languages (Brislin, 1970; Son et al., 2014).

Recontextualization involves interpreting how meaning units relate to the specific research question, often leading to the exclusion of those that are not relevant.

Categorization entails organizing retained meaning units into homogeneous groups based on common characteristics or concepts. This goes beyond synonymizing, as it requires subjective judgment and interpretive analysis. In deductive coding, categories are established beforehand, grounded in existing theories or literature. Conversely, in inductive coding, categorization is a dynamic, iterative process where researchers continuously refine categories. It is important for researchers to balance between overly specific categories, which may be too narrow, and overly broad categories that risk masking significant patterns. In most of the studies referenced in the previous section, only categories that are cited by a minimum percentage of consumers (typically ranging from 5% to 10%) were retained. These categories constitute the final set of features to be used in statistical analyses.

Compilation involves synthesizing the categorized meaning units into a coherent narrative or summary that highlights the key findings and helps address the research questions; this stage is not part of the feature extraction process.

4.2.5 Tools for Data Annotation

Data annotation can be time-consuming and resource-intensive, especially with large datasets. Specialized qualitative data analysis software helps manage and code data while ensuring traceability throughout the annotation process, enhancing scalability, efficiency, and streamlining workflows. Most tools offer features like predefined annotation templates, auto-suggest functions, quality control mechanisms, and support for data visualization, category management, and annotation consolidation. Notably, *RQDA* (http://rqda.r-forge.r-project.org), built on the R programming language, integrates seamlessly with other R packages, broadening its analytical capabilities and allowing users to apply advanced statistical methods during analysis.

5 Using Natural Language Processing for Semi-automated and Automated Feature Extraction

5.1 What Are Semi-automated and Automated Feature Extraction?

Manual feature extraction using content analysis offers maximum flexibility and control; however, as the volume of text data grows, manual procedures become increasingly labour-intensive and impractical. This makes the approach unsuitable for routine use and unscalable for large datasets. To address these challenges, researchers have increasingly turned to computer-aided techniques such as computational content analysis, text mining, text analytics, or more generally, automated text analysis. These methods leverage NLP, a branch of AI focused on enabling machines to analyse, understand, interpret, and generate human language (Khurana et al., 2023). It encompasses a range of techniques that allow computers to process text data for tasks such as classification, translation, summarization, sentiment analysis, and more. One of the primary advantages of NLP for text data preprocessing is its ability to apply consistent procedures across large datasets, thereby reducing the need for manual intervention and minimizing human error. Implementing these methods typically requires proficiency in programming languages such as R or Python, along with a solid understanding of algorithms and data structures.

The level of automation refers to the extent to which automated tools are used in the feature extraction process, as well as the degree of human involvement in guiding, supervising, or refining these tools. Semi-automated feature extraction streamlines content analysis by handling basic text preprocessing tasks, particularly those involved in identifying meaning units during the decontextualization stage. This reduces the time and effort required, while preserving human oversight at critical decision points where human input remains essential for interpreting nuanced patterns and ensuring that extracted features align with the research or application goals. This approach offers a balance between control and scalability.

In contrast, fully automated techniques enable the processing of large volumes of text with minimal human involvement once the system has been configured. This approach is highly efficient and scalable, allowing for the rapid analysis of massive datasets. However, it tends to be less flexible, as any necessary adjustments or refinements often require reconfiguring or retraining the system. Additionally, automated methods may lack contextual awareness or domain-specific expertise, which can limit the quality or relevance of the extracted features. With the increasing availability of pre-trained models and the rise of Large Language Models (LLMs), these limitations are becoming less restrictive, paving the way for more robust, end-to-end solutions in automated feature extraction.

5.2 *What Is a Typical Pipeline for Text Data Feature Extraction?*

5.2.1 Overview

Many operations in NLP mirror those traditionally carried out manually in content analysis. However, humans can navigate complex linguistic structures by leveraging their intuitive understanding of language to make context-specific decisions. They can implicitly perform multiple text preprocessing tasks simultaneously and flexibly, adjusting the process as needed. In contrast, semi-automated and automated feature extraction relies on a more structured and systematic approach, generally involving several key stages: text cleaning, text normalization, text transformation, and feature selection. At each stage, some NLP tasks are widely used, while others are applied more selectively, depending on the particular requirements of the analysis (Antons et al., 2020; Nesca et al., 2022). Indeed, there is no one-size-fits-all approach, as the pipeline of operations is highly dependent on the specific research objectives and the level of granularity chosen for the features list (see Fig. 5.1).

Python is widely used by data scientists due to its versatility and strong integration with machine learning and AI frameworks. Libraries like *SpaCy* and *NLTK* make it especially effective for processing large text datasets. Conversely, R is more commonly utilized in sensory and consumer sciences because of its robust statistical analysis capabilities. It is also popular in the academic community, as it is supported by extensive resources and packages such as *tidyverse* that simplify text manipulation and cleaning. For further reading on NLP techniques, readers can consult technical books on Python (Lee, 2024) or R (Campesato, 2022).

5.2.2 Text Cleaning Operations

Text cleaning aims to remove irrelevant or non-informative content from the raw text data. The specific approach can vary depending on the context of the research, since content considered irrelevant in one context may be important in another. Below are some of the most commonly used operations in text cleaning.

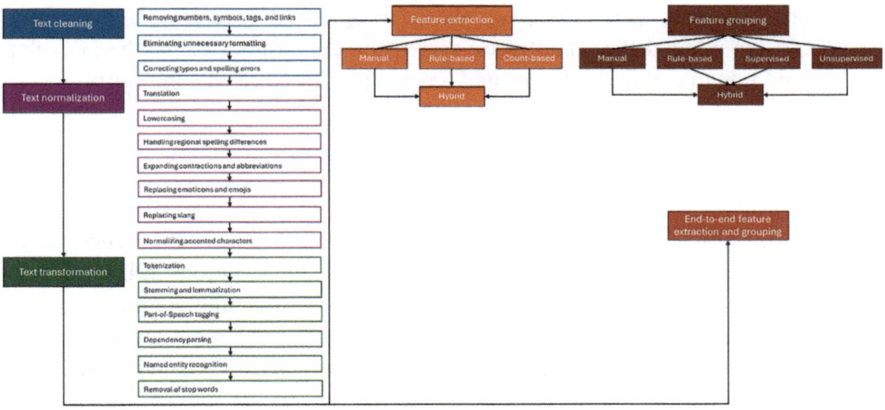

Fig. 5.1 Pipeline of unit NLP operations for text data feature extraction

Removing Numbers, Symbols, Tags, and Links

Numbers and symbols are commonly removed during the text cleaning process, as they are often considered irrelevant or non-informative for most analyses. HTML tags, URLs, and email addresses are also typically eliminated, particularly when the text originates from web pages. For this task (and many others), regular expressions (regex) are widely used as a powerful, versatile, and efficient tool for text cleaning. Regular expression are sequences of characters that defines a search pattern, typically used for pattern matching within strings. They allow the identification, removal, or replacement of specific characters or text fragments (Kaur, 2014).

Eliminating Unnecessary Formatting

Text often contains extra spaces, tabs, line breaks, or duplicated punctuation marks. In general, formatting can be removed; however, in tasks such as dependency parsing (see later), it may be preserved, as it can help in understanding sentence structure and syntactic relationships.

Correcting Typos and Spelling Errors

Text data typed by consumers often contains typos, spelling errors, and other inconsistencies. These errors can arise from various factors such as fast typing, autocorrect issues, or unfamiliarity with correct spelling. To address this issue, text preprocessing often includes a spelling correction step, which aims to identify and correct these errors. Spelling correction can be achieved through various methods, with one widely used tool being *Hunspell* (Al-Hussaini, 2017). *Hunspell* is an open-source spell checker that utilizes a dictionary and algorithmic techniques, such as Levenshtein distance, to compare misspelled words with valid dictionary entries and suggest the closest matches. However, *Hunspell* operates at the word level and does not consider the context surrounding the word. For more advanced spelling correction, sentence-level models that consider contextual information can be used (Hládek et al., 2020). Most programming languages offer functions that integrate online service APIs for sentence-level correction, although they generally require a paid subscription to use. APIs (Application Programming Interfaces) are sets of protocols that allow different software systems to communicate, enabling developers to send text to external services and receive processed results in return.

5.2.3 Text Normalization Operations

Text normalization aims at ensuring that the text data is consistent, coherent, and uniform across the dataset. This process helps eliminate variations in language and format that could impact subsequent analyses. Below are some common text normalization operations. Most of these, when performed at the word level (i.e., without considering context), can be efficiently carried out using semantic resources (such as dictionaries or lists of words, symbols, or expressions) combined with regular expressions, which are used to detect common variants and replace them with standardized forms.

Translation
When working with multilingual datasets, translation is essential for normalization to maintain consistency across various languages. The initial step in automated preprocessing of multilingual datasets is language detection, which determines the language of the text by using algorithms that analyse patterns, word frequencies, and contextual cues (Vuddanti et al., 2024).

Once the language is detected, translation can take place at different levels. Word-level translation typically relies on multilingual dictionaries such as *BabelNet (Bovi & Navigli,* 2017). For sentence-level translation, models that account for contextual information are preferred (Tan et al., 2020). Some models are integrated into APIs which can be accessed within programming languages to facilitate automated translation tasks.

Handling Regional Spelling Differences
Regional spelling differences are variations in the way words are spelled between different dialects. For example, "color" (American English) and "colour" (British English) are semantically equivalent, but they have different spellings depending on the region.

Lowercasing
Converting all text to lowercase helps ensure consistency by treating words with different capitalizations as equivalent. Lowercasing is typically achieved through simple string manipulation functions available in most programming languages. However, there are cases where lowercasing is not relevant. For instance, in Named Entity Recognition (NER, see later), lowercasing could confuse proper nouns like "Apple" (the company) and "apple" (the fruit). Capitalization can be also used to convey emphasis, such as "I LOVE this!" versus "I love this". Acronyms and initialisms should remain capitalized to preserve their meaning. Additionally, in some languages, capitalization can have grammatical significance.

Expanding Contractions and Abbreviations
When consumers provide reviews or feedback, contractions like "can't" or "won't" might appear frequently. Expanding these contractions into their full forms, such as "cannot" or "will not" is crucial to ensure that negation is correctly understood. Similarly, abbreviations like "lol" (laughing out loud) can indicate humour or enjoyment, which may imply positive emotional engagement with a product.

Replacing Emoticons and Emojis
Emoticons (e.g.,:-) or:D) and emojis (e.g., 😊, 😂) are often used by consumers to convey emotions, sentiments, or reactions in text. In tasks like sentiment analysis, it is crucial to replace these symbols with their corresponding words to capture the intended emotion or sentiment. For example, the emoticon ":)" can be replaced with "happy" or "smile", and emojis like 😂 might be converted to "laugh" or "joy". Specialized lexicons and sentiment databases, such as *EmoBase* (Diwersy et al., 2014), *EmoReact* (Nojavanasghari et al., 2016), or *Multidimensional Lexicon of Emojis* (Godard & Holtzman, 2022), can be used to map emojis and emoticons to words that reflect the same sentiment or emotion.

Replacing Slang
Slang terms or informal language may carry specific meanings or emotional connotations, but they might not be understood correctly by NLP models without proper replacement or normalization. For example, "lit" could be replaced with "exciting" or "fun", or "fam" with "family" or "friends". Specialized resources such as *urbandictionary.com* (Smith, 2011) or *slang.net* (Derioh et al., 2023) can be used to map slang terms to their standard or regular word equivalents.

Normalizing Accented Characters
In some languages, such as French, text may include accented characters that are not consistently typed by consumers, resulting in variations such as "é" being typed as "e" or "ç" as "c". Dedicated libraries provide broader solutions for handling accented characters across different languages. However, caution is necessary, as removing accents may change the meaning of words, such as in French where "mais" (but) differs from "maïs" (corn).

5.2.4 Text Transformation Operations

Transformation operations involve converting the cleaned and normalized text into meaningful units for further analysis. Below are some of the most commonly used text transformation techniques.

Tokenization
Tokenization is a fundamental step for many NLP tasks which consists in breaking down text into smaller units called "tokens", such as sentences, words, or subword units.

Sentence tokenization focuses on splitting text into sentences, typically by detecting punctuation marks. This approach is useful when the relationships between sentences are important, such as in tasks like text summarization or machine translation. However, it can be error-prone with ambiguous punctuation or unconventional formatting.

Word tokenization is the most common form of tokenization, where text is split into individual words based on spaces. For example, the sentence "the chocolate is delicious." becomes ["the", "chocolate", "is", "delicious", "."]. While this approach is simple, it can become inefficient when handling compound words like "red fruit", which are treated as two separate tokens despite forming a meaningful unit. To address this, one simple solution is to replace spaces with an underscore in compound words, using a custom list of words and regular expressions.

Rule-based tokenization applies predefined rules or regular expressions to split the text. It can be more sophisticated than simple space-based tokenization, accounting for specific punctuation marks, contractions, and other language-specific features. However, it can be limited by the complexity of language and might not handle all edge cases.

Statistical tokenization leverages machine learning models, such as Hidden Markov Models or Conditional Random Fields can be used to predict token boundaries based on sequences of observed tokens, learning from annotated text data. While these methods are more accurate and adaptable than rule-based approaches, they require a significant amount of data for training and can be computationally intensive. Pre-trained tokenizers, such as those provided by *SpaCy*, can be integrated into R workflows with the help of appropriate packages. These models also support subword tokenization, that can help in handling variations in words more effectively, using methods such as Byte Pair Encoding or WordPiece to break words into smaller units. For example, the word "unhappiness" might be tokenized as ["un", "happiness"].

The reader interested in technical details can refer to Jurish & Würzner, 2013, Tomanek et al., 2007, Song et al., 2020, or Kozma & Voderholzer, 2024.

Stemming and Lemmatization

Stemming is a technique used to reduce words to their root form by stripping off prefixes and suffixes. For example, the word "sourness" may be reduced to "sour". While stemming is computationally efficient, it does not take into account the meaning of the word. As a result, it can produce non-standard or incomplete forms, such as reducing "better" to "bet". Popular stemmers include the Porter stemmer and snowball Stemmer (Jivani, 2011).

Lemmatization reduces a word to its base form (lemma), but it does so based on the meaning and grammatical context of the word. Unlike stemming, lemmatization takes into account the part of speech of a word. For instance, "better" becomes "well" when lemmatized. Lemmatization produces more accurate, context-sensitive forms since it considers the word's role in the sentence, but it is more computationally intensive. Tools like the *WordNet* lemmatizer leverage lexical databases to determine the correct lemma for each word based on its context (Miller et al., 1990). Regular expressions and lexicons are commonly used in simpler or rule-based systems for lemmatization, providing more explicit mappings of inflected forms to their base forms. Morphological analysis involves breaking down a word into its morphological components (prefixes, suffixes, root words) and using morphological rules to derive the base form. Statistical lemmatization uses models trained on large corpora of text to predict the lemma of a word based on patterns observed in the data. Neural Network-based lemmatization uses deep learning techniques to learn complex mappings between words and their lemmas from large annotated datasets. These models can handle ambiguous cases and context better than traditional methods.

The reader interested in technical details can refer to Toporkov and Agerri (2024) or Stanković et al. (2020).

Part-of-Speech Tagging

Part-of-Speech (POS) tagging is the process of assigning a grammatical label (such as noun, verb, adjective, etc.) to each token in a sentence (see Table 5.2).

This step is essential for understanding the syntactic structure of text. POS tagging can be performed using different algorithms and techniques depending on the text complexity and computational resources.

Rule-based POS tagging relies on dictionaries or handcrafted grammatical rules based on context, word endings, and surrounding words. For example, a word ending in "-ly" may be tagged as an adverb. While accurate, it requires manual creation of rules, which can be time-consuming and prone to errors.

Stochastic POS tagging uses probabilistic models like Hidden Markov Models or Maximum Entropy Models to predict the most likely sequence of POS tags. Conditional Random Fields combined with Long Short-Term Memory Networks can also be used to predict the entire sequence of POS tags considering the relationships between adjacent words and capturing long-range dependencies. Popular pretrained models, such as those from *SpaCy*, utilize deep learning architectures.

The reader interested in technical details can refer to Bărbulescu & Morariu, 2020, Ratnaparkhi, 1996, or Veerasekharreddy et al., 2022.

Dependency Parsing

Dependency parsing identifies the grammatical relationships between words in a sentence. The goal of dependency parsing is to break down the sentence into a hierarchical structure that reflects how the words depend on each other. This structure is typically represented as a dependency tree, where each node represents a syntactic unit and edges represent the grammatical connections between them (Fig. 5.2).

Table 5.2 Example of POS tagging applied to the sentence "*The chocolate is delicious.*" (using R package *udpipe*)

doc_id	sentence_id	token_id	token	lemma	pos
text1	1	1	The	the	DET
text1	1	2	chocolate	chocolate	NOUN
text1	1	3	is	be	AUX
text1	1	4	delicious	delicious	ADJ
text1	1	5	.	.	PUNCT

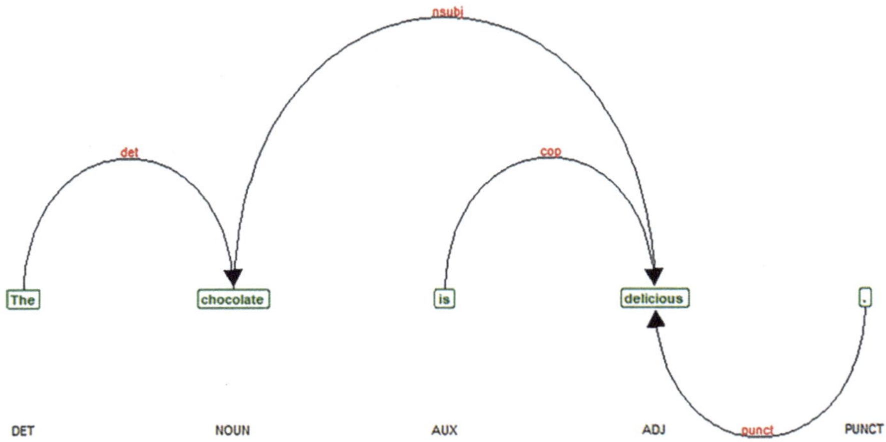

Fig. 5.2 Example of dependency parsing tree applied to the sentence "*The chocolate is delicious.*" (using R packages *udpipe* and *textplot*)

Transition-based parsers are efficient and widely used in systems like the *Stanford NLP*. Neural network models like bidirectional Long Short-Term Memory Networks are increasingly used to improve dependency parsing. Recent advancements integrate pre-trained word embeddings into dependency parsing models, helping to improve the parser's performance, particularly with rare or complex vocabulary. The reader interested in technical details can refer to Zhang et al. (2016) or Smith et al. (2018).

Named Entity Recognition

Named entity recognition (NER) involves identifying and classifying named entities in text into predefined categories such as people, organizations, locations, dates, and more. For example, in the sentence "*Apple was founded by Steve Jobs in Cupertino in 1976*", NER would recognize "Apple" as an organization, "Steve Jobs" as a person, "Cupertino" as a location, and "1976" as a date. NER can be performed using various approaches, including rule-based methods, machine learning models, and deep learning techniques (Jehangir et al., 2023). An example of application if food sensory and consumer science includes the recognition of food names (Cenikj et al., 2022).

Removal of Stop Words

Stop words are common words such as low-content words, connectors, or auxiliary terms (e.g., "the", "and", "is", "in") that do not carry significant meaning in most NLP tasks. These words can be removed during the text cleaning phase using regular expressions applied on predefined stop word lists. However, when techniques like POS tagging, dependency parsing or NER are applied, this step is often unnecessary, as such methods usually focus on retaining only content-rich words, particularly nouns, adjectives or specific entities.

5.2.5 Feature Selection Operations

Once the raw text data has been cleaned, normalized, and transformed, various approaches can be used to select the most relevant features, ranging from simple heuristics to more complex, layered methodologies (Legate et al., 2024). In practice, hybrid approaches are often employed, combining count-based, rule-based, supervised, and unsupervised learning methods. These approaches typically involve varying degrees of human supervision to identify relevant features and, when needed, to group them into broader conceptual categories.

Count-Based Feature Extraction

Count-based feature extraction refers to a data-driven process of deriving features from text data by counting the occurrences of specific text units, either within individual evaluations or across the entire dataset. The most frequently cited units are then used as features for further analysis.

As text data is inherently qualitative, it must be transformed into a numerical format to enable computational analysis. One common approach is numerical vectorization, such as the Bag-of-Words (BoW) model (Qader et al., 2019). In BoW, text is represented as a Document-Term Matrix (DTM), a structured format in which each row corresponds to a document and each column to a unique word (unigram) in the corpus. Each cell in the matrix indicates the frequency of a specific word within a given document (Table 5.3).

This approach can be effective for extracting simple features where no full sentences were constructed (e.g., in sentence completion or free word association tasks) for gathering general information about a product space and deriving sensory attribute lists (Chen et al., 2021). However, BoW ignores word order and thus does not capture syntactic structure or relationships between words, resulting in a loss of context that can be a significant drawback when trying to capture nuanced features.

To incorporate more contextual information, N-grams, which are sequences of N consecutive words (e.g., bigrams, trigrams), can be used instead of unigrams in the BoW model (Sidorov et al., 2014). This method can reveal patterns and associations between words. For example, in the previous example, a DTM based on bigrams would capture the compound word "milk chocolate", the negation "not like", and the associations "very sweet", "too sweet". The reader can refer to Calvert et al. (2023), Espinosa-Ramos et al. (2025) or Jaeger and Rasmussen (2021) for applications of N-grams in sensory and consumer science.

Table 5.3 Example of Document-Term Matrix (DTM) resulting from applying the BoW model to the sentences "*I don't like the taste of milk chocolate in this sample*" and "*The taste of the sample is very sweet, too sweet for me*", following text cleaning, normalization, lemmatization, tokenization, stop words removal

Document	not	like	taste	milk	chocolate	sample	very	sweet	too
Sentence 1	1	1	1	1	1	1	0	0	0
Sentence 2	0	0	1	0	0	1	1	2	1

Another widely used method is Term Frequency-Inverse Document Frequency (TF-IDF), which assigns weights to words or N-grams based on their frequency in a document relative to their frequency across the entire corpus (Das et al., 2023). Term Frequency (TF) measures how frequently a word appears in a specific document. It is calculated by dividing the number of times a word w occurs in a document d by the total number of words in that document. Term Inverse Document Frequency (IDF) measures how rare or unique a word is across the entire corpus of documents. It is calculated as the log of the total number of documents divided by the number of documents containing the word w. The final weight of a word in a document is calculated by multiplying its TF by its IDF, allowing models to focus on the more meaningful words in the corpus, i.e., those the most discriminative. The reader can refer to Laguna et al. (2020), Chen et al. (2020), or Calvert et al. (2023) for applications of TF-IDF in sensory and consumer science.

While count-based methods like BoW and TF-IDF are relatively easy to implement, they can generate irrelevant or excessive features. For instance, lower-order N-grams might miss deeper contextual meanings, but as the value of N increases, the number of possible combinations grows exponentially. This causes the feature space to become very large, leading to sparsity in the DTM, where many N-grams appear only once or a few times throughout the dataset. Consequently, information can become diluted, potentially undermining the validity of subsequent analyses. Typically, features mentioned by at least 5% of consumers are considered meaningful or relevant for analysis, though this threshold may vary depending on the text data source. In addition, human expertise is often necessary to validate the relevance of the extracted features, ensuring they align with the research question.

Experts may also want to group similar features into broader conceptual categories, which reduces redundancy, simplifies the feature space, and highlights core concepts. For instance, in wine sensory evaluation, features such as "strawberry" and "raspberry" represent distinct sensory experiences related to aroma perception. According to standardized wine aroma terminology system such as that of Noble et al. (1987), both can be classified under a broader category like "berry" or "fruity". Grouping features in this way enhances the efficiency and interpretability of the analysis by focusing on overarching themes rather than isolated word variations. This manual refinement of the feature list, whether it involves grouping features or not, is commonly used in sensory and consumer science and can be regarded as a form of semi-automated content analysis (Antille et al., 2024; Bécue-Bertaut et al., 2008; Deneulin & Bavaud, 2016; Galmarini et al., 2013; Grasso & Jaworska, 2020; Symoneaux et al., 2012; ten Kleij & Musters, 2003; Tian et al., 2021; Visalli et al., 2025).

Rule-Based Feature Extraction

Count-based methods, being data-driven, may lack precision in capturing meaningful features, especially when vocabulary is highly diverse or when fine-grained feature extraction is required. In contrast, rule-based methods are knowledge-driven and rely on predefined expert rules to guide feature extraction. This approach ensures domain relevance and can reduce or eliminate the need for post hoc expert review and validation of the feature list. These methods use explicit if-then logic, linguistic patterns, or heuristics, often implemented through tools like regular expressions, to identify features aligned with specific research objectives. The rules are typically established based on prior studies, existing literature, specialized word lists, or domain-specific lexicons (Asseo & Niv, 2022; Gonzaga et al., 2019; Kāle et al., 2021; Mehraliyev et al., 2020). Alternatively, relevant word lists can be initially created through preliminary feature extraction from the dataset, then refined by human experts, and subsequently used to apply rules systematically across the dataset (Espinosa-Ramos et al., 2025; Gonzaga et al., 2019; Lahne et al., 2013; Spinelli et al., 2017).

Rule-based feature extraction can be applied at different levels of granularity, either in a single step, such as directly categorizing words or expressions into broad feature categories, or through multiple steps, for example, by first identifying specific features, then associating them with concept words, and finally grouping them into broader categories. In addition to stemming and lemmatization algorithms, *WordNet* is a widely used resource for synonymization. It functions as a lexical database that organizes words into sets of synonyms (synsets), offering valuable semantic relationships and hierarchies that enhance understanding across various contexts (Miller et al., 1990). However, it is not always suitable for domain-specific contexts (Hamilton & Lahne, 2020).

Rule-based feature extraction is relatively easy to implement and offers strong control and interpretability. It requires minimal human intervention once the rules are established. However, it demands significant effort to define and maintain these rules, and it lacks flexibility and generalizability when the data deviate from the predefined patterns.

Supervised Learning Feature Extraction

Supervised learning methods involve training machine learning classifiers on human-labelled datasets (typically the results of manual or semi-automated content analysis) to predict the most probable feature for any unlabelled text data. In active learning, human oversight is integrated into the training loop to iteratively refine the model by correcting its predictions when ambiguity or complexity arises, thereby improving accuracy over time.

Text data can be analysed using a variety of machine learning classifiers, each suited to different types of linguistic patterns and analytical objectives. Logistic Regression is a simple yet effective classifier for binary classification tasks, such as distinguishing between positive and negative sentiment. K-nearest neighbours works by comparing a piece of text to others in the dataset and assigning it the most common label among its closest matches. Support vector machines are useful for finding the best boundary that separates different categories of text, especially when dealing with a large number of features. Decision trees make classifications by asking a series of yes-or-no questions about the data, forming a logical tree structure. Random forests build on this idea by combining many decision trees to improve accuracy and reduce the risk of errors. Naive Bayes is a fast and effective method based on probabilities, particularly useful when the words in the text are assumed to be independent of each other. Conditional random fields are used for labelling sequences of words where the relationship between words matters. The reader can refer to Maalouf (2011), Cunningham and Delany (2021), Brereton and Lloyd (2010), Breiman (2001), Peretz et al. (2024), or Liliana and Basaruddin (2017) for technical details.

More advanced deep learning models use artificial neural networks with many layers (hence the term "deep learning") to identify complex patterns in large datasets. These networks are inspired by the way the human brain works and are made up of connected units called neurons, which process and pass information through the network. As the model trains, it learns by adjusting internal weights to improve its predictions. Several types of deep learning architectures are commonly used for feature extraction. Feedforward Neural Networks (FNN) are the simplest kind, where information flows in one direction (from input to output) without any feedback loops. Recurrent Neural Networks (RNN) are designed to handle sequential data like sentences or time series. Unlike FNNs, RNNs have a memory mechanism that lets them pass information from one step to the next, making them useful for tasks that depend on word order or context over time. Convolutional Neural Networks (CNN), originally created for image recognition, are also used in text analysis. CNNs scan text using filters that help detect local patterns, such as key phrases or combinations of words, within a sentence. To improve how sequential data is processed, specialized versions of RNNs have been developed. Long Short-Term Memory (LSTM) networks and Gated Recurrent Units (GRU) are designed to better capture long-term dependencies in text, helping models understand relationships between words that are far apart in a sentence or paragraph. The reader can refer to Murat (2006), Indolia et al. (2018), Islam et al. (2019) or Sherstinsky (2020) for technical details, or to Miller et al. (2021) or Hamilton et al. (2023) for applications of deep learning models for feature extraction in sensory and consumer science.

Supervised learning methods can achieve high accuracy and are effective at capturing complex patterns. However, they require substantial amounts of high-quality labelled data. When such datasets are unavailable, transfer learning can be used. This involves taking a model trained on one task and applying it to a different, but related, task. Fine-tuning is a specific form of transfer learning in which a pre-trained model is further trained on a smaller, task-specific dataset to adapt it to the new context. Transformer models are a class of neural network architectures that rely on a self-attention mechanism to process data in parallel rather than sequentially, making them significantly more efficient and capable of handling long-range dependencies in text (Vaswani et al., 2017). Since their introduction, transformer models have evolved into specialized architectures like *BERT*, *GPT*, and *T5*, each optimized for specific tasks and widely available as pre-trained models available through open-source libraries such as *Hugging Face's Transformers*, making them increasingly popular in domain-specific NLP applications.

Unsupervised Learning Feature Extraction

Unsupervised learning methods are used to find patterns in text data without needing any prior labelling or annotations. These techniques work on the assumption that words or attributes that appear in similar contexts likely have similar meanings.

Clustering algorithms like K-means (Ikotun et al., 2023) and hierarchical clustering (Ran et al., 2022) group words or features based on how often they appear together in the dataset and how similar they are mathematically. K-means groups data points into a predefined number of clusters, denoted as K. The algorithm works by iteratively adjusting the cluster centers to minimize the variation within each cluster. This is achieved by reducing the sum of squared distances between each data point and the center of its assigned cluster. Hierarchical clustering, in contrast, organizes data points into a nested hierarchy, allowing for grouping at various levels of granularity. Unlike K-means, it does not require the number of clusters to be specified in advance. Instead, it produces a dendrogram, i.e., a tree-like diagram that visualizes how clusters are formed or divided at different levels. Hierarchical clustering can follow two main strategies: an agglomerative (bottom-up) approach, which starts with individual items and gradually merges them into larger clusters, or a divisive (top-down) approach, which begins with the entire dataset and recursively splits it into smaller clusters.

These methods are relatively quick to implement and are good at detecting hidden patterns in large datasets. However, clusters do not inherently carry a direct semantic meaning; they are formed based purely on mathematical similarity or distance between data points. That is why human validation is often necessary to ensure that the clusters make sense in the context of the domain being studied (Batista et al., 2023; Mahieu et al., 2020a, 2020b, 2022; Polli et al., 2021; Visalli et al., 2023a). A layer of semantic understanding can be added by organizing features into hierarchical structures based on pre-existing expert knowledge. Using this information, clusters combining statistic and semantic criteria can be formed automatically, removing the need for human validation (Visalli et al., 2025).

Clustering algorithms are typically applied to vectors or matrices of data points, but they can also be used on word embeddings. Word embeddings are numeric representations of words that capture their meanings based on the context in which they appear. Unlike treating words as separate, discrete symbols, word embeddings represent each word as a dense vector in a continuous space, which is learned from analysing large collections of text.

This method captures semantic relationships between words by placing similar words close to each other in this vector space (see Table 5.4 and Fig. 5.3). Tools like *Word2Vec, FastText, GloVe*, and *BERT* generate these embeddings and have been used to identify semantically similar word clusters. The reader can refer to Nithyanandan (2019), Sivakumar et al. (2020), Xu and Du (2019), Pennington et al. (2014), or Devlin et al. (2018) for technical details, or to Kim et al. (2018), Reed et al. (2019), Youn et al. (2020), Morales-Garzón et al. (2020), Eftimov et al. (2020), or Yoo et al. (2023) for applications.

Table 5.4 Example of cosine distances between wine sensory attributes computed from *FastText* embeddings (using R packages *text2vec* and *lsa*)

	Fruit	Strawberry	Raspberry	Cherry	Bitter	Astringent	Tannin	Alcohol
Fruit	1.00	0.56	0.49	0.52	0.30	0.33	0.41	0.28
Strawberry	0.56	1.00	0.78	0.63	0.22	0.20	0.24	0.21
Raspberry	0.49	0.78	1.00	0.63	0.23	0.26	0.27	0.21
Cherry	0.52	0.63	0.63	1.00	0.26	0.21	0.29	0.15
Bitter	0.30	0.22	0.23	0.26	1.00	0.47	0.30	0.22
Astringent	0.33	0.20	0.26	0.21	0.47	1.00	0.52	0.30
Tannin	0.41	0.24	0.27	0.29	0.30	0.52	1.00	0.35
Alcohol	0.28	0.21	0.21	0.15	0.22	0.30	0.35	1.00

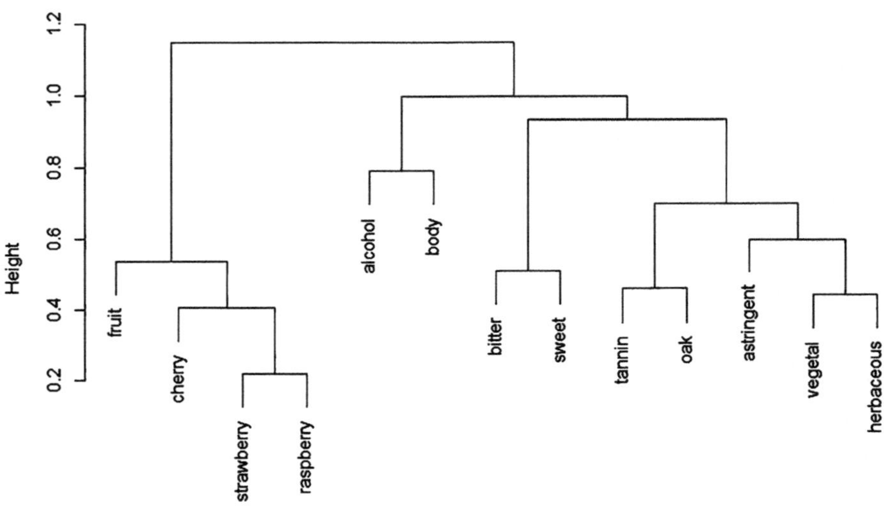

Fig. 5.3 Example of hierarchical clustering dendrogram of wine sensory, based on cosine distances computed from *FastText* embeddings

End-to End Feature Extraction

Large Language Models (LLMs) are sophisticated transformer-based systems pre-trained on extensive datasets, often comprising billions of parameters. Designed as general-purpose tools, LLMs are capable of generating coherent, contextually relevant responses across responses to diverse prompts (Ma et al., 2024). One of the most common and accessible implementations of LLMs is in chatbots, which leverage these models to engage users in natural language conversations, interpreting prompts and generating responses in a human-like manner. Prompt engineering is the process of crafting input prompts to guide LLMs in performing a specific task. The process begins by providing the chatbot with sufficient information to establish the context and clearly define the objectives of the request. Next, a prompt is formulated to clearly articulate the task. To be effective, prompts should be explicit, concise, and free of ambiguity. Proving illustrative examples can further enhance the model's understanding. Once a prompt is designed, it is tested on sample data, and refined iteratively based on the model's output.

Through effective prompt engineering, LLMs like *GPT-4* can perform complex tasks such as attribute extraction and grouping. These end-to-end systems are user-friendly, integrating multiple steps into a single process and eliminating the need for traditional NLP operations. However, this approach has limitations. It often lacks reproducibility due to the opaque nature of evolving pre-trained models and prompt variability. Other challenges include limited domain-specific accuracy, high computational demands, increased operational costs, and privacy concerns, particularly when relying on cloud-based infrastructures (Visalli et al., 2025).

5.3 How to Guarantee High-Quality Text Preprocessing and Feature Extraction?

Thorough documentation of the preprocessing and feature extraction strategy is essential. This includes detailing each preprocessing step, explaining the rationale behind the choice of each operation, and specifying the tools and resources involved. Such comprehensive documentation ensures transparency, explainability, consistency, and reproducibility of the preprocessing and feature extraction process.

Domain knowledge is critical for validating the results. Manual review of a data sample by experts helps determine whether extracted features accurately reflect the meaning and context of the original text. This is especially important when using lemmatization or stemming, where words must be reduced correctly without losing essential nuances. Special attention should be paid to domain-specific vocabulary and study-related concepts.

In count-based feature extraction, it is essential to ensure that the final feature list is complete (i.e., free of missing, redundant, or irrelevant entries). When applying feature selection thresholds, there is a risk of discarding low-frequency synonyms that might have been retained if grouped. In rule-based approaches, quality check involves verifying that the rule set is both exhaustive, relevant for the dataset and correctly applied. For unsupervised methods, clusters must align with the research objectives and remain stable across multiple iterations. In supervised feature extraction, the performance of downstream models serves as an indirect quality check. Good model performance suggests that the features capture relevant patterns. Cross-validation can further test generalizability, ensuring the model does not overfit and performs well on unseen data.

Despite its importance, quality checking prior to analysis is often overlooked (Nesca et al., 2022). Hutchings et al. (2023) note that standard precision measures from sensory and consumer science may not transfer well to unstructured text data. A robust method for evaluating feature extraction quality is comparison against a "ground truth", a manually curated set of features (generally obtained though content analysis) reflecting the true content and context of the text (Jaeger & Rasmussen,

2021; Visalli et al., 2025). Several performance metrics are commonly used. Precision measures the proportion of correctly predicted categories or labels out of all predicted categories (true positives divided by true and false positives). Recall, on the other hand, assesses how well the model identifies all the actual relevant categories or labels (true positives divided true positives and false negatives). The F-measure combines both precision and recall into a single metric, offering a balanced view of both accuracy and completeness (Derczynski, 2016).

6 Summary: Advantages and Limitations of Each Feature Extraction Approaches

Table 5.5 summarizes the main advantages and limitations of each feature extraction approaches in the context of sensory and consumer science. The reader can refer to Hartmann et al. (2019), Kathuria et al. (2021), Chai (2023) or Siino et al. (2024) for technical reviews of tools and techniques for preprocessing of text data.

7 Conclusion and Perspectives

The field of text data processing in sensory and consumer science is rapidly evolving, transitioning from traditional manual methods to more scalable automated approaches. Both manual and automated feature extraction have important roles: manual methods leverage domain expertise and deep contextual understanding suitable for detailed experimental analysis but are impractical for large datasets, while automated methods offer efficiency and scalability but may lack nuance and flexibility. Semi-automated approaches provide a balanced solution, combining human oversight with improved efficiency.

Feature extraction techniques vary widely—count-based, rule-based, supervised, unsupervised, and end-to-end methods (including those based on large language models)—each with distinct advantages and limitations. No single approach fits all scenarios, and despite technological advances, human supervision remains essential. For practitioners and researchers in sensory and consumer science, adopting best practices and staying current with NLP advancements is crucial to harness cutting-edge models, enhance analysis quality, and derive valid insights from text data.

Looking ahead, rapid advances in artificial intelligence, especially the development of increasingly sophisticated LLMs, are poised to transform current practices in sensory and consumer science. As general-purpose models, LLMs may currently lack specialized knowledge tailored to sensory and consumer sciences. To address this limitation, techniques like Retrieval-Augmented Generation (Gao et al., 2023) are gaining traction, providing LLMs with relevant external context to improve their

Table 5.5 Summary of the main advantages and limitations of each feature extraction approaches

Approach	Advantages	Limitations
Manual	Domain-specific insights; highly adaptable; fine-grained feature extraction	Substantial experience and training required; labour-intensive; prone to human error; non-deterministic
Semi-automated	Less time-consuming and more scalable than manual feature extraction; human oversight	Scalability issues with large datasets; non-deterministic
Automated	Most efficient and scalable	Significant computational resources required for some models
Count-based	Data-centric; no need for rules, labelled data or model training; straightforward to implement with basic programming skills; deterministic	Sensitive to sparsity; does not capture semantic relationships between words; expert knowledge required to select relevant features
Rule-based	Knowledge-centric; transparent; easily customizable; relatively straightforward to implement with basic programming skills; interpretable domain-specific features; deterministic	Expert knowledge required to create and maintain rules; challenges with handling ambiguities; potential validity and generalizability issues if data deviate from predefined rules
Supervised	Combination of knowledge and data-centric approaches; deterministic	Large amount of labelled data required; advanced skills in programming and statistics needed; performance heavily dependent on quality of labelled data
Unsupervised	No need for rules, labelled data nor model training; well-suited for exploratory analysis; capable of uncovering latent features	Coarse-grained feature extraction; lack of interpretability; expert validation required
End-to-end	Seamless integration; user friendly	Dependent on pre-trained models not specific to sensory and consumer science; highly dependent on prompts; cost, data security and confidentiality concerns with cloud-based APIs; non-deterministic; lack of interpretability

domain relevance. While the exact trajectory of these technologies is difficult to predict, data collection through chatbots as an alternative to traditional questionnaires, is likely to become increasingly common (Yatawara et al., 2025). Real-time interactions with consumers through these interfaces may help clarify ambiguous responses, enhancing data accuracy and reducing the need for complex, subjective preprocessing. Text data collection is expected to broaden beyond sensory descriptions and affective responses, encompassing attitudes, expectations, and spatial-temporal contexts of consumption. Moreover, multimodal data integration is progressing rapidly; future preprocessing workflows will likely combine text data with video or audio self-recordings from consumers. As these technologies evolve, it will be essential to consider ethical, environmental, and societal implications to ensure responsible and sustainable innovation in sensory and consumer science.

References

Adams, J., Williams, A., Lancaster, N., & Foley, M. (2007). Advantages and uses of check-all-that-apply response compared to traditional scaling of attributes for salty snacks. In *7th Pangborn sensory science symposium. Minneapolis, USA, 12–16 August, 2007.*

Aguirre, A., Borneo, M. T., El Khori, S., & Borneo, R. (2019). Exploring the understanding of the term "ultra-processed foods" by young consumers. *Food Research International, 115*, 535–540. https://doi.org/10.1016/J.FOODRES.2018.09.059

Al-Hussaini, L. (2017). Experience: Insights into the benchmarking data of hunspell and aspell spell checkers. *Journal of Data and Information Quality, 8*(3–4). https://doi.org/10.1145/3092700

Antille, N., Audoubert, F., Camilleri, M., Grain, M., Rytz, A., Pineau, N., & Mahieu, B. (2024). Comparison of open comments and check-all-that-apply to collect reasons for liking and disliking chocolates in preference mapping. *Food Quality and Preference, 117*, 105171. https://doi.org/10.1016/J.FOODQUAL.2024.105171

Antons, D., Grünwald, E., Cichy, P., & Salge, T. O. (2020). The application of text mining methods in innovation research: Current state, evolution patterns, and development priorities. *R and D Management, 50*(3), 329–351. https://doi.org/10.1111/RADM.12408

Arellano-Covarrubias, A., Escalona-Buendía, H. B., Gómez-Corona, C., & Varela, P. (2022). Pairing beer and food in social media: Is it an image worth more than a thousand words? *International Journal of Gastronomy and Food Science, 27*, 100483. https://doi.org/10.1016/J.IJGFS.2022.100483

Ares, G., Giménez, A., & Gámbaro, A. (2008). Understanding consumers' perception of conventional and functional yogurts using word association and hard laddering. *Food Quality and Preference, 19*(7), 636–643. https://doi.org/10.1016/J.FOODQUAL.2008.05.005

Ares, G., Barreiro, C., Deliza, R., Giménez, A., & Gámbaro, A. (2010a). Consumer expectations and perception of chocolate milk desserts enriched with antioxidants. *Journal of Sensory Studies, 25*(SUPPL. 1), 243–260. https://doi.org/10.1111/J.1745-459X.2010.00293.X

Ares, G., Giménez, A., Barreiro, C., & Gámbaro, A. (2010b). Use of an open-ended question to identify drivers of liking of milk desserts. Comparison with preference mapping techniques. *Food Quality and Preference, 21*(3), 286–294. https://doi.org/10.1016/j.foodqual.2009.05.006

Ares, G., Giménez, A., & Deliza, R. (2010c). Influence of three non-sensory factors on consumer choice of functional yogurts over regular ones. *Food Quality and Preference, 21*(4), 361–367. https://doi.org/10.1016/J.FOODQUAL.2009.09.002

Ares, G., Varela, P., Rado, G., & Giménez, A. (2011). Identifying ideal products using three different consumer profiling methodologies. Comparison with external preference mapping. *Food Quality and Preference, 22*(6), 581–591. https://doi.org/10.1016/J.FOODQUAL.2011.04.004

Ares, G., de Saldamando, L., Giménez, A., Claret, A., Cunha, L. M., Guerrero, L., de Moura, A. P., Oliveira, D. C. R., Symoneaux, R., & Deliza, R. (2015). Consumers' associations with wellbeing in a food-related context: A cross-cultural study. *Food Quality and Preference, 40*(PB), 304–315. https://doi.org/10.1016/J.FOODQUAL.2014.06.001

Ares, G., Vidal, L., Allegue, G., Giménez, A., Bandeira, E., Moratorio, X., Molina, V., & Curutchet, M. R. (2016). Consumers' conceptualization of ultra-processed foods. *Appetite, 105*, 611–617. https://doi.org/10.1016/J.APPET.2016.06.028

Asseo, K., & Niv, M. Y. (2022). Harnessing food product reviews for personalizing sweetness levels. *Food, 11*(13), 1872. https://doi.org/10.3390/FOODS11131872/S1

Balbi, S., & Triunfo, N. (2013). Statistical tools in the joint analysis of closed and open-ended questions. In *Survey data collection and integration* (pp. 61–72). Springer Berlin Heidelberg. https://doi.org/10.1007/978-3-642-21308-3_4

Bărbulescu, A., & Morariu, D. I. (2020). Part of speech tagging using hidden Markov models. *International Journal of Advanced Statistics and IT&C for Economics and Life Sciences, 10*(1), 31–42. https://doi.org/10.2478/IJASITELS-2020-0005

Batista, L. F., Rocha, F., dos Dias, M. M. S., dos Pires, A. C. S., & Vidigal, M. C. T. R. (2023). Comfort plant-based food: What do consumers want? – A focus group study with different consumers group. *International Journal of Gastronomy and Food Science, 34*, 100810. https://doi.org/10.1016/J.IJGFS.2023.100810

Bayerl, P. S., & Paul, K. I. (2011). What determines inter-coder agreement in manual annotations? A Meta-Analytic Investigation. *Computational Linguistics, 37*(4), 699–725. https://doi.org/10.1162/COLI_A_00074

Bécue-Bertaut, M. (2014). Tracking verbal-based methods beyond conventional descriptive analysis in food science bibliography. A statistical approach. *Food Quality and Preference, 32*, 2–15. https://doi.org/10.1016/J.FOODQUAL.2013.08.010

Bécue-Bertaut, M., Álvarez-Esteban, R., & Pagès, J. (2008). Rating of products through scores and free-text assertions: Comparing and combining both. *Food Quality and Preference, 19*(1), 122–134. https://doi.org/10.1016/J.FOODQUAL.2007.07.006

Bengtsson, M. (2016). How to plan and perform a qualitative study using content analysis. *NursingPlus Open, 2*, 8–14. https://doi.org/10.1016/J.NPLS.2016.01.001

Bovi, C. D., & Navigli, R. (2017). Multilingual semantic dictionaries for natural language processing: The case of BabelNet. In *Semantic computing* (pp. 149–163). https://doi.org/10.1142/9789813227927_0017

Boyd, R. L., & Schwartz, H. A. (2021). Natural language analysis and the psychology of verbal behavior: The past, present, and future states of the field. *Journal of Language and Social Psychology, 40*(1), 21–41. https://doi.org/10.1177/0261927X20967028

Breiman, L. (2001). Random forests. *Machine Learning, 45*(1), 5–32. https://doi.org/10.1023/A:1010933404324

Brereton, R. G., & Lloyd, G. R. (2010). Support vector machines for classification and regression. *Analyst, 135*(2), 230–267. https://doi.org/10.1039/B918972F

Brislin, R. W. (1970). Back-translation for cross-cultural research. *Journal of Cross-Cultural Psychology, 1*(3), 185–216. https://doi.org/10.1177/135910457000100301

Calvert, M. D., Cole, E., Neill, C. L., Stewart, A. C., Whitehead, S. R., & Lahne, J. (2023). Exploring cider website descriptions using a novel text mining approach. *Journal of Sensory Studies, 38*(5), e12854. https://doi.org/10.1111/JOSS.12854

Campesato, O. (2022). Natural language processing using R pocket primer. In *Natural language processing using R pocket primer*. https://doi.org/10.1515/9781683927297

Cenikj, G., Petelin, G., Korousic Seljak, B., & Eftimov, T. (2022). SciFoodNER: Food named entity recognition for scientific text. In *Proceedings – 2022 IEEE international Conference on Big Data, Big Data 2022* (pp. 4065–4073). https://doi.org/10.1109/BIGDATA55660.2022.10020459

Chai, C. P. (2023). Comparison of text preprocessing methods. *Natural Language Engineering, 29*(3), 509–553. https://doi.org/10.1017/S1351324922000213

Chen, W. K., Riantama, D., & Chen, L. S. (2020). Using a text mining approach to hear voices of customers from social media toward the fast-food restaurant industry. *Sustainability, 13*(1), 268. https://doi.org/10.3390/SU13010268

Chen, Z., Gurdian, C., Sharma, C., Prinyawiwatkul, W., & Torrico, D. D. (2021). Exploring text mining for recent consumer and sensory studies about alternative proteins. *Foods, 10*(11), 2537. https://doi.org/10.3390/FOODS10112537

Choi, B. C. K., & Pak, A. W. P. (2004). A catalog of biases in questionnaires. *Preventing Chronic Disease, 2*(1), A13. https://pmc.ncbi.nlm.nih.gov/articles/PMC1323316/

Cunningham, P., & Delany, S. J. (2021). K-nearest neighbour classifiers-A tutorial. *ACM Computing Surveys, 54*(6). https://doi.org/10.1145/3459665

Das, M., Kamalanathan, S., & Alphonse, P. (2023). A comparative study on TF-IDF feature weighting method and its analysis using unstructured dataset. *CEUR Workshop Proceedings, 2870*, 98–107. https://arxiv.org/abs/2308.04037v1

de Andrade, J. C., de Aguiar Sobral, L., Ares, G., & Deliza, R. (2016). Understanding consumers' perception of lamb meat using free word association. *Meat Science, 117*, 68–74. https://doi.org/10.1016/J.MEATSCI.2016.02.039

Deneulin, P., & Bavaud, F. (2016). Analyses of open-ended questions by renormalized associativities and textual networks: A study of perception of minerality in wine. *Food Quality and Preference, 47*, 34–44. https://doi.org/10.1016/J.FOODQUAL.2015.06.013

Derczynski, L. (2016). *Complementarity, F-score, and NLP evaluation* (pp. 261–266). https://aclanthology.org/L16-1040/

Derioh, M. M. G., Mohd Nasir, P. N. S. N., Roche, C. A., Mohd Nizam, M. A. H., & Syed Abd Rahman, S. H. (2023). How English urban slang has affected verbal and written communication among youths in Selangor. *International Journal of Academic Research in Business and Social Sciences, 13*(7). https://doi.org/10.6007/IJARBSS/V13-I7/17234

Devlin, J., Chang, M. W., Lee, K., & Toutanova, K. (2018). BERT: Pre-training of Deep Bidirectional Transformers for Language Understanding. In *NAACL HLT 2019–2019 Conference of the North American Chapter of the Association for Computational Linguistics: Human Language Technologies – Proceedings of the Conference* (Vol. 1, pp. 4171–4186). https://arxiv.org/abs/1810.04805v2

Diwersy, S., Goossens, V., Grutschus, A., Kern, B., Kraif, O., Melnikova, E., & Novakova, I. (2014). Traitement des lexies d'émotion dans les corpus et les applications d'EmoBase. *Openedition Journals, 13*, 269–293. https://doi.org/10.4000/CORPUS.2537

Eftimov, T., Popovski, G., Valenčič, E., & Seljak, B. K. (2020). FoodEx2vec: New foods' representation for advanced food data analysis. *Food and Chemical Toxicology, 138*, 111169. https://doi.org/10.1016/J.FCT.2020.111169

Erlingsson, C., & Brysiewicz, P. (2017). A hands-on guide to doing content analysis. *African Journal of Emergency Medicine, 7*(3), 93–99. https://doi.org/10.1016/J.AFJEM.2017.08.001

Espinosa-Ramos, J. I., Hutchings, S. C., Dixit, Y., Jin, D., Ryan, G. S., Marsden, M., Jaeger, S. R., & Reis, M. M. (2025). Comparison of free comment online product reviews and central location product testing for sensory product characterisation: A case study with coffee consumers. *Food Quality and Preference, 125*, 105377. https://doi.org/10.1016/J.FOODQUAL.2024.105377

Fife, S. T., & Gossner, J. D. (2024). Deductive qualitative Analy: Evaluating, expanding, and refining theory. *International Journal of Qualitative Methods, 23*. https://doi.org/10.1177/16094069241244856

Galmarini, M. V., Symoneaux, R., Chollet, S., & Zamora, M. C. (2013). Understanding apple consumers' expectations in terms of likes and dislikes. Use of comment analysis in a cross-cultural study. *Appetite, 62*, 27–36. https://doi.org/10.1016/j.appet.2012.11.006

Gao, Y., Xiong, Y., Gao, X., Jia, K., Pan, J., Bi, Y., Dai, Y., Sun, J., Wang, M., & Wang, H. (2023). *Retrieval-augmented generation for large language models: A survey.* https://arxiv.org/abs/2312.10997v5

Glaser, B. G., & Strauss, A. L. (2017). Discovery of grounded theory: Strategies for qualitative research. In *Discovery of grounded theory: Strategies for qualitative research* (pp. 1–271). https://doi.org/10.4324/9780203793206

Godard, R., & Holtzman, S. (2022). The multidimensional lexicon of Emojis: A new tool to assess the emotional content of Emojis. *Frontiers in Psychology, 13*, 921388. https://doi.org/10.3389/FPSYG.2022.921388

Gonzaga, L. S., Capone, D. L., Bastian, S. E. P., Danner, L., & Jeffery, D. W. (2019). Using content analysis to characterise the sensory Typicity and quality judgements of Australian cabernet sauvignon wines. *Foods, 8*(12), 691. https://doi.org/10.3390/FOODS8120691

Grasso, S., & Jaworska, S. (2020). Part meat and part plant: Are hybrid meat products fad or future? *Foods, 9*(12), 1888. https://doi.org/10.3390/FOODS9121888

Hamilton, L. M., & Lahne, J. (2020). Fast and automated sensory analysis: Using natural language processing for descriptive lexicon development. *Food Quality and Preference, 83*, 103926. https://doi.org/10.1016/J.FOODQUAL.2020.103926

Hamilton, L. M., Neill, C. L., & Lahne, J. (2023). Flavor language in expert reviews versus consumer preferences: An application to expensive American whiskeys. *Food Quality and Preference, 109*, 104892. https://doi.org/10.1016/j.foodqual.2023.104892

Hartmann, J., Huppertz, J., Schamp, C., & Heitmann, M. (2019). Comparing automated text classification methods. *International Journal of Research in Marketing, 36*(1), 20–38. https://doi.org/10.1016/J.IJRESMAR.2018.09.009

Hládek, D., Staš, J., & Pleva, M. (2020). Survey of automatic spelling correction. *Electronics, 9*(10), 1670. https://doi.org/10.3390/ELECTRONICS9101670

Hutchings, S. C., Dixit, Y., Al-Sarayreh, M., Torrico, D. D., Realini, C. E., Jaeger, S. R., & Reis, M. M. (2023). A critical review of social media research in sensory-consumer science. *Food Research International, 165*, 112494. https://doi.org/10.1016/J.FOODRES.2023.112494

Ikotun, A. M., Ezugwu, A. E., Abualigah, L., Abuhaija, B., & Heming, J. (2023). K-means clustering algorithms: A comprehensive review, variants analysis, and advances in the era of big data. *Information Sciences, 622*, 178–210. https://doi.org/10.1016/J.INS.2022.11.139

Indolia, S., Goswami, A. K., Mishra, S. P., & Asopa, P. (2018). Conceptual understanding of convolutional neural network- A deep learning approach. *Procedia Computer Science, 132*, 679–688. https://doi.org/10.1016/J.PROCS.2018.05.069

Islam, M., Chen, G., & Jin, S. (2019). An overview of neural network. *American Journal of Neural Networks and Applications, 5*(1), 7. https://doi.org/10.11648/J.AJNNA.20190501.12

Jaeger, S. R., & Rasmussen, M. A. (2021). Importance of data preparation when analysing written responses to open-ended questions: An empirical assessment and comparison with manual coding. *Food Quality and Preference, 93*, 104270. https://doi.org/10.1016/J.FOODQUAL.2021.104270

Jehangir, B., Radhakrishnan, S., & Agarwal, R. (2023). A survey on named entity recognition – Datasets, tools, and methodologies. *Natural Language Processing Journal, 3*, 100017. https://doi.org/10.1016/J.NLP.2023.100017

Jivani, A. (2011). A comparative study of stemming algorithms.

Jurish, B., & Würzner, K.-M. (2013). Word and sentence tokenization with hidden Markov models. *Journal for Language Technology and Computational Linguistics, 28*(2), 61–83. https://doi.org/10.21248/JLCL.28.2013.176

Kāle, M., Šķilters, J., & Rikters, M. (2021). Tracing multisensory food experiences on Twitter. *International Journal of Food Design, 6*(2), 181–212. https://doi.org/10.1386/IJFD_00030_1

Kathuria, A., Gupta, A., & Singla, R. K. (2021). A review of tools and techniques for preprocessing of textual data. *Advances in Intelligent Systems and Computing, 1227*, 407–422. https://doi.org/10.1007/978-981-15-6876-3_31

Kaur, G. (2014). Usage of regular expressions in NLP. *International Journal of Research in Engineering and Technology, 03*(01), 168–174. https://doi.org/10.15623/IJRET.2014.0301026

Khurana, D., Koli, A., Khatter, K., & Singh, S. (2023). Natural language processing: State of the art, current trends and challenges. *Multimedia Tools and Applications, 82*(3), 3713–3744. https://doi.org/10.1007/S11042-022-13428-4

Kim, A. Y., Ha, J. G., Choi, H., & Moon, H. (2018). Automated text analysis based on skip-gram model for food evaluation in predicting consumer acceptance. *Computational Intelligence and Neuroscience, 2018*(1), 9293437. https://doi.org/10.1155/2018/9293437

Kozma, L., & Voderholzer, J. (2024). *Theoretical analysis of byte-pair encoding*. https://arxiv.org/abs/2411.08671v1

Krippendorff, K. (2019). Content analysis: An introduction to its methodology. In *Content analysis: An introduction to its methodology*. https://doi.org/10.4135/9781071878781

Laguna, L., Fiszman, S., Puerta, P., Chaya, C., & Tárrega, A. (2020). The impact of COVID-19 lockdown on food priorities. Results from a preliminary study using social media and an online survey with Spanish consumers. *Food Quality and Preference, 86*, 104028. https://doi.org/10.1016/J.FOODQUAL.2020.104028

Lahne, J., Trubek, A. B., & Pelchat, M. L. (2013). Consumer sensory perception of cheese depends on context: A study using comment analysis and linear mixed models. *Food Quality and Preference, 32*, 184–197. https://doi.org/10.1016/j.foodqual.2013.10.007

Lee, R. S. T. (2024). *Natural language processing: A textbook with python implementation*. Springer Nature Singapore.

Legate, A., Nimon, K., & Noblin, A. (2024). (Semi)automated approaches to data extraction for systematic reviews and meta-analyses in social sciences: A living review. *F1000Research, 13*, 664. https://doi.org/10.12688/F1000RESEARCH.151493.2
Liliana, D. Y., & Basaruddin, C. (2017). A review on conditional random fields as a sequential classifier in machine learning. In *ICECOS 2017 – Proceeding of 2017 international conference on electrical engineering and computer science: Sustaining the cultural heritage toward the smart environment for better future* (pp. 143–148). https://doi.org/10.1109/ICECOS.2017.8167121
Lotfi, C., Srinivasan, S., Ertz, M., & Latrous, I. (2021). Web scraping techniques and applications: A literature review. In *SCRS conference proceedings on intelligent systems* (pp. 381–394). https://doi.org/10.52458/978-93-91842-08-6-38
Ma, P., Tsai, S., He, Y., Jia, X., Zhen, D., Yu, N., Wang, Q., Ahuja, J. K. C., & Wei, C. I. (2024). Large language models in food science: Innovations, applications, and future. *Trends in Food Science & Technology, 148*, 104488. https://doi.org/10.1016/J.TIFS.2024.104488
Maalouf, M. (2011). Logistic regression in data analysis: An overview. *International Journal of Data Analysis Techniques and Strategies, 3*(3), 281–299. https://doi.org/10.1504/IJDATS.2011.041335
Mahieu, B., Visalli, M., Thomas, A., & Schlich, P. (2019). Free comment using speech recognition: An alternative to check-all-that-apply (CATA) for sensory characterization of red wines at home. In *13th Pangborn sensory science symposium "engage with the future" 28 July–1 August 2019*. Edinburgh International Convention Centre, Edinburgh, UK.
Mahieu, B., Visalli, M., Thomas, A., & Schlich, P. (2020a). Free-comment outperformed check-all-that-apply in the sensory characterisation of wines with consumers at home. *Food Quality and Preference, 84*, 103937. https://doi.org/10.1016/J.FOODQUAL.2020.103937
Mahieu, B., Visalli, M., Thomas, A., & Schlich, P. (2020b). Using free comment with consumers to obtain temporal sensory descriptions of products. *Food Quality and Preference, 86*, 104008. https://doi.org/10.1016/J.FOODQUAL.2020.104008
Mahieu, B., Visalli, M., & Schlich, P. (2022). Identifying drivers of liking and characterizing the ideal product thanks to free comment. *Food Quality and Preference, 96*, 104389. https://doi.org/10.1016/J.FOODQUAL.2021.104389
McHugh, M. L. (2012). Interrater reliability: The kappa statistic. *Biochemia Medica, 22*(3), 276. https://doi.org/10.11613/bm.2012.031
Mehraliyev, F., Kirilenko, A. P., & Choi, Y. (2020). From measurement scale to sentiment scale: Examining the effect of sensory experiences on online review rating behavior. *Tourism Management, 79*, 104096. https://doi.org/10.1016/J.TOURMAN.2020.104096
Miller, G. A., Beckwith, R., Fellbaum, C., Gross, D., & Miller, K. J. (1990). Introduction to wordnet: An on-line lexical database. *International Journal of Lexicography, 3*(4), 235–244. https://doi.org/10.1093/IJL/3.4.235
Miller, C., Hamilton, L., & Lahne, J. (2021). Sensory descriptor analysis of whisky lexicons through the use of deep learning. *Foods, 10*(7), 1633. https://doi.org/10.3390/FOODS10071633
Morales-Garzón, A., Gómez-Romero, J., & Martin-Bautista, M. J. (2020). A word embedding model for mapping food composition databases using fuzzy logic. In *Information processing and management of uncertainty in knowledge-based systems* (Vol. 1238, p. 635). Springer International Publishing. https://doi.org/10.1007/978-3-030-50143-3_50
Murat, H. S. (2006). A brief review of feed-forward neural networks. *Communications Faculty Of Science University of Ankara, 50*(1), 11–17. https://doi.org/10.1501/COMMUA1-2_0000000026
Nesca, M., Katz, A., Leung, C. K., & Lix, L. M. (2022). A scoping review of preprocessing methods for unstructured text data to assess data quality. *International Journal of Population Data Science, 7*(1), 1757. https://doi.org/10.23889/IJPDS.V6I1.1757
Nithyanandan, S. (2019). Deep learning models for word sense disambiguation: A comparative study. *SSRN Electronic Journal*. https://doi.org/10.2139/SSRN.3437615
Noble, A. C., Arnold, R. A., Buechsenstein, J., Leach, E. J., Schmidt, J. O., Stern, P. M. (1987). Modification of a Standardized System of Wine Aroma Terminology American Journal of Enology and Viticulture 38(2):143–146. https://doi.org/10.5344/ajev.1987.38.2.143

Nojavanasghari, B., Baltrušaitis, T., Hughes, C. E., & Morency, L. P. (2016). Emo react: A multimodal approach and dataset for recognizing emotional responses in children. In *ICMI 2016 – Proceedings of the 18th ACM international Conference on multimodal interaction* (pp. 137–144). https://doi.org/10.1145/2993148.2993168

Olson, E. L. (2017). The rationalization and persistence of organic food beliefs in the face of contrary evidence. *Journal of Cleaner Production, 140*, 1007–1013. https://doi.org/10.1016/J.JCLEPRO.2016.06.005

Pennington, J., Socher, R., & Manning, C. D. (2014). GloVe: Global vectors for word representation. In *EMNLP 2014 – 2014 conference on empirical methods in natural language processing, proceedings of the conference* (pp. 1532–1543). https://doi.org/10.3115/V1/D14-1162

Peretz, O., Koren, M., & Koren, O. (2024). Naive Bayes classifier—An ensemble procedure for recall and precision enrichment. *Engineering Applications of Artificial Intelligence, 136*, 108972. https://doi.org/10.1016/J.ENGAPPAI.2024.108972

Polli, G. M., Maria da Silveira, F., Magnabosco, F. M., Soares dos Santos, G. H., Stella, P. D., Pinto, P. R., Zibetti, M. R., & Apostolidis, T. (2021). Representations of food among vegetarians in Brazil: A psychosocial approach. *International Journal of Gastronomy and Food Science, 24*, 100317. https://doi.org/10.1016/J.IJGFS.2021.100317

Qader, W. A., Ameen, M. M., & Ahmed, B. I. (2019). An overview of bag of words;importance, implementation, applications, and challenges. In *Proceedings of the 5th international engineering Conference, IEC* (pp. 200–204). https://doi.org/10.1109/IEC47844.2019.8950616

Ran, X., Xi, Y., Lu, Y., Wang, X., & Lu, Z. (2022). Comprehensive survey on hierarchical clustering algorithms and the recent developments. *Artificial Intelligence Review, 56*(8), 8219–8264. https://doi.org/10.1007/S10462-022-10366-3

Ratnaparkhi, A. (1996). A maximum entropy model for part-of-speech tagging. Conference on Empirical Methods in Natural Language Processing.

Reed, D. R., Mainland, J. D., & Arayata, C. J. (2019). Sensory nutrition: The role of taste in the reviews of commercial food products. *Physiology & Behavior, 209*, 112579. https://doi.org/10.1016/J.PHYSBEH.2019.112579

Sherstinsky, A. (2020). Fundamentals of recurrent neural network (RNN) and long short-term memory (LSTM) network. *Physica D: Nonlinear Phenomena, 404*, 132306. https://doi.org/10.1016/J.PHYSD.2019.132306

Sidorov, G., Velasquez, F., Stamatatos, E., Gelbukh, A., & Chanona-Hernández, L. (2014). Syntactic N-grams as machine learning features for natural language processing. *Expert Systems with Applications, 41*(3), 853–860. https://doi.org/10.1016/J.ESWA.2013.08.015

Siino, M., Tinnirello, I., & La Cascia, M. (2024). Is text preprocessing still worth the time? A comparative survey on the influence of popular preprocessing methods on transformers and traditional classifiers. *Information Systems, 121*, 102342. https://doi.org/10.1016/J.IS.2023.102342

Sivakumar, S., Videla, L. S., Rajesh Kumar, T., Nagaraj, J., Itnal, S., & Haritha, D. (2020). Review on Word2Vec word embedding neural net. In *Proceedings – International Conference on Smart Electronics and Communication, ICOSEC* (pp. 282–290). https://doi.org/10.1109/ICOSEC49089.2020.9215319

Smith, R. E. (2011). Urban dictionary: Youth slanguage and the redefining of definition: What's up with meep and other words in the Urban Dictionary. *English Today, 27*(4), 43–48. https://doi.org/10.1017/S0266078411000526

Smith, A., de Lhoneux, M., Stymne, S., & Nivre, J. (2018). An investigation of the interactions between pre-trained word Embeddings, character models and POS tags in dependency parsing. In *Proceedings of the 2018 Conference on empirical methods in natural language processing* (pp. 2711–2720). EMNLP. https://doi.org/10.18653/V1/D18-1291

Son, J. S., Do, V. B., Kim, K. O., Cho, M. S., Suwonsichon, T., & Valentin, D. (2014). Understanding the effect of culture on food representations using word associations: The case of "rice" and "good rice". *Food Quality and Preference, 31*(1), 38–48. https://doi.org/10.1016/J.FOODQUAL.2013.07.001

Song, X., Salcianu, A., Song, Y., Dopson, D., & Zhou, D. (2020). Fast WordPiece Tokenization. In *EMNLP 2021–2021 conference on empirical methods in natural language processing* (pp. 2089–2103). https://doi.org/10.18653/v1/2021.emnlp-main.160

Spinelli, S., Dinnella, C., Masi, C., Zoboli, G. P., Prescott, J., & Monteleone, E. (2017). Investigating preferred coffee consumption contexts using open-ended questions. *Food Quality and Preference, 61*, 63–73. https://doi.org/10.1016/J.FOODQUAL.2017.05.003

Stanković, R. S., Branislavašandrih, B., Krstev, C., Utvi'c, U., Mihailo, M., & Mihailoškori'c, M. (2020). Machine learning and deep neural network-based lemmatization and Morphosyntactic tagging for Serbian. In *Language Resources and Evaluation (LREC 2020)* (pp. 3954–3962) https://aclanthology.org/2020.lrec-1.487/

Symoneaux, R., Galmarini, M. V., & Mehinagic, E. (2012). Comment analysis of consumer's likes and dislikes as an alternative tool to preference mapping. A case study on apples. *Food Quality and Preference, 24*(1), 59–66. https://doi.org/10.1016/J.FOODQUAL.2011.08.013

Tan, Z., Wang, S., Yang, Z., Chen, G., Huang, X., Sun, M., & Liu, Y. (2020). Neural machine translation: A review of methods, resources, and tools. *AI Open, 1*, 5–21. https://doi.org/10.1016/J.AIOPEN.2020.11.001

ten Kleij, F., & Musters, P. A. D. (2003). Text analysis of open-ended survey responses: A complementary method to preference mapping. *Food Quality and Preference, 14*(1), 43–52. https://doi.org/10.1016/S0950-3293(02)00011-3

Thomas, D. R. (2006). A general inductive approach for analyzing qualitative evaluation data. *American Journal of Evaluation, 27*(2), 237–246. https://doi.org/10.1177/1098214005283748

Tian, G., Lu, L., & McIntosh, C. (2021). What factors affect consumers' dining sentiments and their ratings: Evidence from restaurant online review data. *Food Quality and Preference, 88*, 104060. https://doi.org/10.1016/J.FOODQUAL.2020.104060

Tomanek, K., Wermter, J., & Hahn, U. (2007). Sentence and token splitting based on conditional random fields.

Toporkov, O., & Agerri, R. (2024). On the role of morphological information for contextual lemmatization. *Computational Linguistics, 50*(1), 157–191. https://doi.org/10.1162/COLI_A_00497

Valentin, D., Chollet, S., Lelièvre, M., & Abdi, H. (2012). Quick and dirty but still pretty good: A review of new descriptive methods in food science. *International Journal of Food Science & Technology, 47*(8), 1563–1578. https://doi.org/10.1111/J.1365-2621.2012.03022.X

Vaswani, A., Brain, G., Shazeer, N., Parmar, N., Uszkoreit, J., Jones, L., Gomez, A. N., Kaiser, Ł., & Polosukhin, I. (2017). *Attention is all you need.* Advances in neural information processing systems, 30.

Veerasekharreddy, B., Rao, K. S., & Koppula, N. (2022). Parts of speech tagging using conditional random field with long short term memory for English texts. In *2022 13th international Conference on computing communication and networking technologies, ICCCNT 2022*. https://doi.org/10.1109/ICCCNT54827.2022.9984380

Vidal, L., Ares, G., Machín, L., & Jaeger, S. R. (2015). Using twitter data for food-related consumer research: A case study on "what people say when tweeting about different eating situations". *Food Quality and Preference, 45*, 58–69. https://doi.org/10.1016/J.FOODQUAL.2015.05.006

Visalli, M., Dubois, M., Schlich, P., Ric, F., Cardebat, J. M., & Georgantzis, N. (2023a). Relevance of free comment to describe wine temporal sensory perception: An application with panels varying in culture and expertise. *Food Quality and Preference, 105*, 104785. https://doi.org/10.1016/j.foodqual.2022.104785

Visalli, M., Schlich, P., Mahieu, B., Thomas, A., Weber, M., & Guichard, E. (2023b). First steps towards FAIRization of product-focused sensory data. *Food Quality and Preference, 104*, 104765. https://doi.org/10.1016/j.foodqual.2022.104765

Visalli, M., Symoneaux, R., Mursic, C., Touret, M., Lourtioux, F., Coulibaly, K., & Mahieu, B. (2025). Can natural language processing or large language models replace human operators for pre-processing word and sentence-based free comments sensory evaluation data? *Food Quality and Preference, 127*, 105456. https://doi.org/10.1016/J.FOODQUAL.2025.105456

Vuddanti, S., Ariveni, D. S. L., Kethepalli, N. S. S., Manchimi, L., & Yajjavarapu, J. (2024). Multilingual language detection and translation system with multinomial naive Bayes. In *2024 5th International Conference on Image Processing and Capsule Networks (ICIPCN)* (pp. 271–275). https://doi.org/10.1109/ICIPCN63822.2024.00051

Wang, W. (2023). Different natural languages, equal importance. *Patterns, 4*(8), 100821. https://doi.org/10.1016/J.PATTER.2023.100821

Xu, J., & Du, Q. (2019). A deep investigation into fasttext. In *Proceedings –21st IEEE international Conference on high performance computing and communications, 17th IEEE international Conference on Smart City and 5th IEEE international Conference on data science and systems, HPCC/SmartCity/DSS 2019* (pp. 1714–1719). https://doi.org/10.1109/HPCC/SMARTCITY/DSS.2019.00234

Yatawara, K., Sampath, T., Kalupahana, P. L., Rathnayake, S., Jayasuriya, N., & Rathnayake, N. (2025). A systematic review on consumer adoption of AI-driven Chatbots. *Vision*. https://doi.org/10.1177/09722629251332349

Yoo, R., Kim, S. Y., Kim, D. H., Kim, J., Jeon, Y. J., Park, J. H. Y., Lee, K. W., & Yang, H. (2023). Exploring the nexus between food and veg*n lifestyle via text mining-based online community analytics. *Food Quality and Preference, 104*, 104714. https://doi.org/10.1016/J.FOODQUAL.2022.104714

Youn, J., Naravane, T., & Tagkopoulos, I. (2020). Using word Embeddings to learn a better food ontology. *Frontiers in Artificial Intelligence, 3*, 584784. https://doi.org/10.3389/FRAI.2020.584784

Zhang, X., Du, Y., Sun, W., & Wan, X. (2016). Transition-based parsing for deep dependency structures. *Computational Linguistics, 42*(3), 353–389. https://doi.org/10.1162/COLI_A_00252

Part IV
Food Related Text Data Collection: Experimental Aspects

This section focuses on practical and experimental aspects related to the use of food-related text data in sensory and consumer research.

Chapter 6 presents the results of a survey aimed at understanding the perspectives and needs of sensory practitioners regarding various methods of text data collection. It also outlines a step-by-step framework for analysing such data. Chapter 7 introduces text highlighting, a novel evaluation method that offers new possibilities for capturing consumer responses; its implementation and potential applications are illustrated with published case studies. Finally, Chapter 8 explores how affective meaning can be extracted from free comments and online reviews using sentiment analysis, with large-scale applications of these techniques further developed in Part V.

Chapter 6
Collection and Analysis of Free Comments for Sensory and Hedonic Description

Benjamin Mahieu and Ronan Symoneaux

1 Introduction

Based on the reported literature, the Free Comment (FC) method was introduced in sensory and consumer science to describe the sensory characteristics of products by Ten Kleij and Musters (2003). These authors' motivation was that free responses "are not often used for detailed analyses" while they "undoubtedly contain very rich information" with the additional benefit of being "stated in consumer language". Indeed, the essence of FC consists in instructing panellists to describe the products in their own terms answering open-ended questions.

Since then, FC has gained more attention in the sensory and consumer science community and now belongs to the usual toolbox of sensory and consumer scientists. In this toolbox, FC is positioned as a descriptive sensory characterization method considering that sensory characteristics of products can be described by a finite number of identifiable sensory descriptors. To achieve this characterization, FC relies on a principle of presence/absence of sensory descriptors in the products at the panellists' level. The closest sensory characterization method to FC that shares this presence/absence rationale is Check-All-That-Apply (CATA) (Adams et al., 2007). The difference between FC and CATA is the way the sensory descriptors used to describe the products of interest are established. In CATA, sensory descriptors are established a priori by panel leaders based on existing knowledge of the products (literature, previous studies, etc.) or thanks to pre-evaluations. In FC, sensory descriptors are established a posteriori using either Natural Language

B. Mahieu (✉)
Oniris, INRAE, StatSC, Nantes, France
e-mail: benjamin.mahieu@oniris-nantes.fr

R. Symoneaux
GRAPPE, ESA, USC 1422 INRAE, SensoVeg, SFR 4207 QUASAV, Angers, France

M. Visalli, M. V. Galmarini (eds.), *Collection and Analysis of Text Data in Sensory and Consumer Science*, https://doi.org/10.1007/978-3-032-04851-6_6

Processing (NLP) (Hamilton et al., 2023; Hamilton & Delarue, 2025; Hamilton & Lahne, 2020; Jaeger & Rasmussen, 2021; Lahne et al., 2014; Mahieuet al., 2020a), expert coding of the collected free comments (Ares et al., 2010; Hanaei et al., 2015; Rocha et al., 2023; Symoneaux et al., 2012) or Large Language Models (LLM) (Visalli et al., 2025b).

In this context, the main advantage of FC is not to rely on a previous list of sensory descriptors. Indeed, using an a priori list of sensory descriptors leads to several issues that are de facto avoided when using FC. For example, since an a priori list of sensory descriptors contains only a limited number of sensory descriptors, it may result in a loss of information and is likely biased by the dumping effect (Campo et al., 2010; Coulon-Leroy et al., 2017; Krosnick, 1999; Varela et al., 2018) and the acquiescence bias (Callegaro et al., 2015; Kim et al., 2017; Klayman & Ha, 1987; Koriat et al., 1980; Yzerbyt & Leyens, 1991). Another issue is that the a priori list of sensory descriptors may suggest attributes to panellists they would not have thought otherwise (Coulon-Leroy et al., 2017; Kim et al., 2017; Krosnick, 1999; Reja et al., 2003; Schuman & Presser, 1979; Züll, 2016) and the first words presented (in a list) are likely to receive more attention from the evaluators (Castura, 2009; Kim et al., 2017; Krosnick, 1999). In addition, FC is straightforward for the panellists and easy to implement for the panel leaders, which constitute non-negligible practical benefits.

Despite its benefits, FC also has some drawbacks. The main drawback of FC is the pretreatment of text data to establish the list of sensory descriptors. Indeed, refined pretreatment by expert coding takes time, is cumbersome and non-deterministic while automated systems, either based on NLP or LLM, are faster, less cumbersome and deterministic (for NLP but not for LLM) but at the price of a rougher information in some situations (Visalli et al., 2025b). Therefore, overall, no optimal pretreatment exists for FC. Other drawbacks of FC are that some panellists may encounter some difficulties to verbalize properly their perception (Krosnick, 1999; Reja et al., 2003) while instructions require high precision to avoid uninformative descriptions from the panellists (Reja et al., 2003; Züll, 2016).

Different versions of the FC method have been described in the literature. They differ in how much they restrict panellists in terms of what to describe and how to do it. The less restrictive version allows panellists to describe any sensory modality (e.g., visual aspect, flavour) with as many terms as they wish without any imposed format or limitations on the length of the response (Ten Kleij & Musters, 2003). Another version of FC imposes panellists to describe the products with respect to different sensory modalities (Hanaei et al., 2015; Mahieu et al., 2020a, 2022). With the same rationale of restricting panellists to somewhat channel their thoughts, a version of FC restricts them to provide their descriptions with a limited number of terms (Ares et al., 2010) which is probably borrowed from the word association task (Ares et al., 2008). A third version imposes panellists to describe products through a hedonic point of view by providing separate descriptions of what they like and of what they dislike (Lahne et al., 2014; Symoneaux et al., 2012). In the same vein (i.e.

hedonic oriented), some version of FC introduces the notion of an ideal product: Free-JAR (Luc et al., 2020) and Ideal-Free Comment (IFC) (Mahieu et al., 2022). Finally, FC descriptions may be imposed to panellists to be provided including a temporal intra-intake component (Mahieu et al., 2020b; Visalli et al., 2023a). Note that in all previous studies, regardless of the FC version used, FC is at the core of the study and not a subsidiary element.

Despite of how FC data are collected and pretreated, it is usually submitted to statistical analyses pertaining the chi-square framework, either usual (Mahieu et al., 2020) or multiple-response one (Mahieu et al., 2021), the multiple-response one being better suited to the nature of FC data. While the FC literature concentrates on chi-square analyses, statistical analyses undertaken for CATA data analysis (Meyners et al., 2013; Meyners & Hasted, 2021, 2022) can be considered as an alternative since, after pretreatment, FC data have the exact same nature and structure as CATA data.

Beyond what is reported in the literature, little is known about how practitioners actually use the FC method. The specific ways in which it is applied—in terms of the FC version, its placement within questionnaires, pretreatment procedures, and the (statistical) analysis of the pretreated data—remain undocumented. The first aim of this chapter is to present the use of FC from a day-to-day perspective as used by professionals in 2024–2025. This provides an opportunity to gather their perception of the advantages, limitations, and perspectives associated with this method. This first part especially highlights the challenges associated with the pretreatment of comments and the lack of knowledge regarding appropriate statistical analyses.

The second aim of the chapter is to outline the current state of statistical analyses that can be performed on pretreated FC data, drawing from both published literature and the day-to-day practitioners' uses. A case example based on hedonic FC data (i.e., descriptions of likes and dislike) is also included, following by a conclusion section.

2 Background on the Current Use of FC for Sensory and Hedonic Description: A Survey

As an introduction to this chapter, rather than conducting an exhaustive literature review, a study was carried out among sensory analysis professionals in France and Europe in Autumn 2024. The objectives of this survey were to examine the use of open-ended questions in sensory evaluation, identify the motivations and barriers associated with their use, explore their application contexts, document methodological practices related to data acquisition and processing, and highlight current or future avenues for exploration.

2.1 Methodology

2.1.1 Survey Design

A comprehensive online survey was developed and distributed to members of the sensory and consumer science community. The survey targeted professionals affiliated with organizations such as the Société Française d'Analyse Sensorielle (SFAS) and the European Sensory Science Society (E3S). Entitled "*Use of Free Comments and Open-Ended Questions in Sensory Evaluation*", the questionnaire was distributed through the SFAS member list and the E3S newsletter.

The survey primarily consisted of open-ended prompts designed to elicit detailed qualitative responses. Respondents were asked to describe the various contexts in which they use open-ended questions for evaluating products and concepts (e.g., hedonic, descriptive, or other applications) and to explain their reasons for incorporating such questions into their tests. Additionally, they were asked about the integration of open-ended questions with closed-ended formats such as CATA, JAR, overall hedonic ratings, and hedonic ratings by sub-category (e.g., appearance, texture).

From a methodological perspective, respondents were questioned about the data acquisition tools they used (e.g., paper-based, digital, oral), the types of question formulations employed, and how they solicit comments (e.g., through a single question or multiple prompts). They also described the response formats they use, including sentence-based responses with or without length constraints and shorter responses in limited text fields. Furthermore, they were asked about experimental constraints affecting the use of FC.

Participants also provided insights on their practices for correcting and preprocessing responses before statistical analysis and interpretation. This included methods such as manual processing, computerized tools, software solutions, artificial intelligence, or a combination of several approaches. They were further questioned about the statistical methods and software packages they used to analyse open-ended responses.

Finally, respondents shared their remarks, questions, doubts, and expectations regarding the use of open-ended questions in product evaluation.

2.1.2 Participants

The survey received 39 responses, including 31 fully completed questionnaires and eight partially completed ones. Respondents represented a diverse cross-section of the sensory analysis community, including 16 professionals from the industrial sector, 10 service providers, three academics, and two students. The majority had experience using open-ended questions, with 34 indicating they incorporate them into their practices, while five reported that they do not use them. Among the participants, 25 were affiliated with SFAS, while six belonged to the European Sensory

Science Society. This diversity provided a solid foundation for analysing the various applications and perceptions of open-ended questions across different professional contexts.

2.1.3 Data Analysis

Responses underwent thematic content analysis to identify recurring patterns and insights. This process involved manually categorizing responses into key themes, including motivations for using FC, application contexts, methodological practices, perceived limitations, and potential directions for future research.

2.2 Results

2.2.1 The Main Motivations for Using FC

The primary motivations for using FC in sensory evaluation are multifaceted, reflecting the advantages of open-ended questions over their closed-ended counterparts. FC are valued for their ability to provide richer, more expansive, and exhaustive responses, complementing the insights gained from closed-ended questions. Respondents emphasize that open-ended questions allow for a broader capture of diverse opinions and expressions, ensuring that no critical information is forgotten.

The spontaneity of responses is a key advantage, as it allows for uninfluenced consumer feedback. FC capture spontaneous verbatim, providing natural and less biased insights into consumer perceptions. This approach is particularly useful for gathering immediate and varied responses that are not constrained by predefined answer choices. It allows consumers to express their thoughts in their own words, offering a more authentic reflection of their experiences and preferences.

Some respondents also mentioned that FC are also instrumental in identifying weak signals that might be missed by traditional closed-ended questions. They help detect subtle or unexpected insights that can differentiate products in ways that standard questions, such as JAR or CATA, cannot. This ability to uncover hidden aspects of consumer perception makes FC a valuable tool for comprehensive product evaluation.

While not necessarily a pertinent justification, some individuals explain that they use FC out of habit or because it is mandated by others. External demands, such as client requirements and adherence to norms and standards, also drive the use of FC. Often, clients or project teams request this kind of responses to meet specific needs or to follow established practices.

One respondent specifies a particular use of FC in the context of trained panels. They are used to validate the coherence of panellists' responses and to integrate new attributes into evaluations.

2.2.2 Fairly Consensual Uses of These Open Questions

The use of FC is primarily associated with product or concept evaluation. Respondents frequently report employing it in product testing to gather insights after consumers taste or experience a product. FC can be collected in both home-use tests and laboratory settings. In some instances, FC extends beyond products to encompass broader concepts or offers. They are used to gather customer feedback when presenting concepts, assessing product-concept fit, and exploring usage scenarios. This broader application allows for a more comprehensive understanding of consumer perceptions and preferences beyond the confines of specific product attributes.

FC are extensively used in hedonic testing contexts. They are employed to justify liking and preference ratings, helping to identify what consumers appreciate or dislike about a product. Questions are often framed to capture detailed reasons behind consumer preferences, focusing on sensory dimensions such as appearance, flavour and texture and others. The use of FC is often coupled with other questions. They are asked following a global appreciation question in hedonic tests to delve deeper into what pleases or displeases consumers. Other respondents use them after several hedonic questions, one for each sensory dimension. This approach allows for a deeper understanding of consumer preferences and perceptions, complementing the quantitative data obtained from closed-ended questions.

Very often, FC are associated with Just-About-Right (JAR) or Check-All-That-Apply (CATA) questions to provide a more detailed understanding of consumer acceptability and preferences. They are used to identify drivers of liking that may not be captured by JAR scales or to complement and refine consumer feedback in hedonic tests.

Additionally, FC are used, by some, to justify purchase intentions. Following questions about purchase intent, respondents are asked to explain their reasoning, particularly when the intent is negative. This helps in understanding the barriers to purchase and provides insights into consumer decision-making processes.

FC method is also used in product evaluation context, such as profiling, triangular tests, and sorting tasks. They are used to collect panellist comments that fall outside predefined descriptors, characterizing differences between products in discriminative tests or to describe freely groups or products in a sorting task. This application helps identifying and understanding the sensory characteristics that differentiate products. In descriptive sensory tests, FC help identify specific off-odours or off-flavours, missing attributes or other comments from trained panellists. Some respondents use them with trained panels to generate descriptors.

Beyond traditional uses, FC are employed, by some respondents, to capture emotions, usage patterns, and consumer knowledge. For example, one respondent describes the use of FC in perceptual experiments to evaluate tactile experience. Another measures consumer knowledge about product attributes, such as defining organic products using FC. This versatile application of FC allows for a holistic understanding of consumer interactions with products and concepts, enriching the data collected through structured questionnaires.

FC is rarely used on its own, without accompanying closed-ended questions. This contrasts with the scientific literature, which often presents FC as a standalone alternative. It appears that practitioners have embraced this method for its complementarity with closed-ended questions and other descriptive approaches.

2.2.3 Methodological Practices

Data Acquisition

The majority of practitioners use computerized systems to collect FC, leveraging various software and platforms. These digital tools include sensory software, online questionnaires, and applications accessible via smartphones, tablets, or computers. Platforms like *Fizz* (Biosystèmes, France) are frequently mentioned. The reliance on digital methods is evident, with many indicating that data acquisition is almost entirely computerized.

The use of paper for collecting FC is becoming less common, though it is still employed in specific contexts. Paper-based methods are occasionally used, particularly in profiling or free sorting tasks involving experts. Some practitioners prefer this approach when working with elderly individuals who are less comfortable with computerized systems. However, there is a noticeable decline in paper usage. Paper questionnaires are sometimes used for interviews or note-taking, but digital methods are generally preferred for their efficiency and ease of data management. Some respondents mentioned resorting to paper as a temporary switch in rare instances when computerized systems fail.

Oral methods are also used to gather FC, providing a more spontaneous and detailed form of feedback. Oral interviews, whether conducted in person or recorded via platforms like *Microsoft Teams* (Microsoft Corp., USA), allow participants to express their thoughts more extensively than they might in written responses. This approach often leads to richer descriptions of perceptions and experiences, as participants tend to elaborate more when speaking. In some cases, oral responses are transcribed by an interviewer to ensure that all feedback is accurately captured.

A few practitioners incorporate advanced technologies such as chatbots and voicebots to enhance the collection of FC. These tools can prompt respondents for additional details, ensuring that feedback is comprehensive and detailed. While not yet widely adopted, these technologies offer innovative ways to gather and analyse qualitative data, potentially improving the depth and breadth of insights obtained from FC.

Question Formulation

The formulation of open-ended questions in sensory evaluations often involves segmenting usage based on quality and defects, particularly in hedonic tests. A common approach is to separate questions into two categories: one focusing on what consumers liked (qualities or positive aspects) and another on what they disliked (defects or negative aspects). This dual-question category format allows for a clear distinction between positive and negative feedback, providing a structured way to gather detailed insights. Questions such as "*What did you like about this product?*" and "*What did you dislike about this product?*" are frequently used to prompt respondents to elaborate on their experiences.

In some cases, a single, more global question is employed to capture overall impressions or comments about the product. This approach is often used to gather general feedback or to justify preferences, such as asking consumers which product they preferred and why. Open-ended questions like "*Do you have any remarks or comments?*" or "*What are your comments on the proposed flavour/concept?*" encourage respondents to share their thoughts freely, without the constraint of specific categories. These questions are typically optional and allow for a broader range of responses, capturing both positive and negative aspects in a single prompt.

Other open-ended questions are generally unique and tailored to specific contexts, such as justifying purchase intentions or addressing discomfort. For example, following a closed question about purchase intent, respondents might be asked why they would not buy a product, providing insights into potential barriers to purchase.

Response Constraints

The constraints imposed on respondents when answering open-ended questions vary, with some imposing no limitations and others restricting responses to a limited number of words or phrases.

Many surveys provide open text areas without limitations, allowing respondents to express their thoughts freely using phrases or keywords. This approach encourages spontaneous and detailed responses, capturing the nuances of consumer perceptions. However, some software tools impose character limits, which can restrict the length of responses. In other cases, respondents are asked to provide shorter answers, using a limited number of words or phrases to describe their experiences.

Among those who recommend this approach, many request responses to consist of only three words or expressions. This method is used to focus responses and highlight key attributes or impressions, making it easier to analyse and interpret the data. By limiting responses to a few words or ideas, researchers can quickly identify the most salient aspects of a product, although this may sacrifice some depth in the feedback received.

2.2.4 Analysis of Practices

Practices Are Still Dominated by Manual Processing Despite the Emergence of AI

The processing of free comments remains predominantly manual, with respondents almost universally mentioning the use of spreadsheets like *Excel* (Microsoft Corp., USA) to organize and analyse responses. Some refer to a combination of manual and computerized approaches, where software is used to classify comments, but the overall process remains largely dependent on human effort. Even in cases where there is a semi-automated cleaning process (using macro), a significant portion of the work is still done manually.

However, the introduction of AI is gradually gaining traction, and several respondents stated they are experimenting with AI tools. Some are using a mix of manual and AI-based approaches, with AI being employed for certain tasks like summarizing or analysing comments. There are mentions of specific AI tools currently in testing, but results are still unsatisfactory. These tools often struggle with tasks like recognizing the difference between positive and negative comments, misinterpreting certain terms or synonymizing. Despite these challenges, many respondents continue to explore AI's potential in improving the processing of comments, such as using AI for summary generation or product-specific analysis. Some also report using a combination of AI with other solutions like dictionaries, synonyms, antonyms, and automatic selection of word types.

In addition to AI, some respondents rely on statistical software with specific modules for handling open-ended questions. Tools like *SPSS* (IBM Corp., USA) and *LeSphinx* (Le Sphinx Développement, France) are used to process free comments from consumers, offering an alternative to the more manual methods.

A Principle of Pretreatment Quite Consensual Among Users

The pretreatment of verbatim responses involves several stages of processing. For the vast majority of respondents, this includes a phase of manual (or sometimes AI-assisted) pre-cleaning, which involves correcting spelling errors, removing plurals, and other refinements. This is followed by lemmatization, synonymisation, and categorization.

For those who do not impose constraints on responses, the process begins by dividing the overall comment into multiple sub-comments based on the ideas expressed. Each sub-comment is then assigned a category, such as aspect, taste, or texture, and further grouped into subcategories depending on the required level of precision (e.g., "acid," "sour," and "very sour" may be consolidated under "acid"). Words with similar meanings and synonyms are also grouped to streamline analysis.

When practitioners code this data, they typically apply between one and three levels of precision, though this is not always explicitly defined. The macro level is the most apparent, involving a synthesis by major sensory categories such as aspect, odour, texture, and taste. However, the level of granularity beyond this is not always clear. Synonymous comments are grouped under a single term, with lemmatization and human synonymisation playing a crucial role in the analysis. However, the criteria used by practitioners to determine the level of grouping between items and to integrate intensity information (e.g., "slight," "very much," "a little bit") into subcategories remain unclear. It appears that these decisions are often made arbitrarily. Some respondents mention an iterative approach that involves conducting an initial frequency analysis to determine the need to group certain subcategories.

Statistical Analysis Often Limited to Frequency Calculation

The majority of analyses focus solely on frequency calculations, counting the number of comments by category and product. Comments that reach a certain citation threshold are highlighted, a process often referred to as text analytics, which is typically performed manually using tools like *Excel* (Microsoft Corp., USA). The results are commonly visualized through histograms and word clouds, emphasizing the recurrence of specific terms.

However, due to the often small dataset sizes, the results may not be widely used, and many respondents admit that no advanced statistical analysis is applied. The analysis is generally limited to identifying and interpreting the most frequently cited terms.

Some users incorporate inferential statistics into their analyses, employing methods such as chi-square tests and MR-Chi2 (Mahieu et al., 2021). These techniques are used to compare response percentages and analyse the frequency of certain categories across products. Some users also apply Z-tests for proportions. However, there is no consensus on whether to use one-tailed or two-tailed tests to determine significance. These statistical methods help in identifying relationships and differences within the data.

The use of multidimensional analysis techniques, such as Correspondence Analysis (CA) and Multiple Correspondence Analysis (MCA) was mentioned. For example, in free sorting tests, these analyses can help in understanding the verbalization of sorting criteria. Other techniques mentioned include Multiple-Response Correspondence Analysis (MR-CA) (Mahieu et al., 2021). These advanced statistical methods provide deeper insights into the data, beyond simple frequency counts.

A reflection on the attributes and thresholds to consider for statistical analyses emerges from the responses. Some apply a minimum citation threshold, often removing low-frequency occurrences. Common practices include keeping only terms cited at least once or twice for a given product or setting stricter filters at 5–10% of the panel. The approach varies with dataset size, with significance levels generally assessed at 5% and 10%.

Statistical Analysis and Software

Various software tools are used for data processing. Excel remains a common choice, either directly or through *XLSTAT* (Addinsoft, France), with several respondents mentioning manual analysis in spreadsheets. *XLSTAT* is frequently cited as the primary tool for statistical analysis.

Generalist statistical software and sensory analysis tools are also used. Respondents mention *SPSS, LeSphinx*, and *Thematic* (Thematic Ltd., USA), as well as specialized tools such as *Tastelweb*® (ABT Informatique, France). *FIZZ* is another commonly used software for data processing in sensory analysis.

Some rely on internally developed solutions, including custom macros in *Excel*, proprietary software, or tools built in R. Specific R packages such as *FactoMineR*,

SensoMineR, and *MultiResponseR* are frequently mentioned. Additionally, packages like *tm*, *stringr*, and *tidyr* are referenced for statistical analysis and text processing.

Confidentiality concerns are noted, with some respondents specifying that certain tools or methods remain undisclosed. AI-based applications are emerging and are still in the testing phase.

2.2.5 Limitation Behind the Use of Free Comment

The main limitation of open-ended questions lies in the time-consuming and labour-intensive nature of manual analysis. Many respondents highlight the difficulty of processing these responses, leading some to avoid open-ended questions in surveys due to the extensive effort required.

Concerns also arise regarding the quality and value of the insights gained. Some believe that open-ended responses offer limited additional value compared to structured questions like JAR or CATA, often merely reinforcing findings already evident in closed-ended quantitative data. While these responses may reassure clients, they rarely provide unexpected insights. Extracting genuinely novel or actionable inputs from consumer comments is difficult.

The variability in responses and the influence of consumer background further complicate analysis. Some respondents mention excessive variation in open-ended data, while others note that responses can be strongly influenced by factors such as country of origin and social class.

Another challenge is the robustness of the data. Some respondents point out that statistical reliability is often weak, making interpretation difficult. Additionally, consumer verbatim feedback is frequently vague and generic—comments like "pleasant taste", "nice texture", or "good appearance" provide little actionable insight for further product improvement.

A lack of knowledge about tools, analyses, and statistical methods associated with open-ended data processing is also a recurring issue. Some respondents express interest in training opportunities and clearer guidance on appropriate statistical analyses. Others seek information on automated preprocessing methods, such as those developed in academic research but not fully documented for practical implementation. Questions also remain about which tools to use for collecting and analysing free-text responses outside traditional consumer test questionnaires and how to efficiently compile and interpret such feedback, particularly in linking keywords to positive or negative product perceptions.

2.2.6 Expectations and Future Directions

There is a strong desire to improve the processing, analysis, and reporting of FC, with users seeking to save time and have access to easier-to-use tools. The need for automated solutions is growing, as many express interest in software or algorithms that can handle tasks such as categorization, lemmatization, and synonymisation more efficiently. The goal is to reduce the time spent on preparation, adjustment, elimination, and grouping. There is a demand for tools that can create a taxonomy of words, making the process more manageable and less dependent on manual effort. This is especially important as the complexity of products and the nuances in language can make categorization difficult.

Handling quantifiers and concepts that are more complex remains a challenge, as automation often struggles with understanding the context and meaning when words need to be associated with adjacent terms. The categorization of aspects like odour, texture, and flavour can be complicated, especially in French. Determining the extent of synonymisation is another area that requires careful consideration to ensure accurate and meaningful analysis.

In terms of quality of analysis, there is an expectation to present findings in a more modern, design-oriented, and marketable way in reports and presentations. This involves not just improving the efficiency of data processing but also enhancing the visual and communicative aspects of the results.

The introduction of AI raises concerns about confidentiality and the reproducibility of results. Users are cautious about using AI due to the sensitive nature of the data and the need for consistent outcomes. Ensuring that AI tools can handle data securely and produce reliable results is crucial for their adoption.

Chatbots are seen as a potential future direction for consumer questioning, offering the possibility of more interactive and engaging experiences. They allow consumers to feel like they are communicating with someone directly, which could replace traditional closed-ended questions and encourage more detailed and thoughtful responses, even from those who initially have "nothing to say".

2.2.7 Some Missing Methodologies in this Panorama

This questionnaire administered to the professional community of sensory analysis highlights the main uses of French practitioners and some Europeans. We found the approaches classically described in literature (Ares et al., 2010; Hanaei et al., 2015; Lahne et al., 2014; Mahieu et al., 2020a; Symoneaux et al., 2012; Ten Kleij & Musters, 2003). However, some current applications were not reported in the survey.

The Free-JAR was recently proposed by Luc et al. (2020). This method involves asking consumers to categorize products into three hedonic groups ("I don't like it," "I like it a bit," and "I like it a lot"). Once each product is classified, consumers describe them using their own descriptors within a JAR framework, such as "*This product is too…*", "*This product is not… enough*" or "*This product is just right*". Mahieu et al. (2022) proposed an extension of Free Comment called Ideal-Free Comment (IFC). IFC instructs consumers to describe first actual products and then their ideal product using FC. The benefit comes from introducing the ideal product description, which enables positioning this ideal relative to other products. Finally, Mahieu et al. (2020b) introduced FC-AEF (Free Comment Attack-Evolution-Finish) as an alternative to temporal technics such as TDS and TCATA. This method combines the strengths of Free Comment (FC) and a temporal sensory analysis based on three steps Attack-Evolution-Finish (AEF) (Visalli et al., 2020). Consumers describe their sensory experiences for each phase (Attack, Evolution, Finish) without being limited by a fixed list of descriptors but by using free comments.

3 Analysis of Free Comment

In the presented questionnaire, respondents reported methodological difficulties related to the use of Free Comment. Among these, data preprocessing is frequently mentioned but will only be briefly discussed in this chapter as it is covered in Chap. 5 of this book.

Another difficulty encountered concerns the application of appropriate statistical tests to this type of data. Many respondents rely on frequency calculations and basic statistics due to limited knowledge on suitable analytical tools.

The remainder of this chapter aims to capture the current state of statistical analyses that can be performed on pretreated FC data, drawing on both the literature and the reported practitioners' uses.

3.1 Pretreatment

The raw free comments need to be pretreated for the information they contain to be cleaned and standardized with the final aim of building a list of sensory descriptors a posteriori from the free comments. The pretreatment of free comments can be performed manually by human operators (Ares et al., 2010; Hanaei et al., 2015; Rocha et al., 2023; Symoneaux et al., 2012; Ten Kleij & Musters, 2003). It can also be automatized through existing non-sensory specific NLP libraries from some computer software (Hamilton et al., 2023; Hamilton & Delarue, 2025; Hamilton & Lahne, 2020; Jaeger & Rasmussen, 2021) or through dedicated expert system crossing NLP tools, statistics, and domain knowledge (Visalli et al., 2025b). Some pretreatment methods lie in between full automatized and full manual treatment: some steps of the pretreatment are done using computer software automatically and some manual interventions are performed for some specific steps (Lahne et al., 2014; Mahieu et al., 2020a, 2022; Visalli et al., 2023b). Large Language Models (LLM) such as ChatGPT (Open AI) may also be useful in pre-treating FC (Visalli et al., 2025b). The main benefits of (fully) automated approaches are that they are relatively fast, more reproducible and more deterministic but they may be less performant than proficient human operators in terms of discrimination and repeatability (Visalli et al., 2025b). Central steps to build a list of sensory descriptors a posteriori from the free comments are the following:

- Check for and correct typing and spelling errors.

- Remove uninformative information (e.g. punctuations, stop words, name of the product, etc.).
- Standardize the different forms of a term based on lemmatization and/or stemming.
- Take into account negations and/or quantifiers.
- Take into account compound terms and/or any relevant n-grams.
- Apply a filter based on part-of-speech tagging (e.g., keeping only nouns, adjectives and ad hoc n-grams).
- Group terms that convey similar sensory information.
- Apply a frequency filtering (e.g. keeping only terms cited by at least one same product by at least 5% of the panel).

For more information on the pretreatment of free comments, the interested reader is invited to read part III of the present book.

The terms resulting from the pretreatment constitute the list of sensory descriptors shared by the panel that will be used to characterize the products from a sensory point of view in the same fashion as CATA experiments. In the typical structure of the dataset, each row corresponds to one evaluation of one product by one panellist. The first two columns identify to which product and panellist each row corresponds. Subsequent columns correspond to sensory descriptors resulting from the pretreatment of free comments and encode whether the corresponding sensory descriptor has been cited within the evaluation (coded as 1) or not cited (coded as 0).

3.2 Basic Analyses

The simplest (but still interesting) analysis that can be performed with FC data is to count the number of citations of each descriptor by product. These counts can then be displayed in a barplot or a word cloud by product. Counting of citations can, for example, be used as a first contact with the data and allows investigating how panellists most frequently describe each product.

One drawback that simply counting citations may have is that descriptors highly cited across the product range may overlook important differences between individual products. For example, if the product space under investigation is chocolate, it is likely that descriptors such as *cocoa* or *chocolate taste* will be frequently cited for all products. To circumvent this issue, instead of simply investigating the number of citations for each descriptor and product, one can examine the chi-square contribution of each cell in the contingency table crossing products and descriptors. Each cell contains the number of citations of the descriptor (column) for the corresponding product (row). Like basic counting of citations, these chi-square contributions can then be displayed either on a barplot or a word cloud by product. This approach focuses more on the differences between products rather than on their absolute descriptions. As it will be explained in the next sections, using the chi-square contribution from the multiple-response chi-square framework (Mahieu et al., 2021) should be preferred to the usual chi-square framework for FC data.

3.3 Statistical Analyses for Products Sensory Characterization

3.3.1 Test for Overall Differences Between Products

One of the first questions that generally arises in the field of sensory and consumer science is whether or not there is a product effect, i.e. that at least two products in the product space under study are different from a sensory point of view. This may be investigated using an omnibus test of overall differences between products.

Since after pretreatment FC data have the same nature and structure as CATA data, analyses that are applied to CATA data are also applicable to FC data. In the context of CATA data analysis, it has been proposed to test for overall differences between products using the sum of descriptor-wise Cochran's Q (Cochran, 1950) test statistics using random permutations to estimate the null distribution of the test statistic (Meyners et al., 2013).

In the context of FC data analysis that historically focused more on the contingency table crossing products and descriptors, where each cell contains the number of citations of the descriptor of the column for the product of the row, testing for overall differences between products was performed using chi-square test of independence (Lahne et al., 2014; Symoneaux et al., 2012). More recently, it has been shown that the usual chi-square test of independence is not suited to the nature of FC (and CATA) data because several citations come from the same evaluation, i.e. one pair of panelist and product (Mahieu et al., 2021). These authors developed a multiple-response chi-square test that is better suited to FC data and thus should be preferred over the usual chi-square test. Indeed, the multiple-response chi-square test considers an entire evaluation as a statistical unit, which is more in line with the FC data nature. As the sum of Cochran's Q, the null distribution of the multiple-response chi-square statistic is estimated using random permutations.

3.3.2 Sensory Mapping of Products and Descriptors

When the test of overall differences between products turns out significant, the main questions that arise are: how are differences between products structured, which products are closer/further from a sensory point of view and with respect to which sensory dimensions. These questions can be answered by mapping the sensory space of the products with respect to their FC sensory descriptors.

Originally, the sensory mapping of products and descriptors from FC data has been performed using Correspondence Analysis (CA, Benzécri, 1973) of the contingency table crossing product in rows, sensory descriptors in columns, and the frequency of citation of a sensory descriptor for each product at the panel level as cell values (Ten Kleij & Musters, 2003). As for the chi-square test, it has been shown that the usual CA is not suited to the nature of FC (and CATA) data because several citations come from the same evaluation, i.e. one pair of panellist and product, and that the Multiple-Response Correspondence Analysis (MR-CA) is more suited to FC data than the usual CA (Mahieu et al., 2021). Indeed, MR-CA considers an entire evaluation as a statistical unit, which is more in line with the FC data nature. For this reason, MR-CA should be preferred over usual CA for the analysis of FC data.

Another approach that may be applicable to FC data to provide the sensory map of products and descriptors comes from Temporal-Check-All-That-Apply (TCATA) data analysis. It consists in performing a Principal Components Analysis (PCA) of the matrix that crosses product in rows, sensory descriptors in columns, and the average proportion of citation of a sensory descriptor for a product at the panel level as cell values (Castura et al., 2022).

There is close connection between PCA of average proportions of citations and MR-CA. Let us denote by n_{pd} the number of time descriptor d was cited for product p at the panel level and by E_p the number of times product p was evaluated at the panel level where E_p equals the number of panelists if the experimental design is balanced. Then PCA of average proportions of citations is the PCA of the matrix whose general term is:

$$\left(\frac{n_{pd}}{E_p}\right)_{1\le p\le P,1\le d\le D}$$

and MR-CA is the (E_p weighted) PCA of the matrix whose general term is:

$$\left(\frac{n_{pd}}{E_p\sqrt{\frac{\sum_p n_{pd}}{\sum_p E_p}}}\right)_{1\le p\le P,1\le d\le D}$$

Where $\frac{\sum_p n_{pd}}{\sum_p E_p}$ is nothing more than the estimation of the probability of a descriptor to be cited within an evaluation of any product. As one can see, the only difference between the two methods is that MR-CA weights descriptors inversely proportional to their estimated probability of citation within any evaluation, something PCA of average proportions of citations does not. As a result, the MR-CA places more emphasis on the specific characteristics of products than the PCA of average citation proportions. In other words, if a few products are frequently described with certain descriptors, while the majority of other products are rarely described with these descriptors, the MR-CA will clearly highlight this.

Whatever the method performed to map products and descriptors from FC data, investigating the stability of the product configuration and the degree of discrimination of the products is crucial to avoid overinterpretations. In sensory and consumer science, this stability is usually investigated by constructing a large number of virtual panels (say 1000) of the same size as the original by bootstrapping (random sampling with replacement) on the panellists of the original panel. The same method as the one used on the original panel is used on the virtual panels data, providing a new product configuration for each virtual panel that is superimposed on the real panel configuration. To superimpose virtual panels configurations, two rationales exist: partial bootstrap and total bootstrap (Cadoret & Husson, 2013). Partial bootstrap consists of projecting the virtual panels' configurations onto the actual panel map as supplementary individuals. Total bootstrap consists in matching the virtual panels' configurations to the actual panel's by mean of Procrustes rotations. Cadoret and Husson (2013) shown that both partial and total bootstrap provide too optimistic confidence ellipses. They advocated using truncated total bootstrap that corresponds to the total bootstrap method but relying only on the first sensory dimensions to perform the Procrustes rotations to avoid overfitting.

Determining a suitable number of "first sensory dimensions" still remains a difficult question. When using MR-CA, one can rely on the associated test of dimensionality (Mahieu et al., 2021). To avoid overloaded maps, the superimposition of virtual panels configurations is usually summarized by confidence ellipses containing $1 - \alpha$ of the bootstrap replicates of each product.

Besides enabling the investigation of the stability of the product configuration and the degree of discrimination among products, the bootstrap replicates can be used to test the pairwise discrimination of products more formally (Castura et al., 2023). The principle of the proposed test is, for each pair of products to be tested, to estimate the probability that the origin (multivariate zero) belongs to the distribution of differences in matched bootstrap replicates of the two products tested. If the origin has an estimated probability to belong to this distribution lower than α, then the products are considered as significantly different. In other words, what is estimated is the probability of the difference between the products to include zero.

3.3.3 Test of Differences Between Products by Descriptor

Descriptor Tests

A common question that arises in sensory and consumer science is whether there exists a product effect by descriptor, this to investigate which descriptors enable discriminating products and which not. To the best of our knowledge, no test was proposed in the literature in the context of FC data analysis to investigate whether there is a product effect at the descriptor level. However, several tests were proposed in the context of CATA data analysis and again, as stated before, these tests may be considered to investigate the same question on FC data.

For this purpose, the most common test (up to now) is Cochran's Q test (Cochran, 1950; Meyners et al., 2013). Cochran's Q test is a non-parametric test that evaluates whether several treatments have the same probability of occurrence of a binary outcome in block designs. Here the panellists are considered the blocks and the products the treatments. Cochran's Q test perfectly matches a typical FC sensory data design but is limited in more complex situations such as: incomplete designs, experimental designs behind the product effect and designs with repetitions and/or sessions where interactions and/or session effects are to be investigated or at least controlled. Usual ANalysis Of VAriance (ANOVA) conducted on the binary CATA data has been studied as an alternative to Cochran's Q test (Meyners & Hasted, 2021). These authors showed that, in this context, ANOVA has good statistical properties, notably holding the nominal α risk, and providing very close p-values to those from Cochran's Q test. As an alternative to usual ANOVA, logistic regression has been proposed to analyse CATA data (Bi & Kuesten, 2022). Logistic regression is theoretically more suited to CATA data than usual ANOVA, but it has been shown that usual ANOVA has better statistical properties than logistic regression is the context of typical CATA data, notably holding the nominal α risk while logistic regression does not (Meyners & Hasted, 2022). Therefore, usual ANOVA should be preferred to logistic regression.

When the omnibus test turns out significant, it is common to perform post-hoc analyses to investigate which pairs of products can be considered as having a different proportion of citations with respect to the descriptor for which the omnibus test turns out significant. For consistency of the analysis, the sign test is preferred if the omnibus test was Cochran's Q test while traditional Least-Square-Difference (LSD) tests are preferred if the omnibus test was ANOVA. Note that LSD tests have also been studied by Meyners and Hasted (2021) and shown to have the same good statistical properties as ANOVA in the context of CATA data analysis. Whatever the post-hoc test, it is recommended to perform p-value corrections for multiple testing to avoid inflation of empirical α risk due to multiple testing, for example using the False Discovery Rate (FDR) correction (Benjamini & Hochberg, 1995).

Descriptor by Product Tests

Another question that arises in sensory and consumer science is testing whether a product can be said to be more or less intense than the average with respect to each descriptor. In the context of FC data, "more or less" refers to whether a descriptor is mentioned more or less frequently compared to the average.

Historically, to test whether a product was significantly more or less described by a descriptor, so-called chi-square test per cell were performed (Lahne et al., 2014; Symoneaux et al., 2012). Some authors used Fisher's exact test per cell (Mahieu et al., 2020a) that is essentially an exact version of the chi-square test for 2 by 2 contingency tables. As for the overall chi-square test, the usual chi-square test and/or Fisher's exact test per cell is not suited to the nature of FC (and CATA) data because several citations come from the same evaluation, i.e. one pair of panellist and product. To circumvent this issue, in the multiple-response chi-square framework, the multiple-response hypergeometric test has been developed (Mahieu et al., 2021). This considers an entire evaluation as a statistical unit, which is more in line with the nature of FC data. Multiple-response hypergeometric test relies on Monte-Carlo simulations to estimate the null distribution of the descriptor by product proportion of citations.

As for post-hoc analyses of descriptors tests, it is recommended to perform p-value corrections to avoid inflation of empirical α risk due to multiple testing, for example using the False Discovery Rate (FDR) correction (Benjamini & Hochberg, 1995). Note that after p-value correction, the lowest corrected p-value by descriptor can be used in a similar fashion as the p-value of the descriptor test of the previous section to investigate whether or not there is a product effect on descriptors (Simes, 1986).

3.3.4 Overall Consistency of Statistical Analyses

As seen in the previous sections, some analyses are naturally related and form a comprehensive and coherent framework. The chi-square frameworks are the most comprehensive, as they can be used from basic wordclouds to descriptor-by-product tests, including overall differences testing and mapping products and descriptors—while maintaining a consistent rationale throughout. Since the introduction of the multiple-response chi-square framework (Mahieu et al., 2021) in the sensometric toolbox, there appears to be little reason to continue using the traditional chi-square framework, as the latter is flawed for FC data, whereas the former is better suited to this type of analysis.

Alternatively, to the multiple-response chi-square framework, Cochran's Q tests-based analysis of FC data may be performed while staying in the same rationales for testing overall differences and performing by descriptor tests. However, mapping products and descriptors notably requires using other rationales than Cochran's Q tests and thus chi-square-based methods are more comprehensive frameworks.

3.4 Pairing Free Comment Data with Liking Data (When Measured Alongside)

It is commonplace in sensory and consumer science that when consumers are asked to describe some products with respect to their sensory characteristics, they are also asked to provide a liking score and FC does not make exception. For this reason, this section deals with such situation and in particular it presents some analyses to pair FC data with liking data to investigate sensory drivers of liking and disliking.

In sensory and consumer science, the analysis of hedonic ratings and product description result in drivers of liking (and disliking) which can be represented in (internal or external) preference maps (Carroll, 1972; Danzart, 2009; Greenhoff & MacFie, 1994; McEwan, 1996). The external preference mapping assumes that sensory characteristics of products are the principal reasons for liking or disliking a product and thus investigates product acceptance through the prism of their sensory characteristics. The internal preference mapping first focuses on the liking itself and then investigates if this can be understood by the sensory characteristics of the products. For sake of conciseness, the external preference mapping point of view is adopted in this section since it appears to be the most used in practice with FC and CATA data (Ares et al., 2010; Mahieu et al., 2022; Rocha et al., 2023; Visalli et al., 2023b; Wakihira et al., 2023). For an example of the internal preference map with FC data, the reader can refer to Antille et al. (2024). Furthermore, it should be noted that the developments in this section are made considering a non-hedonic FC version, however the same analyses can be performed with hedonic FC considering each descriptor with a "like version" and "dislike version". The example will be performed on hedonic FC to illustrate this point.

3.4.1 Consumer Clusters

While sensory data can be assumed objective, liking data is by nature subjective. For this reason, the existence of consumer segments in the population that like and dislike different sensory characteristics should be considered to study the drivers of liking and disliking of homogeneous consumer clusters (in terms of liking). In fact, the average consumer may not represent any of them.

Analyzing whether clusters of consumers exist in terms of liking can be performed using MDPREF (Carroll, 1972). If MDPREF reveals distinct preference directions among different consumer groups, it is more appropriate to first conduct a clustering analysis before investigating the drivers of liking. Indeed, the factors driving liking are likely to differ across clusters, making it unjustified to analyze them jointly.

To perform consumer clustering—since consumers are considered as variables in this context—the Clustering Around Latent Variables (CLV) framework (Vigneau & Qannari, 2003) is highly appropriate. The CLV framework is essentially an adaptation to the variable case of the usual ascendant hierarchical clustering followed by k-means consolidation devoted to units.

3.4.2 Computation of Drivers of Liking at the Panel Level

To pair FC data with liking data at the panel level, the liking data can be projected on the sensory map of the products and descriptors (coming for example from MR-CA). Projecting hedonic data on the sensory map is usually performed by regressing liking data on the sensory components. Depending on assumptions, statistical criterion and/or the nature of the data, different regression models can be considered: from the simplest vectorial model to the complete quadratic model (Danzart, 2009), and either individual or average liking by cluster may be modelled as response.

For the projection of liking data on FC sensory map, using the vectorial model appears more aligned with the nature of data that are intrinsically presence/absence data where average proportions of citations are used to map products. Indeed, using some form of quadratic model would implicitly assume that an optimal proportion of citations exists for each sensory descriptor to maximize liking, which sounds questionable. On the contrary, a vectorial model shows the direction of increased liking with increasing or decreasing proportions of citations. This implicitly assumes that the more likely/unlikely a sensory descriptor is cited within an evaluation the more liking increases depending on the valence of the sensory descriptor for the liking. This assumption is closer to the intrinsic presence/absence nature of the individual data. To simplify the interpretation of the map, it is recommended to project only average liking by cluster and possibly overall average liking to report implicitly which clusters are more important in terms of proportion of consumers (Ares et al., 2010; Mahieu et al., 2022; Rocha et al., 2023).

3.4.3 Computation of Drivers of Liking at the Individual Level Within Clusters

One alternative way to investigate drivers of liking is to relate the presence/absence of the sensory descriptors within an evaluation to the liking of the corresponding evaluation to investigate whether when a sensory descriptor is cited, it increases or decreases liking in average. Obviously, a sensory descriptor that increases liking in average is likely a driver of liking and a sensory descriptor that decreases liking in average is likely a driver of disliking. Note that this relation between the presence/absence of the sensory descriptors within an evaluation with the corresponding liking better be investigated within clusters of consumers rather than overall since different clusters of consumers likely like and dislike different sensory characteristics.

At least two methods exist in the literature to investigate the presence/absence of the sensory descriptors within an evaluation with the corresponding liking. The first one, called penalty-lift analysis (Meyners et al., 2013), was developed on the context of CATA data analysis. It consists in computing the difference between the average liking score of evaluations where the sensory descriptor is cited with the average liking score of evaluations where the sensory descriptor is not cited and consider this difference as the impact of the sensory descriptor on liking. The second one consists in regressing liking scores on the sensory descriptors effect (0: absence; 1: presence) together with the consumer and possibly the product effect to control for these two last factors (Mahieu et al., 2022; Rocha et al., 2023; Visalli et al., 2023b; Wakihira et al., 2023). The regression loading of each sensory descriptor is then considered as its impact on liking. In both methods, if the impact of a sensory descriptor is positive the sensory descriptor is likely a driver of liking while if the impact is negative the sensory descriptor is likely a driver of disliking. These impacts of sensory descriptors on liking are typically displayed as horizontal barplot. As compared to penalty-lift analysis, the regression-based method shows three benefits. First, it controls for the subject effect. Second, it controls for correlations between citations of sensory descriptors to some extent. Third, it enables investigating the significance of a sensory descriptor's impact on liking straightforwardly. Note the two first benefits correct two limitations of penalty-lift analysis mentioned by Meyners et al. (2013). For these three benefits, the regression-based method should be preferred.

3.4.4 Panel Versus Individual Level

The investigation the relation between FC and hedonic data at the panel and the individual levels are rather complementary tools than competitors. The panel level allows for quick visual investigation of liked/disliked products in conjunction with how these products are relatively positioned with respect to the sensory dimensions. It allows investigating general trends between the sensory and the hedonic data where sensory data represent probabilities of citation. On the contrary, the individual level is less appropriate to investigate whether or not a specific sensory descriptor is effectively a driver of liking/disliking. Indeed, at the panel level, both the sensory and the hedonic data are averaged and thus sensory data from some consumers may be interpreted alongside hedonic data of some others. The individual level is more appropriate to investigate the specific impact of sensory descriptors on liking. Indeed, the effective presence or absence is set in regards of liking variations directly without averaging, which is a finer grain than the panel level. Furthermore, the individual level brings an information on the magnitude and the significance of the impact of sensory descriptors on liking, something impossible to read at the panel level.

3.5 *When Descriptors Matter More Than Products*

In some experimental settings, it can happen that the relation between descriptors and the extent to which they are cited together or not is of paramount importance, at least more than describing products. This can be the case when, for example, one wants to investigate sensory congruencies.

The distance between two descriptors can be computed at the evaluation level using the phi correlation (Cramer, 1946; Meyners et al., 2013) by taking the square root of one minus the phi correlation coefficient (Gower & Legendre, 1986). The resulting pairwise descriptors distances can be encoded in square symmetric distance matrix and this matrix submitted to MultiDimensional Scaling (MDS, Borg & Groenen, 2005) or Principal Coordinates Analysis (PCoA) (Torgerson, 1958) to map the proximities between descriptors. In such a map, the closer two descriptors are, the more they are cited (or not cited) together and the less one of them is cited without the other within an evaluation. Note that when the matrix is submitted to PCoA, the square root is essential in the computation of the distance between the descriptors since it renders this distance Euclidian resulting in the eigenvalues of PCoA to be non-negative.

To investigate whether two descriptors are significantly cited together within an evaluation, a Monte-Carlo test has been proposed (Heussen et al., 2023). This test considers as a null hypothesis that the two descriptors are cited independently while the alternative hypothesis is that when one is cited the other is significantly likely to be cited also. The test statistic is the number of times they have been cited together within an evaluation and the null distribution of the test statistic is estimated using random permutations.

4 Example

This final section presents a case study on madeleines, allowing for the visualization of some of the previously suggested statistical analyses.

4.1 Dataset

The dataset used in this example is part of published data (Visalli et al., 2025a) used in a published paper (Visalli et al., 2025b). It contains the pretreated FC data of five distinct plus one replicate madeleines described by 100 consumers using a hedonic FC, with like (L) and dislike (D) descriptions, without any further restrictions. In addition to the hedonic FC, the consumers also provided a liking score for each madeleine using a 11-point discrete scale. From the published data, only the data pretreated by the expert system (operator "ES") and collected with sentences (collection mode "S") were used in the following example. The so-called concepts resulting from the pretreatment of the expert system have been filtered before data analysis so that retained concepts are cited by at least 5% of the consumers for at least one same product. The remaining concepts were considered as sensory descriptors. For more information on data collection and pretreatment, the reader is invited to read the corresponding publications (Visalli et al., 2025a, b).

4.2 Data Analysis

To illustrate basic analyses that are common in day-to-day practitioners' uses, wordclouds and barplots of the two most different products in terms of liking are presented based on raw citation counts as presented in Sect. 3.2.

The analyses from the multiple-response chi-square framework were performed to test for overall differences between products (MR-Chi^2 test), map products and descriptors (MR-CA) and test differences between products by descriptor (MR-hypergeometric tests) as presented in Sect. 3.3. Analyses in the presence of liking presented in Sect. 3.4 were performed at both the panel and individual level considering no clusters for simplicity. Finally, the proximity between descriptors were mapped as presented in Sect. 3.5.

4.3 Results

4.3.1 Basic Analyses

Figure 6.1 shows how products are frequently described in absolute terms. P605 was the most appreciated and the most cited terms are cited as liked. The only term highly cited as disliked is *nothing* suggesting that many consumers could not find something they disliked in this product. The opposite occurs for P108: the most cited terms are disliked and *nothing* is mainly cited as liked. Overall, the most cited terms for both products are rather generic. What is mainly liked for P605 is its taste, visual aspect and size while the main defect for P108 appears to be its texture and taste.

Fig. 6.1 Wordclouds based on raw citation rate (first row) and their corresponding barplot (second row) of products P108 (first column) and P605 (second column). Liked (L) terms are represented in green while disliked (D) terms are coloured terms are presented in red

4.3.2 Statistical Analyses Using the Multiple-response Chi-Square Framework

Following the multiple-response chi-square test, it turns out that there are differences between products with respect to their average proportions of citations for some descriptors ($MRChi^2 = 971.70$, $p < 0.001$).

Figure 6.2 shows the sensory dimensions of the product space and the relative position and proximity of the products with respect to these sensory dimensions. It can be seen that the first dimension opposes products that are mostly disliked for their visual aspect with an overall liked texture with some specific texture defects, on the positive part to products with liked visual aspect but disliked texture on the negative part. The second dimension mostly carries liking information and opposes products that have liked characteristics on its negative part to products that have disliked characteristics on the positive part notably for taste and texture and slightly for appearance.

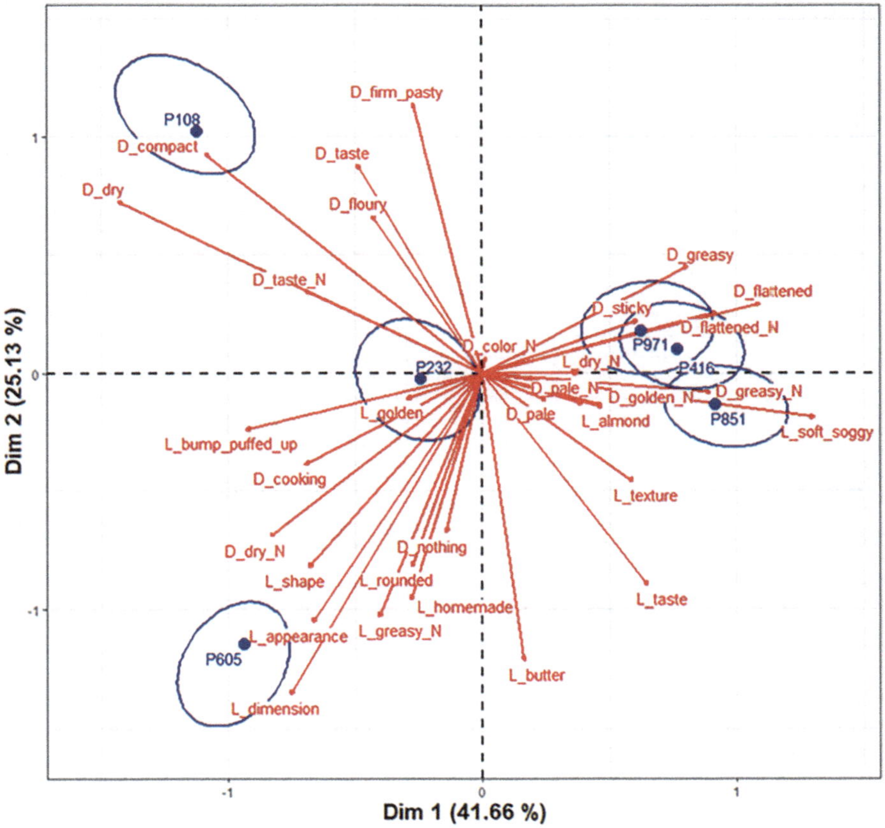

Fig. 6.2 Biplot of the first two dimensions of MR-CA. Confidence ellipses are built at $\alpha = 5\%$ from truncated total bootstrap using the significant dimensions in the Procrustes Rotations. Only descriptors having at least one significant association (see Fig. 6.3) are displayed

	P108	P232	P416	P605	P851	P971
D_aftertaste	6	2	0	0	0	4
D_airy_N	7	2	2	2	1	3
D_appearance	8	9	9	2	3	8
D_butter_N	4	5	0	1	0	2
D_color_N	1	6	0	0	3	0
D_compact	20	9	2	6	2	1
D_cooking	5	2	3	8	0	0
D_cooking_N	0	0	2	2	6	2
D_dimension_N	9	8	8	1	3	5
D_dry	33	13	7	18	0	6
D_dry_N	6	11	0	14	1	4
D_firm_pasty	14	4	4	0	5	5
D_flattened	1	2	14	0	8	17
D_flattened_N	1	0	10	0	6	10
D_floury	6	1	2	1	1	0
D_golden_N	0	3	0	0	7	0
D_greasy	5	0	9	0	12	10
D_greasy_N	0	0	9	1	9	3
D_nothing	22	19	29	39	26	21
D_odor_fragrant	1	0	3	3	3	6
D_odor_N	2	6	1	3	6	5
D_pale	0	8	0	0	6	0
D_pale_N	0	5	0	0	7	0
D_persistent	4	1	5	0	0	3
D_shape	2	1	3	0	3	6
D_soft_N	6	2	1	7	1	2
D_sticky	4	1	2	0	15	1
D_sweet	4	12	11	4	10	13
D_sweet_N	3	1	0	5	2	2
D_taste	25	13	7	7	14	6
D_taste_N	21	16	3	13	10	9
D_texture	7	4	2	4	5	0
L_airy_light	3	5	5	7	9	12
L_almond	0	4	9	2	3	2
L_appearance	12	23	8	31	14	7
L_appetizing	2	0	1	5	1	1
L_bump_puffed_up	10	6	1	12	1	3
L_butter	0	4	14	19	9	7
L_color	15	18	21	28	14	22
L_cooking	2	3	3	8	5	5
L_dimension	6	7	2	25	7	5
L_dry_N	2	4	2	1	9	1
L_golden	15	2	13	18	6	15
L_greasy_N	0	4	2	8	1	0
L_homemade	0	1	0	6	2	1
L_nothing	27	12	16	21	18	17
L_odor_fragrant	18	20	27	14	17	26
L_rounded	1	0	1	7	2	2
L_shape	9	22	6	22	7	5
L_soft_soggy	2	23	35	10	31	31
L_sticky_melting_N	3	8	12	6	12	11
L_sweet	6	9	14	10	9	8
L_taste	11	38	42	40	43	38
L_texture	3	14	12	11	21	13
L_vanilla	1	4	4	4	8	6

Fig. 6.3 Average proportion of citations of each descriptor (in rows) for each product (in columns). Highlighted cells have a significant multiple-response hypergeometric test at $\alpha = 5\%$, with green being for positive significant association and red for negative significant association

Figure 6.3 shows which descriptors are significantly positively or negatively associated with each product. It is a natural complement to the MR-CA map adding the benefit of taking into account all significant dimensions at the same time and not two at a time like the map. For example, it shows here that P232 appears to have a defect of color but its shape was appreciated, something difficult to see on the MR-CA map due to its central position on the two first axes.

4.3.3 Pairing Free Comment Data with Liking Data

Panel Level

Figure 6.4 presents which sensory dimensions drive the liking and which products are mostly liked or disliked. The most appreciated product appears to be P605 while the least appreciated one is P108, while liking is mostly related to the second dimension of the MR-CA map. This is consistent with comments made in Fig. 6.2 and

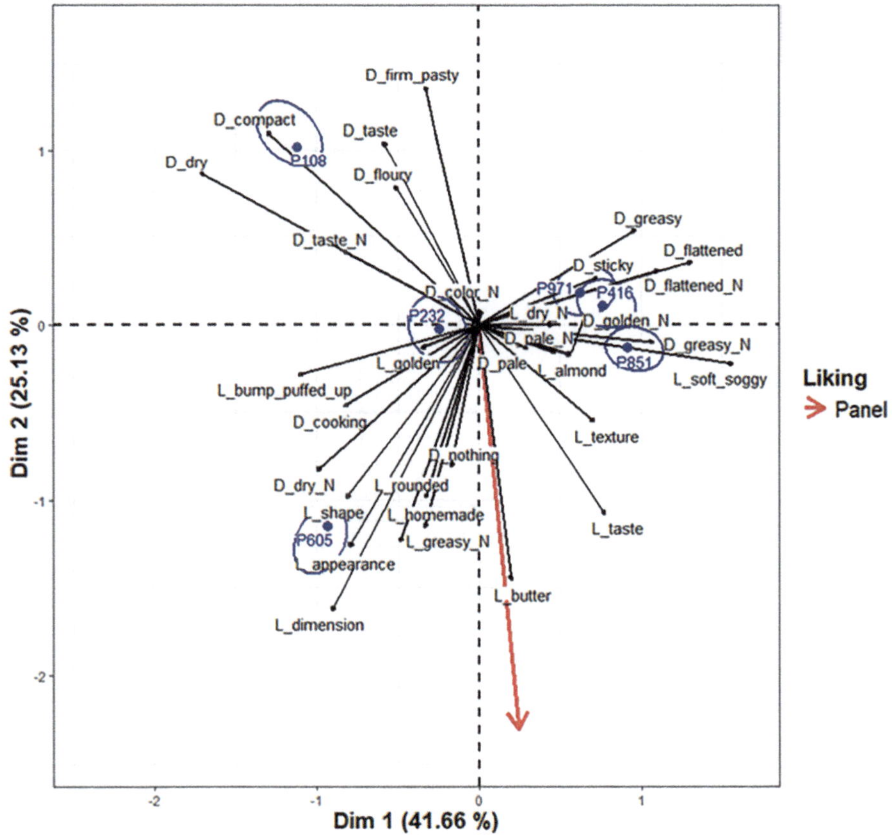

Fig. 6.4 Biplot of the first two dimensions of MR-CA with the panel average liking projected. Confidence ellipses are built at $\alpha = 5\%$ from truncated total bootstrap using the significant dimensions in the Procrustes Rotations. Only descriptors having at least one significant association (see Fig. 6.3) are displayed

shows that overall liking data are consistent with hedonic FC data as liking increases with the probability of liked descriptors to be cited.

Individual Level

Figure 6.5 shows the average impact on liking of each descriptor when cited within the corresponding evaluation, allowing the understanding drivers of liking and disliking. It shows here that liking data are consistent with hedonic FC data as only disliked descriptors (liking nothing being a disliked descriptor) are identified as drivers of disliking and liked descriptors as drivers of liking. Figure 6.5 further highlights that, in order to be appreciated, a product should have a good appearance, taste and texture and odour and that tasting butter, being golden and not dry contributes to convey these characteristics.

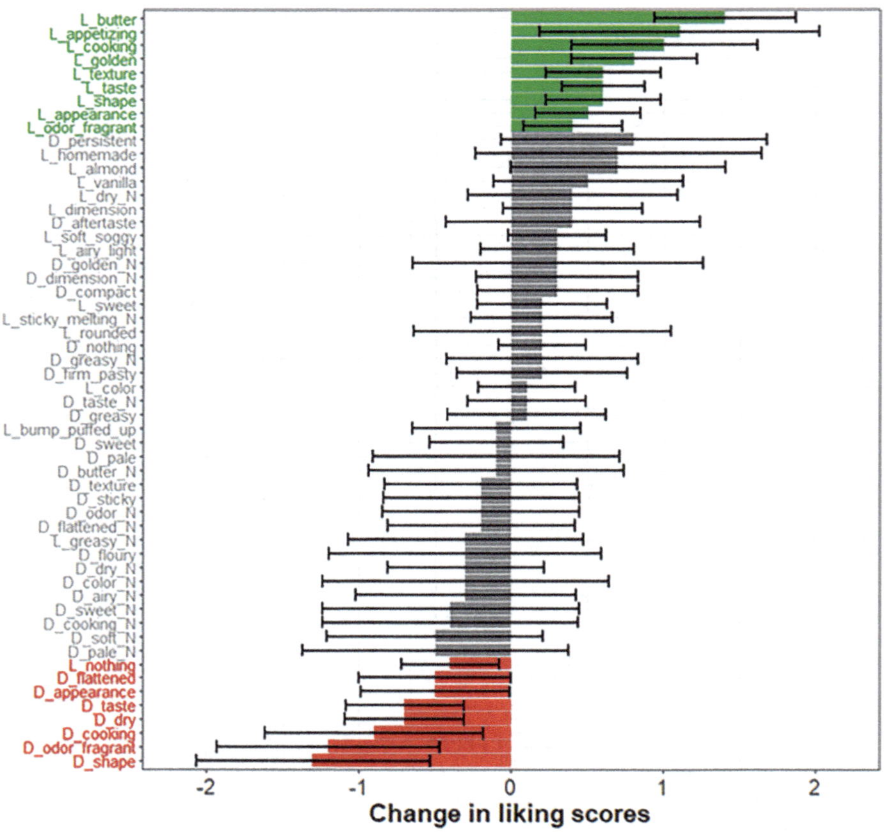

Fig. 6.5 Regression loadings of each sensory descriptor with their respective confidence intervals built at α = 5% or the entire panel

4.3.4 Proximity Between Descriptors

Figure 6.6 shows the relative proximity of descriptors, where two close descriptors tend to be cited or not cited together and the less one of them is cited without the other within an evaluation. Liked descriptors were associated with each other and not with disliked descriptors (and inversely). The first dimension shows mostly an opposition between positive texture and taste descriptors and negative ones while the second dimension shows mostly an opposition between positive and negative aspect descriptors.

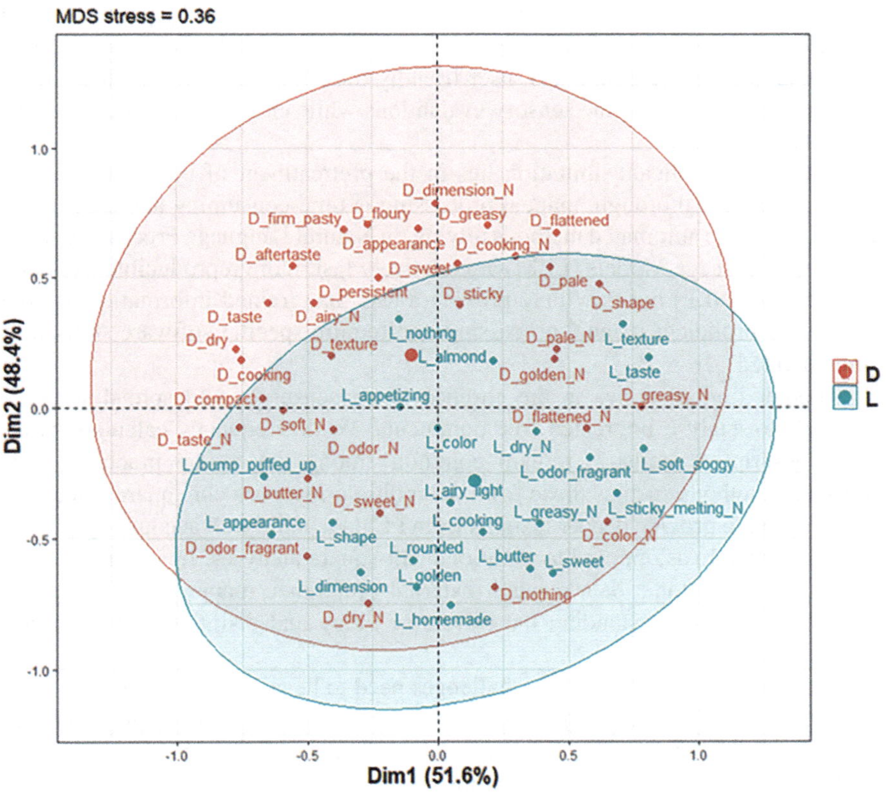

Fig. 6.6 Principal components rotated bi-dimensional MDS solution (Stress = 0.36) of the pair-wise phi-correlation distance between descriptors

5 Conclusion

In this chapter, we explored the use of the Free Comment (FC) method in sensory and consumer science, highlighting its benefits, limitations, and practical applications through a detailed survey done among users of the method.

The FC method, rooted in open-ended questioning, provides a rich and spontaneous capture of consumer perceptions by allowing them to express their thoughts freely, without being constrained by predefined lists of sensory descriptors. This flexibility makes FC particularly valuable in detecting subtle or unexpected insights that might be overlooked by traditional closed-ended methods like Check-All-That-Apply or Just-About-Right scales.

The survey among practitioners of sensory analysis revealed that despite the potential of FC, its application in industry remains constrained by practical issues and a lack of expertise in data pretreatment and statistical analysis. There is a growing demand for more automated, user-friendly tools that can facilitate the integration of FC data into routine sensory evaluations while ensuring robust and reliable insights.

The most significant limitation lies in the pretreatment of the collected comments. Although thorough, manual processing is time-consuming and lacks reproducibility, while automated methods relying on Natural Language Processing (NLP) or Large Language Models (LLM) are relatively fast, more reproducible and deterministic (for NLP) but they may provide a less fine-grained information as compared to a proficient manual processing and require specific software skills to be implemented.

Another limitation lies in the absence of consensus or widespread statistical methods to analyse pretreated free comments. While frequency calculations and basic descriptive statistics remain common among day-to-day practices, more advanced methods such as those from the multiple-response chi-square framework have proven to provide deeper insights from FC data while well accounting for their specific structure as opposed to traditional chi-square analyses. In addition, pairing FC data with hedonic data through external preference mapping and regression models enables understanding the drivers of liking and disliking and the related sensory dimensions.

Looking forward, several key challenges need to be addressed to enable a widespread use of FC with homogeneous practices. These include developing standardized, fast and reproducible pretreatment methods with accessible software to perform them. It also includes providing practitioners with clear guidance on what statistical analyses should be performed, again with accessible software. Overall, this bridges the gap between academic advancements and industrial practices and needs. By doing so, the full potential of FC for generating rich, reliable and easy to analyse product insights will be unleashed, ultimately driving more informed product development and innovation without constrained list of sensory descriptors inducing some biases and constraints.

References

Adams, J., Williams, A., Lancaster, B., & Foley, M. (2007). Advantages and uses of check-all-that-apply response compared to traditional scaling of attributes for salty snacks. In *7th Pangborn sensory science symposium, Minneapolis, USA*.

Antille, N., Audoubert, F., Camilleri, M., Grain, M., Rytz, A., Pineau, N., & Mahieu, B. (2024). Comparison of open comments and check-all-that-apply to collect reasons for liking and disliking chocolates in preference mapping. *Food Quality and Preference*, 105171. https://doi.org/10.1016/j.foodqual.2024.105171

Ares, G., Giménez, A., & Gámbaro, A. (2008). Understanding consumers' perception of conventional and functional yogurts using word association and hard laddering. *Food Quality and Preference, 19*(7), 636–643. https://doi.org/10.1016/j.foodqual.2008.05.005

Ares, G., Giménez, A., Barreiro, C., & Gámbaro, A. (2010). Use of an open-ended question to identify drivers of liking of milk desserts. Comparison with preference mapping techniques. *Sensometrics 2008 (TBC), 21*(3), 286–294. https://doi.org/10.1016/j.foodqual.2009.05.006

Benjamini, Y., & Hochberg, Y. (1995). Controlling the false discovery rate: A practical and powerful approach to multiple testing. *Journal of the Royal Statistical Society. Series B (Methodological), 57*(1), 289–300.

Benzécri, J. P. (1973). *L'Analyse des Données: Vol. 2: L'Analyse des Correspondances*. Dunod.

Bi, J., & Kuesten, C. (2022). Commentary on Meyners and Hasted (2021): On the applicability of ANOVA models for CATA data, food quality and preference, 92. *Food Quality and Preference, 95*, 104340. https://doi.org/10.1016/j.foodqual.2021.104340

Borg, I., & Groenen, J. F. (2005). *Modern multidimensional scaling*. Springer. https://doi.org/10.1007/0-387-28981-X

Cadoret, M., & Husson, F. (2013). Construction and evaluation of confidence ellipses applied at sensory data. *Food Quality and Preference, 28*(1), 106–115. https://doi.org/10.1016/j.foodqual.2012.09.005

Callegaro, M., Murakami, M. H., Tepman, Z., & Henderson, V. (2015). Yes–no answers versus check-all in self-administered modes: A systematic review and analyses. *International Journal of Market Research, 57*(2), 203–224. https://doi.org/10.2501/IJMR-2015-014a

Campo, E., Ballester, J., Langlois, J., Dacremont, C., & Valentin, D. (2010). Comparison of conventional descriptive analysis and a citation frequency-based descriptive method for odor profiling: An application to Burgundy pinot noir wines. *Food Quality and Preference, 21*(1), 44–55. https://doi.org/10.1016/j.foodqual.2009.08.001

Carroll, J. D. (1972). Individual differences and multidimensional scaling. In R. N. Shepard, A. K. Romney, & S. I. Nerloves (Eds.), *Multidimensional scaling: Theory and applications in the behavioral sciences*. Academic.

Castura, J. C. (2009). Do panellists donkey vote in sensory choose-all-that-apply questions? In *8th Pangborn sensory science symposium, Florence, Italy*.

Castura, J. C., Rutledge, D. N., Ross, C. F., & Næs, T. (2022). Discriminability and uncertainty in principal component analysis (PCA) of temporal check-all-that-apply (TCATA) data. *Food Quality and Preference, 96*, 104370. https://doi.org/10.1016/j.foodqual.2021.104370

Castura, J. C., Varela, P., & Næs, T. (2023). Evaluation of complementary numerical and visual approaches for investigating pairwise comparisons after principal component analysis. *Food Quality and Preference, 107*, 104843. https://doi.org/10.1016/j.foodqual.2023.104843

Cochran, W. G. (1950). The comparison of percentages in matched samples. *Biometrika, 37*(3–4), 256–266. https://doi.org/10.1093/biomet/37.3-4.256

Coulon-Leroy, C., Symoneaux, R., Lawrence, G., Mehinagic, E., & Maitre, I. (2017). Mixed profiling: A new tool of sensory analysis in a professional context. Application to wines. *Food Quality and Preference, 57*, 8–16. https://doi.org/10.1016/j.foodqual.2016.11.005

Cramer, H. (1946). *Mathematical methods of statistics*. Princeton University Press.

Danzart, M. (2009). *Evaluation sensorielle. Manuel méthodologique* (SSHA 3eme ed.). Lavoisier.

Gower, J. C., & Legendre, P. (1986). Metric and Euclidean properties of dissimilarity coefficients. *Journal of Classification, 3*(1), 5–48. https://doi.org/10.1007/BF01896809

Greenhoff, K., & MacFie, H. J. H. (1994). Preference mapping in practice. In H. J. H. MacFie & D. M. H. Thompson (Eds.), *Measurement of food preferences*. Blackie Academic & Professionals.

Hamilton, L. M., & Delarue, J. (2025). Understanding perceptions of unfamiliar drinks using natural language in simulated drinking contexts. *Food Quality and Preference, 123*, 105340. https://doi.org/10.1016/j.foodqual.2024.105340

Hamilton, L. M., & Lahne, J. (2020). Fast and automated sensory analysis: Using natural language processing for descriptive lexicon development. *Food Quality and Preference, 83*, 103926. https://doi.org/10.1016/j.foodqual.2020.103926

Hamilton, L. M., Neill, C. L., & Lahne, J. (2023). Flavor language in expert reviews versus consumer preferences: An application to expensive American whiskeys. *Food Quality and Preference, 109*, 104892. https://doi.org/10.1016/j.foodqual.2023.104892

Hanaei, F., Cuvelier, G., & Sieffermann, J. M. (2015). Consumer texture descriptions of a set of processed cheese. *Food Quality and Preference, 40*, 316–325. https://doi.org/10.1016/j.foodqual.2014.05.018

Heussen, F., Holthuysen, N., Kremer, S., Rason, J., & Worch, T. (2023). Beyond liking: Innovative approach using CATA to better understand consumer's associations to products. *Food Quality and Preference, 108*, 104860. https://doi.org/10.1016/j.foodqual.2023.104860

Jaeger, S. R., & Rasmussen, M. A. (2021). Importance of data preparation when analysing written responses to open-ended questions: An empirical assessment and comparison with manual coding. *Food Quality and Preference, 93*, 104270. https://doi.org/10.1016/j.foodqual.2021.104270

Kim, I.-A., Hopkinson, A., Van Hout, D., & Lee, H.-S. (2017). A novel two-step rating-based 'double-faced applicability' test. Part 1: Its performance in sample discrimination in comparison to simple one-step applicability rating. *Food Quality and Preference, 56*, 189–200. https://doi.org/10.1016/j.foodqual.2016.10.010

Klayman, J., & Ha, Y. (1987). Confirmation, disconfirmation, and information in hypothesis testing. *Psychological Review, 94*(2), 211–228. https://doi.org/10.1037/0033-295X.94.2.211

Koriat, A., Lichtenstein, S., & Fischhoff, B. (1980). Reasons for confidence. *Journal of Experimental Psychology: Human Learning and Memory, 6*(2), 107–118. https://doi.org/10.1037/0278-7393.6.2.107

Krosnick, J. A. (1999). Survey research. *Annual Review of Psychology, 50*(1), 537–567. https://doi.org/10.1146/annurev.psych.50.1.537

Lahne, J., Trubek, A. B., & Pelchat, M. L. (2014). Consumer sensory perception of cheese depends on context: A study using comment analysis and linear mixed models. *Food Quality and Preference, 32*, 184–197. https://doi.org/10.1016/j.foodqual.2013.10.007

Luc, A., Lê, S., & Philippe, M. (2020). Nudging consumers for relevant data using free JAR profiling: An application to product development. *Food Quality and Preference, 79*, 103751. https://doi.org/10.1016/j.foodqual.2019.103751

Mahieu, B., Visalli, M., & Schlich, P. (2020). Accounting for the dimensionality of the dependence in analyses of contingency tables obtained with check-all-that-apply and free comment. *Food Quality and Preference, 83*, 103924. https://doi.org/10.1016/j.foodqual.2020.103924

Mahieu, B., Visalli, M., Thomas, A., & Schlich, P. (2020a). Free-comment outperformed check-all-that-apply in the sensory characterisation of wines with consumers at home. *Food Quality and Preference, 84*, 103937. https://doi.org/10.1016/j.foodqual.2020.103937

Mahieu, B., Visalli, M., Thomas, A., & Schlich, P. (2020b). Using free comment with consumers to obtain temporal sensory descriptions of products. *Food Quality and Preference, 86*, 104008. https://doi.org/10.1016/j.foodqual.2020.104008

Mahieu, B., Schlich, P., Visalli, M., & Cardot, H. (2021). A multiple-response chi-square framework for the analysis of free comment and check-all-that-apply data. *Food Quality and Preference, 93*, 104256. https://doi.org/10.1016/j.foodqual.2021.104256

Mahieu, B., Visalli, M., & Schlich, P. (2022). Identifying drivers of liking and characterizing the ideal product thanks to free comment. *Food Quality and Preference, 96*, 104389. https://doi.org/10.1016/j.foodqual.2021.104389

McEwan, J. A. (1996). Preference mapping for product optimization. In T. Naes & E. Risvik (Eds.), *Multivariate analysis of data in sensory science*. Elsevier.

Meyners, M., & Hasted, A. (2021). On the applicability of ANOVA models for CATA data. *Food Quality and Preference, 92*, 104219. https://doi.org/10.1016/j.foodqual.2021.104219

Meyners, M., & Hasted, A. (2022). Reply to Bi and Kuesten: ANOVA outperforms logistic regression for the analysis of CATA data. *Food Quality and Preference, 95*, 104339. https://doi.org/10.1016/j.foodqual.2021.104339

Meyners, M., Castura, J. C., & Carr, B. T. (2013). Existing and new approaches for the analysis of CATA data. *Food Quality and Preference, 30*(2), 309–319. https://doi.org/10.1016/j.foodqual.2013.06.010

Reja, U., Manfreda, K., Hlebec, V., & Vehovar, V. (2003). Open-ended vs. Close-ended questions in web questionnaires. *Developments in Applied Statistics, 19*(1), 159.

Rocha, R. S., Mahieu, B., Tavares Filho, E. R., Zacarchenco, P. B., Freitas, M. Q., Mársico, E. T., Pimentel, T. C., Esmerino, E. A., & Cruz, A. G. (2023). Free comment as a valuable approach to characterize and identify the drivers of liking of high-protein flavored milk drink submitted to ohmic heating. *Food Research International, 165*, 112517. https://doi.org/10.1016/j.foodres.2023.112517

Schuman, H., & Presser, S. (1979). The open and closed question. *American Sociological Review, 44*(5), 692. https://doi.org/10.2307/2094521

Simes, R. J. (1986). An improved Bonferroni procedure for multiple tests of significance. *Biometrika, 73*(3), 751–754. https://doi.org/10.1093/biomet/73.3.751

Symoneaux, R., Galmarini, M. V., & Mehinagic, E. (2012). Comment analysis of consumer's likes and dislikes as an alternative tool to preference mapping. A case study on apples. *Food Quality and Preference, 24*(1), 59–66. https://doi.org/10.1016/j.foodqual.2011.08.013

Ten Kleij, F., & Musters, P. A. D. (2003). Text analysis of open-ended survey responses: A complementary method to preference mapping. *Food Quality and Preference, 14*(1), 43–52. https://doi.org/10.1016/S0950-3293(02)00011-3

Torgerson, W. S. (1958). *Theory & methods of scaling*. Wiley.

Varela, P., Antúnez, L., Carlehög, M., Alcaire, F., Castura, J. C., Berget, I., Giménez, A., Næs, T., & Ares, G. (2018). What is dominance? An exploration of the concept in TDS tests with trained assessors and consumers. *Food Quality and Preference, 64*, 72–81. https://doi.org/10.1016/j.foodqual.2017.10.014

Vigneau, E., & Qannari, E. M. (2003). Clustering of variables around latent components. *Communications in Statistics – Simulation and Computation, 32*(4), 1131–1150. https://doi.org/10.1081/SAC-120023882

Visalli, M., Mahieu, B., Thomas, A., & Schlich, P. (2020). Concurrent vs. retrospective temporal data collection: Attack-evolution-finish as a simplification of Temporal Dominance of Sensations? *Food Quality and Preference, 85*, 103956. https://doi.org/10.1016/j.foodqual.2020.103956

Visalli, M., Dubois, M., Schlich, P., Ric, F., Cardebat, J.-M., & Georgantzis, N. (2023a). Relevance of free comment to describe wine temporal sensory perception: An application with panels varying in culture and expertise. *Food Quality and Preference, 105*, 104785. https://doi.org/10.1016/j.foodqual.2022.104785

Visalli, M., Mahieu, B., Dubois, M., & Schlich, P. (2023b). Hedonic valence of descriptive sensory terms as an indirect measure of liking: A preliminary study with red wines. *Food Quality and Preference*, 104861. https://doi.org/10.1016/j.foodqual.2023.104861

Visalli, M., Symoneaux, R., Mursic, C., Touret, M., Lourtioux, F., Coulibaly, K., & Mahieu, B. (2025a). A dataset of annotated free comments on the sensory perception of madeleines for benchmarking text mining techniques. *Data in Brief, 58*, 111250. https://doi.org/10.1016/j.dib.2024.111250

Visalli, M., Symoneaux, R., Mursic, C., Touret, M., Lourtioux, F., Coulibaly, K., & Mahieu, B. (2025b). Can natural language processing or large language models replace human operators for pre-processing wordand sentence-based free comments sensory evaluation data? *Food Quality and Preference, 127*, 105456. https://doi.org/10.1016/j.foodqual.2025.105456

Wakihira, T., Visalli, M., & Schlich, P. (2023). Temporal drivers of liking by period: A case study on lemon-flavored carbonated alcoholic drinks with consumers in natural settings. *Food Quality and Preference, 106*, 104793. https://doi.org/10.1016/j.foodqual.2022.104793

Yzerbyt, V. Y., & Leyens, J.-P. (1991). Requesting information to form an impression: The influence of valence and confirmatory status. *Journal of Experimental Social Psychology, 27*(4), 337–356. https://doi.org/10.1016/0022-1031(91)90030-A

Züll, C. (2016). Open-ended questions. In *GESIS survey guidelines*. Leibniz Institute for the Social Sciences.

Chapter 7
Text Highlighting Method: Using Textual Data for Attitude Measurement and Other Consumer Behaviour Responses

Sara R. Jaeger, Sok L. Chheang, and Gastón Ares

1 Introduction

This chapter introduces text highlighting as a new method for explicit attitude measurement. It was developed by Jaeger, Ares and colleagues in 2022 and has since been applied in several case studies with adult consumers in different countries. In parallel with applying this novel method, methodological investigations have been ongoing. We summarize our knowledge about the method at this point. Readers are referred to the individual papers for detailed accounts of the individual studies and investigations.

Explicit methods are the most common approach for attitude measurement (Fazio, 2007). These methods rely on participants' self-reported attitudes towards a target concept, typically by indicating their degree of agreement with written statements or reporting how frequently they engage in a certain behaviour. This approach requires the use of scales (Willits et al., 2016), which are susceptible to different types of response bias (Orne, 2009; Podsakoff et al., 2003), particularly in cross-cultural research (Ares, 2018).

S. R. Jaeger (✉)
Department of Food Science, Aarhus University, Aarhus, Denmark
e-mail: sara.jaeger@food.au.dk

S. L. Chheang
The New Zealand Institute for Plant and Food Research Limited, Auckland, New Zealand
e-mail: sokleang.chheang@plantandfood.co.nz

G. Ares
Department of Sensometrics & Consumer Science, Instituto Polo Tecnológico de Pando, Facultad de Química, Universidad de la República, Canelones, Uruguay
e-mail: gares@fq.edu.uy

M. Visalli, M. V. Galmarini (eds.), *Collection and Analysis of Text Data in Sensory and Consumer Science*, https://doi.org/10.1007/978-3-032-04851-6_7

To overcome the challenges associated with using scales, text highlighting was introduced as a new explicit method for attitude measurement by Jaeger et al. (2022a). In text highlighting, participants are presented with a text focused on the target concept/idea/theme. They are asked to read it and to highlight parts they like/feel positive about and parts they dislike/feel negative about. The task is typically implemented using a software interface, as shown in Fig. 7.1. Participants can highlight as much or as little text as they want. Responses to the text highlighting allow

Fig. 7.1 Illustration of text highlighting for attitude research (exemplar text in Latin). Participants are instructed to read the text and use the highlight functions to indicate the content in the text that they like/feel positive about and the content that they dislike / feel negative about. (**a**) Task as it appears at the outset; (**b**) Task as it appears at the end where highlights have been added by a participant who has completed the task

researchers to identify those parts of the text that participants perceived as positive, negative or neutral, which can be regarded as a measure of their attitude towards the text.

Text highlighting is expected to be an easy and familiar task for most participants. Highlighting relevant sections of a text is a popular learning technique (Dunlosky et al., 2013) and is also frequently used in crowd sourcing platforms (Strobelt et al., 2015). Considering that highlighting has been reported to increase engagement and understanding (Latini et al., 2019; Leroy et al., 2021), it is expected to be less tedious than repeatedly answering Likert scales. These characteristics make text highlighting an appropriate task for online surveys, where complexity, survey length, and boredom influence completion rates and data quality (e.g., Galesic & Bosnjak, 2009; Ganassali, 2008; Liu & Wronski, 2018; Roßmann et al., 2018).

2 Developing the Text to Conduct a Text Highlighting Study

Any text can be used for text highlighting, although one developed by a research team specifically for the focal project seems most probable. In such instances, the task of writing the text should receive careful attention. Just like the question wording is paramount when developing a survey, the text is at the center of the text highlighting method.

2.1 Text Content and Sentiment

Text content should flow directly from the research aim/s. For example, if the aim is to explore consumer attitudes to a novel food technology like indoor vertical farming, then the text needs to describe this way of producing food. How this description happens depends on the research aims. The text may be a short, factual, and neutrally worded description, or it could be longer and extend to different advantages (positively worded) and disadvantages (negatively worded) of vertical farming. Alternatively, multiple texts could be used if the research aimed to determine how consumers responded to varying types of information about vertical farming (say, exclusively positive or exclusively negative narratives). Thus, there are no *a priori* rules for what the text should be about or how it should be written. If relevant to the research aims, there are no methodological hindrances to applying text highlighting to existing texts (e.g., blogs).

Besides ensuring that the text is free of spelling errors and grammatically correct, other aspects of the text do, however, need to be considered as they can influence consumers' responses and the outcome of the empirical study. We have found that the interpretation of results can be more direct when the sentences are constructed such that they relate to a single topic only and do not include both positive and negative information about the focal topic.

If the intent is to write a factual and neutrally worded text, then pilot testing should confirm that the text is perceived as such. If the text intends to be persuasive, the emotional tone behind the words needs to be of primary concern. In both instances, sentiment analysis should be performed. The *SÉANCE* tool (Crossley et al., 2017) can be used to evaluate the sentiment of the text. It calculates 20 different indices with input from eight databases/word lists. The tool is free and easy to install and use. Other free alternatives include *MonkeyLearn* and *Daniel Soper's Free Sentiment Analyzer*. The latter allows users to conduct sentiment analysis on any text written in English by computing a sentiment score that reflects the overall sentiment, tone, or emotional feeling, which can range from −100 (very negative or serious tone) to +100 (very positive or enthusiastic tone).

2.2 Text Difficulty and Length

It is important to develop text explicitly with the study participants in mind. For a text on the same topic, different writing styles and language may be used when seeking the opinions of experts and laypeople. A general recommendation is to use simple and familiar language that is suited to the groups of participants taking part in the study. Besides being fairly intuitive, this recommendation is supported by the findings of Jaeger et al. (2022b) in a case study on biodynamic agriculture.

For texts written in English, different free-to-use online tools exist for determining reading difficulty, including *Readability Formulas*, *WordCalc's Readability Calculator* and *Readable Pro*. We have used readabilityformulas.com, which also exists in a Spanish language version (SpanishReadability.com). Tools for other languages may also exist. Depending on the tool, different established measures of readability are used, for example, Flesch Reading Ease (Farr et al., 1951), Gunning Fog (Gunning, 1952), Flesch-Kincaid Grade Level (Kincaid et al., 1975), Coleman-Liau Index (Coleman & Liau, 1975), SMOG Index (McLaughlin, 1969), Automated Readability Index (E. A. Smith & Senter, 1967), and Linsear Write Formula (Klare, 1974). In the Readability Formulas online tool, we have relied on an integrated index that calculates an average reading level and relates that to the age of the reader. For example, a text with a reading level corresponding to that expected of 14–15 year-olds is marked as "fairly difficult", while a "difficult" to read text is associated with the expected reading level of 17–18-year-olds.

The use of technical jargon should also be considered, especially whether or not it is appropriate for the intended participants in the main study. For example, we have made conscious decisions not to use the terms "vertical farming" or "indoor vertical farming" and instead use "indoor farming" and "indoor farms". Other researchers have used the latter two (e.g., Beacham et al., 2019; Gobble & Headrick, 2019; O'Sullivan et al., 2020), and we regarded them as more consumer-friendly, especially if study participants have never heard about controlled environment agriculture. Rather than suggesting that technical terms and jargon should be avoided, we encourage decisions about writing style, word familiarity, etc., should be taken in the context of the stated research aim. This way they can be explicitly justified, and underpinning assumptions made available to readers of the ensuing scientific report/paper.

Regarding text length, we recommend using consecutive highlighting tasks on shorter texts (125–250 words) rather than a single highlighting task on a longer text because the former increases highlighting frequency and is expected to be associated with greater participant task engagement. We found evidence of text length effects in Jaeger and Ares (2022) and Jaeger et al. (2022b). The latter of these papers demonstrated that the overall frequency of highlighting, the frequency of "like" highlighting, the frequency of "dislike" highlighting, and sentiment scores were all influenced by text difficulty and length, although the effects were larger for variations in text length than variations in reading difficulty.

2.3 Pilot Testing and Revising the Text Before Data Collection

We have experienced that arriving at the final text to be used in a text highlighting study requires a longer time than expected at the outset. There needs to be sufficient time for several rounds of revisions that consider input from colleagues in the research team. Professional writers and editors can be approached for input and suggested improvements, and there may need to be several rounds of revision to ensure that factual correctness (if desired) is not traded against simplified language and accessible writing style, etc. As generative AI continues to evolve, this may provide another avenue for text ideation/generation. Before data collection, formal pilot testing with people representative of the main study can be performed. For example, a set of attitude statements that can be rated on 7-point Likert scales (e.g., the text was easy to understand, the text was too long, the text tried to persuade me to have a certain opinion about <topic>, etc.).

When a final text has been agreed upon, it is important to ensure that it is correctly displayed in the data collection platform. This requires that the test links are trialed and approved. Ideally, a final round of pilot testing should occur at this stage, to obtain responses from the interface that will be used during the main study. Actual highlighting data should be evaluated to see that they fit expectations. For example, if the text includes information about products selling at a premium price, or that a product is very unhealthy, unsustainable etc., it can reasonably be expected that some study participants will respond negatively. If no negative highlights are found in the data file, it may be necessary to check that the data collection interface is functioning correctly.

If the study is to be conducted in multiple languages, translation of the text needs to be carefully approached, and research methodology literature that advises on how to undertake such a task should be consulted as needed (e.g., Harkness et al., 2010; McGorry, 2000). For example, in a four-country study aimed at measuring consumer attitudes to vertical farming (Ares et al., 2021), we initially developed the text in English. The final version of the text was translated to German by a bi-lingual researcher assisted by Google Translate, followed by further review and revision by another German speaker before a final version was agreed between the two German-speaking researchers. We also adapted the spelling of the English-language text to American (US) or British (UK) spelling, depending on the country of data collection.

Studies by Ares, Jaeger and colleagues cited in this chapter serve as a guide for the type of information about the text that should be included in scientific reporting (methodology section). These include:

- The full text, where individual sentences are given a unique identifier.
- Length in terms of number of words, sentences and paragraphs.
- The sources used to develop the text.
- Who wrote the text, and who revised it.
- The pilot testing procedure.
- Sentiment of the final text.
- Reading difficulty of the final text.
- Translation procedures if text developed in a language different to that used for data collection.
- If images supplemented the text, these should be reproduced.

3 Collecting Data in a Text Highlighting Study

3.1 Participants

Anyone with the appropriate reading and language skills can complete a text highlighting task. The only method-specific exclusion criterion we have implemented to date relates to red-green colour blindness. This condition can potentially introduce errors during data collection if the highlighter functions are identified by these two colours. A simple self-report question can be used, for example, "*Are you red-green colour blind*?" Those who answer "yes" can be excluded. Existing data suggest that red-green colour vision deficiency (i.e., where the two colours cannot be distinguished) affects 2–8% of men and 0.5% of women, with the highest rates among Euro-Caucasians (Delpero et al., 2005). The reason to focus on red-green colour blindness is that we have used these colours in the data collection interface as per Fig. 7.1.

As applies to all empirical consumer research, standard protocols for ethical conduct should be followed, and human ethics approval obtained from an institutional review board.

3.2 *Text Highlighting Task*

3.2.1 Text Highlighting Responses

Text highlighting was introduced as a dual-response task where participants when completing the highlighting task were asked to select both "like" and "dislike" aspects of the text. That is, they would highlight in green the content that they agreed with, and in red the content that they disagreed with.

However, other implementations are possible. In Jaeger, Chheang, et al. (2023b), as part of a series of methodological investigations, we compared the dual-response approach to implementations of text highlighting where participants selected only "like" highlights or only "dislike" highlights (i.e., single-response text highlighting). The results from a single case study showed no differences for "like" highlights when comparing dual- and single-response highlighting. However, for "dislike" highlighting, significantly fewer words were selected in the dual-response approach, although the difference in frequency of selection was less than 10%. On this basis it seems reasonable to infer that the risk of bias is low. That is, performing a dual-response task is unlikely to lead to largely different results than would be obtained by completing two single-response tasks.

On this, admittedly, limited empirical evidence, we currently consider dual-response text highlighting to be appropriate. Asking the same participant to do two single-response tasks on the same text does not seem warranted and rather could contribute to fatigue and boredom among participants. The only instance where single-response has merit is if the research aim/question mandates a particular focus on negative attitudes and text content that participants disagree with.

For completeness, we note that highlighter functions other than "like" and "dislike" can be used according to research aims and objectives. This requires further investigation, but in principle, any highlighter labels could be used as long as they apply to the text and the task is clear to participants. One attempt was made by Jaeger, Chheang, et al. (2023a), who studied plant-based alternatives to yoghurt and eggs. Participants were instructed to "read the text and use the highlight functions to select the content in the text that makes you want to buy plant-based yoghurt (eggs), and the content that does not make you want to buy plant-based yoghurt (eggs)." The highlighter functions were: "*Makes me want to buy plant-based yoghurt (eggs)*" (green colour) and "*Does not make me want to buy plant-based yoghurt (eggs)*" (red colour). The obtained data had face validity but were less efficient than purchase intent ratings at identifying the product concepts that were regarded most positively by consumers.

3.2.2 Instructions to Participants

Clear directions should be provided to participants about how to perform the highlighting task. Our approach has been to instruct participants to read the text and use the highlighter functions in the survey software to select text content they liked/felt positive about (i.e., "like" highlighting) and text content that they disliked/felt negative about (i.e., "dislike" highlighting). The use of a short video that instructs participants how to complete the task could be considered, but we have not attempted that to date. A very short practice exercise could also be considered.

At the end of the task, if a participant has not highlighted any text, they have been prompted to do so. Further, if at the end of the task, a participant has only used "like" highlights, we have used the following message screen prompt: "*We noticed you only selected text which you like/feel positive about. Please also select text which you dislike/feel negative about, or press continue if you do not dislike/feel negative about any of it.*" In a similar vein, a prompt was given if only "dislike" highlights had been used. It is also possible to implement the task such that each participant must provide both positive and negative highlights, and although we did this initially, we now prefer the less prescriptive approach described above.

3.2.3 Data Capture

To date, we have used a bespoke digital solution for data collection provided by LightSpeed Research (https://www.lightspeedresearch.com.au/), a commercial research provider. We have not explored if text highlighting is offered by other commercial operators but considering that the principle behind the method is simple, it seems probable that it can be made available upon request.

In the studies we have conducted until now with text highlighting, we have restricted completion by mobile phone for the reason that the screen size may be too small to present, in a clear manner, the instructions, the highlighter functions and the text. Moreover, we have customized the implementation such that the instructions, the highlighter functions and the complete text were visible without the need to scroll down. If the text was long enough to require it, we opted to split the task across two screens, presenting it as two separate highlighting tasks—each with full instructions and highlighter options.

In principle, text highlighting can be implemented without digital tools by using a pen-and-paper approach. Considering the rapid expansion of digital access and connectivity, it may be more relevant to develop low-cost options for electronic data collection than pursuing non-digital alternatives.

4 Analysing Data in a Text Highlighting Study

4.1 Word Clouds

For preliminary results, responses to the text highlighting task can be graphically represented using word clouds which represent all the words of the text according to the percentage of participants who highlighted them as "like" and "dislike". The larger the word, the higher the percentage of participants using "like" (or "dislike") highlighting in the task. Although word clouds can be easily interpreted, it is important to consider that they are frequently constructed on a relative basis. That is, they do not provide absolute information about the percentage of participants who used "like" and "dislike" highlighting.

4.2 *Converting Text Highlights to Numbers and Data Properties*

The first step of the analysis of text highlighting is converting participants' responses to numbers. For this purpose, variables are created to indicate whether each of the participants highlighted each words as "like" (+1), "dislike" (−1), or did not highlight it at all (0). To date, we have instructed the online provider to do this conversion, such that the raw data file provided is numerical.

Using the coded data, the percentage of participants that highlight each of the words as "like" and "dislike" can be calculated. However, data analysis at the level of individual words does not lead to a full understanding of participants attitudes towards the target concept. A more comprehensive understanding can be obtained by analysing their responses to sentences or parts of sentences communicating specific ideas (e.g., "small vertical farms" or "yellow kiwifruit"). For this purpose, variables are created to indicate whether each participant highlighted at least one of the words of each of the sentences as "like" (+1), "dislike" (−1), or did not highlight any word (0). The percentages of participants highlighting each of the sentences as "like" and "dislike" are calculated and presented using tables or bar charts. Sentences with the larger percentage of participants highlighting words as "like" are those statements that are positively perceived. Conversely, sentences with the largest percentage of participants highlighting words as "dislike" are the most negatively perceived.

4.3 *Sentiment Scores*

Text highlighting enables us to obtain an overall measure of consumers' attitude towards each of the sentences included in the text through the calculation of sentiment scores. They can be calculated at the level of individual participants or at the aggregate level for each of the sentences (see Chap. 8).

4.3.1 Participants' Sentiment Scores

Categorical coding is used to analyse the text highlighting responses. The value "1" indicates that a word was highlighted as "like", and the value "-1" indicates that a word was highlighted as "dislike". The value "0" is used if a word had not been selected/highlighted.

Overall sentiment scores for each participant can be calculated by summing up the sentiment scores across all the sentences in the text and expressed as absolute numbers or percentages. Thus, scores can range between (−) the number of sentences and (+) the number of sentences, or between −100% and +100%, according to whether they are expressed in absolute numbers or percentages, respectively. A score higher than 0 means that the participant has an overall positive attitude towards the text, whereas a score lower than 0 indicates an overall negative attitude.

Based on these scores, segments of participants with different overall sentiment towards the text can be identified using manual segmentation or hierarchical cluster analysis. For example, Jaeger et al. (2022a) manually created three groups of participants: (i) Positive Group, composed of participants with sentiment scores higher than 0%, (ii) Neutral or Ambivalent Group, composed of participants with sentiment scores equal to 0%, and (iii) Negative Group, composed of participants with sentiment scores lower than 0%.

4.3.2 Sentiment Scores for Sentences

Aggregate sentiment scores can also be calculated for each sentence by averaging the scores of all participants. Positive scores indicate a generally positive sentiment towards the information conveyed by the sentence, whereas negative scores indicate generally negative sentiment towards the information. When expressed as percentages, sentiment scores for individual sentences can range between −100% and +100%. Sentiment scores of −100% indicate that all participants only highlighted words as "dislike" within the sentence, whereas scores of +100% indicate that all participants only highlighted words as "like".

4.4 Penalty-Lift Analysis

Jaeger, Chheang, et al. (2023c) explored the use of text highlighting in food product research. The method was used to explore how consumers perceived specific aspects of green and gold kiwifruit. In addition, consumers' expected liking for both products was measured using hedonic scales. The relationship between responses to text highlighting and expected liking was analysed using penalty-lift analysis, following the approach developed by Meyners et al. (2013) for CATA (check-all-that-apply) questions.

A series of 43 characteristics of kiwifruit were described in the text, which was presented as a text highlighting task per Fig. 7.1. For sections of the text describing each of the characteristics, variables were created to indicate whether participants highlighted words as "like" (+1), "dislike" (−1), or did not highlight any word or highlighted words as both "like" and "dislike" (0). The difference between the average expected liking of consumers who highlighted a characteristic as "like" and those who did not highlight the characteristic or used both "like" and "dislike" highlighting was calculated. The statistical significance of the differences was assessed using a t-test, only for characteristics highlighted as "like" or "dislike" by at least 50 consumers (4.4% of the participants involved in the study). Differences higher than 0 indicate characteristics positively associated with expected liking, whereas differences lower than 0 indicate characteristics negatively associated with expected liking. The same approach can be used for "dislike" highlighting. Results were represented using scatter plots with the percentage of participants highlighting each characteristic as "like" and "dislike" in the x-axis and the difference in expected liking in the y-axis.

5 When to Use Text Highlighting Methodology

5.1 Attitude Measurement

When developing text highlighting, we sought to create a method that could be used as an alternative to direct attitude rating. But what is the evidence that text highlighting measures attitudes and how do we know that it is a "good" measure of attitudes? In other words, how has the method been validated?

In the paper that first introduced the text highlighting method—Jaeger et al. (2022a)—we began to address key questions regarding validity. An important first observation was that consumers systematically approached the task such that positive/desirable content in the text was frequently highlighted as "like", whereas negative/undesirable text content was more frequently highlighted as "dislike". The case study in the paper was about indoor vertical farming, a novel technology for growing herbs, salad and some vegetables (see Jaeger, 2024, for an introduction to vertical farming and a concise review on consumer attitudes), and consumers (adults from the United Kingdom) responded positively to sentences mentioning, for example, reduced carbon emissions, nutrient-rich growth environment and agricultural land being returned to nature. Conversely, responses were negative to text mentioning the use of robots, high energy requirements of indoor vertical farms and premium prices. When overall sentiment scores for participants were calculated and used as the basis for consumer segmentation, the group of people with a positive attitude to vertical farming based on sentiment from text highlighting also rated their attitude as more positive on a later rating task. Similarly, the group of people that, based on text highlighting, were identified to have a negative attitude to vertical farming, on average, rated this technology as less positive. In other words, there was a strong concordance between rated attitude and attitude based on text highlighting responses. The comparison with direct attitude questions also held when the attitude statements related to specific aspects of the text (e.g., climate change, premium prices) and were compared to sentiment scores for "matching" sentences.

The sentiment scores also aligned with expectations based on the influence of food neophobia. This trait, which captures reluctance to eat novel foods (Pliner & Hobden, 1992), and is strongly correlated to food technology neophobia (Cox & Evans, 2008), is a reliable predictor of attitude to vertical farming (Jaeger, 2024). Accordingly, the average degree of food neophobia was higher in the group of consumers who expressed a negative sentiment towards vertical farming in the text highlighting task. Similarly, those people who were positive towards vertical farming based on text highlighting were less food neophobic, on average.

The risk of bias must also be evaluated when establishing a new method, and while this work is still ongoing, some key results have already been presented. Jaeger and Ares (2022), who again used vertical farming as a case study, showed that the characteristics about vertical farming that consumers responded positively and negatively to were minimally influenced by length and content of the text used in text highlighting. Thus, if consumers respond positively to a certain characteristic of a novel food or production method, it would not be expected, based on these results, that another study where the same information was presented in longer/shorter text about the same food/method would produce a result which indicated that consumers were dominantly negative about the focal characteristic. However, what was observed in Jaeger and Ares (2022) was that highlighting responses are not independent of text length and content. Word highlighting frequency decreased for longer texts and sentence sentiment scores were also altered. This means that numerical results should not be directly compared because they are textually dependent. A different way of saying this is that the results from text highlighting are specific to the text from which they were derived.

Also important in establishing the methodology is evidence that it does not introduce bias. In Jaeger and Ares (2022) we asked whether consumers' attitudes were altered by the process of completing the highlighting task and found that they were not, on average. In other words, bias was not introduced by the task of adding highlights to a text while reading it. In a later study (Jaeger, Chheang, et al., 2023b) the use of concurrent thought-listing (Ericsson & Simon, 1980) confirmed this result. Qualitative analysis of responses from ~125 participants in a text highlighting task found that no one expressed thoughts indicative of response bias.

In relation to the inclusion in longer surveys with tasks other than text highlighting, our current evidence is also that bias is not a concern. Specifically, Study 2 of Jaeger, Chheang, et al. (2023b) was dedicated to the question of whether the process of completing the text highlighting task influenced consumers' topic- and product-related responses. We used a between-subjects design with three experimental conditions (EC). The participants assigned to EC1 completed a dual-response text highlighting task ("like" and "dislike" functionality) while the participants assigned to EC2 read the text used in EC1 without completing a text highlighting task. The third group (EC3) read no text and completed no text highlighting task. Next, all groups completed a product evaluation task (expected liking, purchase intent, willingness to pay, product familiarity). There was minimal evidence that these responses were influenced by EC, a finding that was repeated across four substudies. Thus, we concluded that that text highlighting can reasonably be used together with other types of consumer responses without much risk of biasing the latter.

5.2 Cross-Cultural Research

The ability to be used in multiple languages and/or cultures is a desirable characteristic for any research method. Text highlighting lends itself to such applications since it is not restricted to the Latin (or Roman) alphabet but can be used with texts written in any alphabet (Cyrillic, Arabic, Sanskrit, Hangul, Hanzi, etc.). Besides texts in English (British or American spelling), we have used the method with German consumers on a text written in German (e.g., Jaeger, Harker, et al., 2023) and with Chinese consumers on a text written with simplified Chinese characters (e.g., Ares et al., 2021). We have also run studies in Singapore and Malaysia with texts written in English (e.g., Jaeger, Chheang, et al., 2023c), although the need to recruit participants with high English language proficiency hinders the ability to obtain a nationally representative consumer sample.

The remainder of this section is devoted to the knowledge we currently have regarding the use of text highlighting methodology in cross-cultural research, which is primarily based on the work published in Ares and Jaeger (2022). In this paper, the case study was indoor vertical farming, and we included four countries in the research. These were the United Kingdom, the United States, Singapore, and China, and from each country ~650 adult consumers took part. Quota sampling was used such that the samples had the same age and gender profiles. For the text highlighting task, we used the dual-response method, meaning that participants would highlight in green the text content that they liked/agreed with, and in red the text content that they disliked/disagreed with. Our analyses of the highlighting responses did not point to systematic differences in response style between countries. Further, we showed that groups of consumers with different sentiment (Positive, Neutral/Ambivalent, or Negative) towards different sentences in the text were predictably different in their average ratings for related attitudinal statements.

This was a welcome finding, considering that text highlighting has primarily been developed as a method for explicit attitude measurement and is positioned as an alternative to attitude rating. As also outlined in Sect. 1, rating scales are ubiquitous in attitude measurement but subject to response styles, which is a hindrance in cross-cultural research since mean ratings can be biased due to cultural influences. Direct comparison of such means should, therefore, be done with much caution. Two types of response styles (and sources of bias) with particular relevance in cross-cultural research are extreme responding and acquiescence bias. Briefly, the former refers to the tendency to respond using extreme endpoints on rating scales, and the latter is the tendency to respond with agreement and affirmation. There is considerable evidence to suggest that both of these response styles are more common among people from Asian countries, and they have been linked to collectivist cultural orientation in this part of the world (e.g., Johnson et al., 2005; Smith & Fischer, 2008).

Together with ranking and choice-based methods, text highlighting offers a scale-free and less bias-prone alternative to attitude scales. This is a notable method characteristic, and in Ares and Jaeger (2022) we demonstrated the ability of text highlighting to capture specific attitudes towards indoor vertical farming across countries and languages, as well as its ability to overcome response style differences.

6 Combining Text Highlighting with Other Methodologies

6.1 Open-Ended Questions About Selected Highlights

Ares et al. (2023) proposed that text highlighting methodology be combined with open-ended questions to aid interpretation of the text highlighting results. This modification of the method, which transforms it to a quantitative-dominant mixed method approach, does not alter how the text highlighting task is conducted. The add-on happens immediately after the text highlighting task is completed and participants are asked to explain in their own words why they highlighted certain text.

Based on our experiences so far, vague instructions do not work optimally. If participants are asked to "*tell us about the text that you selected as 'like'*", it can be difficult to connect their written responses to an actual section of highlighted text. Therefore, we recommend that when open-ended questions are used, participants be shown a screenshot of the highlights they just added to the text. With this visual input, they are then asked to explain why they made these highlights. For example, the instruction can be: "*To better understand your responses, please tell us about the parts of the text that you highlighted. If you selected text as 'like', what is it about the text that makes you feel positive? Similarly, if you selected text as 'dislike', what is it about that text that makes you feel negative?*" To simplify the process, the shown screenshot may only display one type of highlights, either "like" or "dislike". If the text comprises several paragraphs of text, only one of these can be shown, so as to reduce the response burden for participants and increase the ease of connecting written responses to the text in the highlighting task. As part of the instructions, we also suggest that participants be directed to "*Please explain using specific examples and in as much detail as you can.*" This is to prevent responses that have limited value such as "*I liked the text*" or "*I disagree*".

We showed in Ares et al. (2023) that the combination of text highlighting and open-ended questions led to more comprehensive insights on the topic of the research. The supplementary qualitative component of the mixed method design (i.e., the open-ended questions) helped to elaborate the insights from text highlighting (i.e., the quantitative core component). Thus, we advocate inclusion of open-ended questions, but note that they increase the time required for data analysis as the open-ended responses need to be content coded. This should be done in accordance with established practices for such data (e.g., Krippendorff, 2013; see also Chap. 5). Specifically, for each question, a coding frame has to be developed, and if done manually, typically one of the members of the research team (with appropriate skills and experience) will do this. Typically, we have used inductive coding and developed the coding frame based on approximatively 10% of the written answers. Next, two experienced researchers have revised the coding frame and the final set of codes has been agreed by consensus. Then the final coding began — performed by one of the researchers and verified afterwards by a senior researcher (for greater rigor, inter-coder reliability should be calculated). We have allowed each open-ended answer to be assigned to more than one content category.

For data analysis, the frequency of mention of each category is calculated by counting the number of participants who provided responses to that specific category. Then in accordance with standard practice for such data, examples of responses within each category are selected to illustrate their meaning (see Chap. 6). The rapid development in semi-automated analysis of textual data will probably mean that the time needed to content code textual data will reduce in the future, but there will always be a need to sense-check the data and interpret it manually.

6.2 Other Questions

Directed by the research aim, questions and hypotheses, additional questions and tasks may be included in the study. For example, including participant demographic and socio-economic characteristics would be typical. It could also be relevant to measure psychographic traits and behavior-related variables.

We have also demonstrated how text highlighting can be part of multi-method investigations, for example, Ares et al. (2021) and Jaeger, Harker, et al. (2023). In both instances, research participants required information about the topic of research, respectively, vertical indoor farming and biodynamic agriculture. As they read this information, they completed a text highlighting task. Then they proceeded to other tasks also about the focal topic, for example, rating of attitude statements, evaluation of product concepts, choice tasks, etc. The responses obtained by text highlighting can then be compared to other responses, for example, within consumer segments or by country. We investigated in Jaeger, Chheang, et al. (2023b) whether the process of completing the text highlighting task influenced consumers' topic- and product-related responses and found that there was minimal evidence of this. In other words, researchers can reasonably use text highlighting together with other types of consumer responses without much risk of biasing the latter.

Because the text highlighting methodology is likely to be implemented online, data quality should be evaluated before data analysis. We have not developed specific indices to highlight selection but encourage the application of indices that can be applied to online surveys in general. Readers are referred to Jaeger et al. (2024) and Jaeger and Cardello (2022) for recommendations and guidelines that are specific to sensory and consumer science researchers. Increasingly, reasearchers need to be aware that some survey responses may come from bots, presenting fraudulent data that should be excluded.

7 Case Study

To illustrate some of the analyses mentioned earlier in the chapter, we draw on data from Jaeger, Harker, et al. (2023), specifically those obtained from Australian consumers (n = 311). The focus of the study was on consumer attitudes towards "beyond organic" (or biodynamic) agriculture. Participants were presented with a test and performed a dual highlighting task with "like" and "dislike" functionality.

Figure 7.2 shows word clouds, where Fig. 7.2a represents "like" responses and Fig. 7.2b represents "dislike" responses. The 10 words most frequently selected as "like" were "ecosystems", "healthy", "groundwater", "wildlife", "biodiversity", "soil", "rich", "care", "build" and "farmers." Conversely, the 10 words most frequently selected as "dislike" were "thinking", "magical", "dung", "cow", "genetically", "non-scientific", "modified", "links", "moon" and "sun." The insights from the word clouds identify the most positive aspects of biodynamic agriculture which are key described benefits of biodynamic agriculture for the planet and its inhabitants. The "dislike" responses focused on the descriptions related to magical thinking, as well as genetic modification and animal excrements. In Table 7.1, the sentence sentiment scores are shown for the overall sample, with values for each of the 20 sentences (S1-S20) in the text. Most positively received (highest sentiment score) was sentence S7 "*Beyond organic farmers care for the soil, groundwater, and wildlife to build healthy ecosystems and rich biodiversity*". Least positively received (most negative sentiment score) was sentence S18 "*But it can also be seen as non-scientific with links to magical thinking*". The overall results from the word clouds and the sentiment scores were aligned, but greater detail and insight were gained from the latter. Segmentation based on individuals with positive and negative sentiment scores is also seen in Table 7.1. For simplicity, participants with neutral/ambivalent overall sentiment were excluded (i.e., not shown in Table 7.1). For some sentences, both segments had positive scores, or both had negative scores and, in some cases, the sentence sentiment score was positive in one segment and negative in the other. The pattern of differences in the sentiment scores fitted expectations since sentiment scores in the positive segment were more positive / less negative than in the negative segment. The sentiment difference between the two segments varied by sentence and was larger in some instances (e.g., S17) and smaller in other instances (e.g., S18). This indicates that some aspects of the information in the text were similarly perceived in the two segments, whereas attitudes to the provided information at times differed majorly.

8 Conclusion

The purpose of the present chapter was to present the methodology of text highlighting, which was introduced in 2022 and has since been used to explore consumer attitudes to a range of topics linked to consumer-related food behaviour.

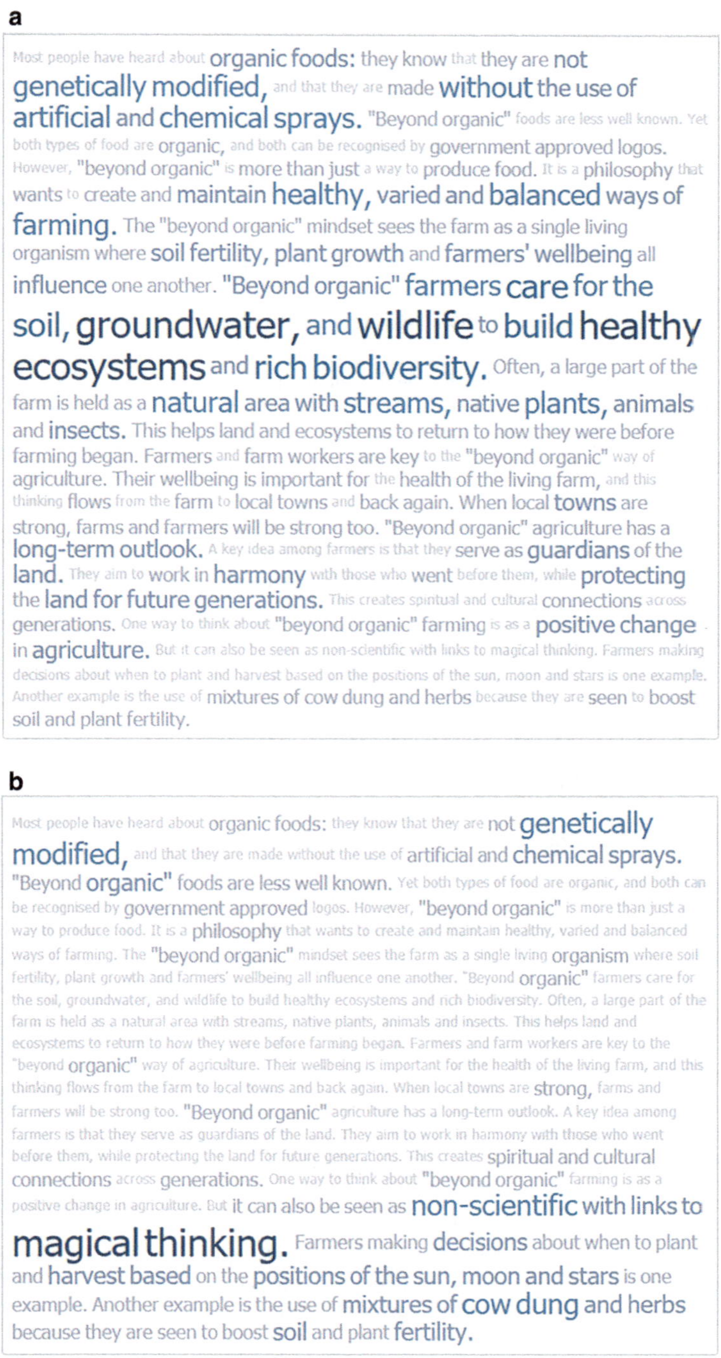

Fig. 7.2 Word clouds from text highlighting data obtained from Australian consumers ($n = 311$): (**a**) "Like" responses; (**b**) "Dislike" responses. Varying colour and font size indicate frequency of highlighting with larger and darker blue font for more frequently highlighted words, and smaller and lighter blue font for less frequently highlighted words

Table 7.1 Average sentiment scores (−100 to 100) for each of the 20 sentences in the text highlighting task for the overall sample (n = 311), segments with positive (n = 208) and negative (n = 39) sentiment towards biodynamic agriculture

Sentences	Overall	Positive	Negative
S1. Most people have heard about organic foods: They know that they are not genetically modified, and that they are made without the use of artificial and chemical sprays.	18.3	29.8	0.0
S2. "Beyond organic" foods are less well known.	3.2	12.0	−30.8
S3. Yet both types of food are organic, and both can be recognised by government approved logos.	10.6	17.8	−12.8
S4. However, "beyond organic" is more than just a way to produce food.	11.9	19.2	−23.1
S5. It is a philosophy that wants to create and maintain healthy, varied and balanced ways of farming.	22.8	36.5	−12.8
S6. The "beyond organic" mindset sees the farm as a single living organism where soil fertility, plant growth and farmers' wellbeing all influence one another.	19.6	33.2	−23.1
S7. "Beyond organic" farmers care for the soil, groundwater, and wildlife to build healthy ecosystems and rich biodiversity.	38.9	53.8	−2.6
S8. Often, a large part of the farm is held as a natural area with streams, native plants, animals and insects.	22.8	33.7	7.7
S9. This helps land and ecosystems to return to how they were before farming began.	13.5	20.7	−5.1
S10. Farmers and farm workers are key to the "beyond organic" way of agriculture.	12.5	19.7	−17.9
S11. Their wellbeing is important for the health of the living farm, and this thinking flows from the farm to local towns and back again.	16.1	22.6	−2.6
S12. When local towns are strong, farms and farmers will be strong too.	13.2	19.2	0.0
S13. "Beyond organic" agriculture has a long-term outlook.	12.9	24.0	−25.6
S14. A key idea among farmers is that they serve as guardians of the land.	14.5	23.6	−2.6
S15. They aim to work in harmony with those who went before them, while protecting the land for future generations.	21.5	33.2	−7.7
S16. This creates spiritual and cultural connections across generations.	3.2	10.1	−20.5
S17. One way to think about "beyond organic" farming is as a positive change in agriculture.	12.2	24.5	−30.8
S18. But it can also be seen as non-scientific with links to magical thinking.	−13.2	−13.0	−17.9
S19. Farmers making decisions about when to plant and harvest based on the positions of the sun, moon and stars is one example.	−2.3	2.9	−30.8
S20. Another example is the use of mixtures of cow dung and herbs because they are seen to boost soil and plant fertility.	0.0	7.2	−25.6

Several methodological studies have also been conducted, but the method is still novel, and the recommendations put forward regarding implementation are based on the best of our knowledge at the time of writing. We hope others will find the approach useful in attitude research and contribute to its further methodological characterization. Ample scope exists for further development (including text development and data analysis), validation and application.

References

Ares, G. (2018). Methodological issues in cross-cultural sensory and consumer research. *Food Quality and Preference, 64*, 253–263.

Ares, G., Ha, B., & Jaeger, S. R. (2021). Consumer attitudes to vertical farming (indoor plant factory with artificial lighting) in China, Singapore, UK, and USA: A multi-method study. *Food Research International, 150*(Part B), 110811. https://doi.org/10.1016/j.foodres.2021.110811

Ares, G., & Jaeger, S. R. (2022). Text highlighting for attitude measurement in cross-cultural consumer research: A methodological study. *Journal of Sensory Studies, 37*(2), e12728. https://doi.org/10.1111/joss.12728

Ares, G., Ryan, G. S., & Jaeger, S. R. (2023). Text highlighting combined with open-ended questions: A methodological extension. *Journal of Sensory Studies, 38*(3), e12816.

Beacham, A. M., Vickers, L. H., & Monaghan, J. M. (2019). Vertical farming: A summary of approaches to growing skywards. *The Journal of Horticultural Science and Biotechnology, 94*(3), 277–283.

Coleman, M., & Liau, T. L. (1975). A computer readability formula designed for machine scoring. *Journal of Applied Psychology, 60*(2), 283–284.

Cox, D. N., & Evans, G. (2008). Construction and validation of a psychometric scale to measure consumers' fears of novel food technologies: The food technology neophobia scale. *Food Quality and Preference, 19*(8), 704–710.

Crossley, S. A., Kyle, K., & McNamara, D. S. (2017). Sentiment analysis and social cognition engine (SEANCE): An automatic tool for sentiment, social cognition, and social-order analysis. *Behavior Research Methods, 49*(3), 803–821.

Delpero, W. T., O'Neill, H., Casson, E., & Hovis, J. (2005). Aviation-relevent epidemiology of color vision deficiency. *Aviation, Space, and Environmental Medicine, 76*(2), 127–133.

Dunlosky, J., Rawson, K. A., Marsh, E. J., Nathan, M. J., & Willingham, D. T. (2013). Improving students' learning with effective learning techniques: Promising directions from cognitive and educational psychology. *Psychological Science in the Public Interest, 14*(1), 4–58.

Ericsson, K. A., & Simon, H. A. (1980). Verbal reports as data. *Psychological Review, 87*(3), 215–251.

Farr, J. N., Jenkins, J. J., & Paterson, D. G. (1951). Simplification of Flesch reading ease formula. *Journal of Applied Psychology, 35*(5), 333–337.

Fazio, R. H. (2007). Attitudes as object–evaluation associations of varying strength. *Social Cognition, 25*(5), 603–637.

Galesic, M., & Bosnjak, M. (2009). Effects of questionnaire length on participation and indicators of response quality in a web survey. *Public Opinion Quarterly, 73*(2), 349–360.

Ganassali, S. (2008). The influence of the design of web survey questionnaires on the quality of responses. *Survey Research Methods, 2*(1), 21–32.

Gobble, M. M., & Headrick, D. (2019). Converging innovations nourish growth of indoor farming. *Research-Technology Management, 62*(2), 7–8.

Gunning, R. (1952). *The technique of clear writing*. McGraw-Hill.

Harkness, J. A., Villar, A., & Edwards, B. (2010). Translation, adaptation, and design. In J. A. Harkness, M. Braun, B. Edwards, T. P. Johnson, L. Lyberg, P. P. Mohler, B.-E. Pennell, & T. W. Smith (Eds.), *Survey methods in multinational, multiregional, and multicultural contexts* (pp. 115–140). Wiley.

Jaeger, S. R. (2024). Vertical farming (plant factory with artificial lighting) and its produce: Consumer insights. *Current Opinion in Food Science, 10*, 101145.

Jaeger, S. R., & Ares, G. (2022). Measuring consumer attitudes using text highlighting: Methodological considerations. *Food Quality and Preference, 96*, 104422.

Jaeger, S. R., & Cardello, A. V. (2022). Factors affecting data quality of online questionnaires: Issues and metrics for sensory and consumer research. *Food Quality and Preference, 102*, 104676. https://doi.org/10.1016/j.foodqual.2022.104676

Jaeger, S. R., Chheang, S. L., & Ares, G. (2022a). Text highlighting as a new way of measuring consumers' attitudes: A case study on vertical farming. *Food Quality and Preference, 95*, 104356.

Jaeger, S. R., Chheang, S. L., & Ares, G. (2022b). Text highlighting for consumer insights: Influence of text length and difficulty. *Food Quality and Preference, 97*, 104492. https://doi.org/10.1016/j.foodqual.2021.104492

Jaeger, S. R., Chheang, S. L., & Ares, G. (2023a). Beyond plant-based alternatives to milk and meat: Product and individual variables influence purchase intention for plant-based yoghurt and eggs. *Food Quality and Preference, 112*, 105019.

Jaeger, S. R., Chheang, S. L., & Ares, G. (2023b). Text highlighting: Three methodological studies to inform guidelines for implementation. *Food Quality and Preference, 109*, 104904.

Jaeger, S. R., Chheang, S. L., & Ares, G. (2023c). Using text highlighting in product research: Case study with kiwifruit in Singapore and Malaysia. *Food Quality and Preference, 104*, 104741.

Jaeger, S. R., Harker, F. R., & Ares, G. (2023). Consumer insights about sustainable and 'beyond organic'agriculture: A study of biodynamics in the United Kingdom, Australia, Singapore, and Germany. *Journal of Cleaner Production, 401*, 136744.

Jaeger, S. R., Rasmussen, M. A., & Cardello, A. V. (2024). Factors affecting data quality of online questionnaires with consumers in sensory and consumer research: Metrics from the literature and empirical insights. *Science Talks, 9*, 100307.

Johnson, T., Kulesa, P., Cho, Y. I., & Shavitt, S. (2005). The relation between culture and response styles: Evidence from 19 countries. *Journal of Cross-Cultural Psychology, 36*(2), 264–277.

Kincaid, J. P., Fishburne, R. P., Jr., Rogers, R. L., & Chissom, B. S. (1975). Derivation of new readability formulas (automated readability index, fog count and flesch reading ease formula) for navy enlisted personnel. *Institute for Simulation and Training, 56*.

Klare, G. R. (1974). Assessing readability. *Reading Research Quarterly, 10*, 62–102.

Krippendorff, K. H. (2013). *Content analysis: An introduction to its methodology* (3rd ed.). Sage Publications.

Latini, N., Bråten, I., Anmarkrud, Ø., & Salmerón, L. (2019). Investigating effects of reading medium and reading purpose on behavioral engagement and textual integration in a multiple text context. *Contemporary Educational Psychology, 59*, 101797.

Leroy, C., Gerjets, P., Oestermeier, U., & Kammerer, Y. (2021). Simultaneous presentation of multiple documents and text-highlighting: Online integrative processes and offline integrated understanding. *Scientific Studies of Reading, 25*(2), 179–192.

Liu, M., & Wronski, L. (2018). Examining completion rates in web surveys via over 25,000 real-world surveys. *Social Science Computer Review, 36*(1), 116–124.

McGorry, S. Y. (2000). Measurement in a cross-cultural environment: Survey translation issues. *Qualitative Market Research: An International Journal, 3*(2), 74–81.

McLaughlin, G. H. (1969). SMOG grading: A new readability formula. *Journal of Reading, 12*(8), 639–646.

Meyners, M., Castura, J. C., & Carr, B. T. (2013). Existing and new approaches for the analysis of CATA data. *Food Quality and Preference, 30*(2), 309–319.

Orne, M. T. (2009). Demand characteristics and the concept of quasi-controls. In R. Rosenthal & R. L. Rosnow (Eds.), *Artifacts in behavioral research* (Vol. 110, pp. 110–137). Oxford University Press.

O'Sullivan, C. A., McIntyre, C. L., Dry, I. B., Hani, S. M., Hochman, Z., & Bonnett, G. D. (2020). Vertical farms bear fruit. *Nature Biotechnology, 38*(2), 160–162.

Pliner, P., & Hobden, K. (1992). Development of a scale to measure the trait of food neophobia in humans. *Appetite, 19*(2), 105–120.

Podsakoff, P. M., MacKenzie, S. B., Lee, J.-Y., & Podsakoff, N. P. (2003). Common method biases in behavioral research: A critical review of the literature and recommended remedies. *Journal of Applied Psychology, 88*(5), 879–903.

Roßmann, J., Gummer, T., & Silber, H. (2018). Mitigating satisficing in cognitively demanding grid questions: Evidence from two web-based experiments. *Journal of Survey Statistics and Methodology, 6*(3), 376–400.

Smith, E. A., & Senter, R. (1967). *Automated readability index* (Vol. 66, Issue 220). Aerospace Medical Research Laboratories, Aerospace Medical Division, Air ....

Smith, P. B., & Fischer, R. (2008). Acquiescence, extreme response bias and culture: A multilevel analysis. In F. J. R. van de Vijver, D. A. van Hemert, & Y. Poortinga (Eds.), *Individuals and cultures in multilevel analysis* (pp. 285–314). Erlbaum.

Strobelt, H., Oelke, D., Kwon, B. C., Schreck, T., & Pfister, H. (2015). Guidelines for effective usage of text highlighting techniques. *IEEE Transactions on Visualization and Computer Graphics, 22*(1), 489–498.

Willits, F. K., Theodori, G. L., & Luloff, A. (2016). Another look at Likert scales. *Journal of Rural Social Sciences, 31*(3), 126–139.

Chapter 8
Extracting Affective Meaning from Text Data Using Sentiment Analysis: Applications in Sensory and Consumer Science

Michel Visalli

1 Introduction

1.1 Limitations of Direct Assessments of Liking and Preferences

Sensory and consumer sciences play an important role in understanding how individuals perceive and evaluate products. At the heart of this field lie the measurements of consumer liking, preferences, and satisfaction, generally assessed through structured tools such as hedonic rating scales (Lim, 2011). These instruments have long provided valuable insights for product development, quality control, and marketing. However, despite their widespread use and practical value, they present several notable limitations, both conceptual and methodological, which have spurred interest in complementary, indirect approaches.

A fundamental limitation of direct assessments is that they require individuals to consciously quantify their hedonic response, often on a discrete numerical or categorical scale. Yet, human affective experience is rarely so easily reducible (Andersen et al., 2021). Feelings of liking or disliking are often nuanced, dynamic, and contextually shaped, involving dimensions that extend beyond a single evaluative judgment (Gutjar et al., 2015). When participants are asked to summarize their hedonic response in a single rating (e.g., "*How much did you like this product on a scale from 1 to 9?*"), they are compelled to condense a potentially multifaceted experience into a simplified format. This can lead to a loss of granularity, masking ambivalence, mixed reactions, or conditional statements (Norris et al., 2019).

M. Visalli (✉)
Centre for Taste and Feeding Behavior (CSGA), National Research Institute for Agriculture, Food and the Environment (INRAE), Dijon, France
e-mail: michel.visalli@inrae.fr

M. Visalli, M. V. Galmarini (eds.), *Collection and Analysis of Text Data in Sensory and Consumer Science*, https://doi.org/10.1007/978-3-032-04851-6_8

Moreover, the act of asking for a direct judgment can itself alter the way people reflect on and report their experience (Wilson & Schooler, 1991). Asking participants to evaluate liking may introduce cognitive biases or lead them to over-rationalize an inherently affective, sometimes pre-conscious response (Fernqvist & Ekelund, 2014). In some cases, participants may not have a stable or well-formed opinion at the time of assessment and are instead constructing a response on the spot (Betsch, 2011). This issue is compounded in laboratory contexts, where the artificiality of the setting and the awareness of being observed can influence the way individuals report their preferences (Boutrolle et al., 2007).

Social desirability bias also plays a significant role (Cerri et al., 2019). Especially in experimental or formal evaluation contexts, participants may feel pressure to conform to perceived expectations, whether those of the researcher, the brand, or their peer group (Robinson & Higgs, 2012). This can result in over-reporting of positive evaluations or under-reporting of negative reactions, thereby obscuring their true preferences and compromising the validity of the data.

Context effects further complicate direct assessments. Liking is not a fixed trait but a context-sensitive judgment. Responses can vary depending on the presence of others, recent experiences, and even the order in which products are presented (Cardello, 2017). For example, a moderately liked product may receive a higher rating if it follows a poorly liked one, or a lower rating if it is evaluated after a highly liked product: a contrast effect that is difficult to control fully in structured assessments (Lim, 2011).

Cultural and individual differences in scale use present challenges for comparability (Kaneko et al., 2021). Some individuals may be more prone to use the extremes of a scale, while others prefer central values. These differences do not necessarily reflect differences in liking but rather in response style. While statistical techniques can help mitigate these effects (Romano et al., 2008), they do not fully resolve the underlying issue: the abstraction and artificiality of asking people to assign numerical values to inherently subjective experiences.

Ultimately, liking alone is an insufficient predictor of food acceptance (Cardello et al., 2000; Dalenberg et al., 2014), and recent research has supported the call for "beyond liking" measures that enhance behavioural prediction beyond what is possible with hedonic responses alone (Giacalone et al., 2022).

1.2 Expressions of Liking and Affective Meaning in Natural Language and Text Data

Natural language is far more than just a collection of explicit opinions: it serves as a powerful medium for individuals to express their subjective experiences, feelings, and evaluations (Hallikainen et al., 2025; Seyeditabari et al., 2018). It functions as a complex system through which liking and affective meaning are conveyed both directly and indirectly, bridging the gap between subjective and objective measurements. In the context of liking and affective meaning, language offers a uniquely rich and nuanced channel that transcends the limitations of numerical ratings or categorical scales (Wen & Laporte, 2025). Unlike direct assessments, which require consumers to consciously quantify their preferences, natural language captures spontaneous, often multifaceted expressions of affect, revealing subtle layers of meaning that structured tools frequently overlook (Hamby et al., 2015).

People tend to describe their sensory and emotional experiences using a diverse array of linguistic resources (e.g., adjectives, metaphors, narratives, and qualifiers) that convey the subtleties of their subjective responses (Caballero, 2019). For example, a consumer reviewing a food product might say "*The flavor was surprisingly bold, with just the right balance of sweetness and spice*", or "*It reminded me of childhood summers, which made the experience feel comforting*." These statements not only indicate liking but also embed emotional associations, memories, and contextual details that enrich the interpretation of affective response (Li et al., 2023a, b).

Moreover, language captures the dynamic nature of liking. While direct measures tend to freeze affect at a single point in time, verbal expressions can reflect shifts in sentiment, ambivalence, or conditional preferences. Consumers may express mixed feelings ("*I loved the texture but found the aroma overpowering*") or conditional attitudes ("*It's great for summer but too light for winter*"). This depth of expression is critical for understanding the complexity of consumer responses, particularly in sensory science where hedonic evaluations are intertwined with multisensory integration and personal context.

The richness of natural language also opens the door to exploring affective meaning beyond simple positive or negative valence (Ullah et al., 2016). Linguistic expressions can reveal emotions such as surprise, nostalgia, excitement, or disappointment, which often underlie liking but are not captured by traditional hedonic scales. Psycholinguistic research highlights that emotional states are frequently encoded in subtle lexical choices, sentence structure, and even stylistic elements, all of which contribute to the overall affective tone of a message (Mairano et al., 2019).

1.3 Inferring Affective Meaning from Sentiment Analysis

Emerging from advancements in natural language processing (NLP), sentiment analysis (or opinion mining) encompasses a range of techniques for the automated detection and quantification of sentiments (defined as affective evaluations expressed toward an object, experience, or event, including emotions and affective states) within text content (Mohammad, 2016). When sentiments are directly expressed through clear emotional language in the text, the analysis is termed "explicit sentiment analysis"; if sentiments are inferred from context or subtle cues without overt sentiment words, it is referred to as "implicit sentiment analysis" (Balahur et al., 2012; Tubishat et al., 2018).

The theoretical foundation of sentiment analysis is not limited to computational techniques. It is deeply rooted in linguistic, cognitive, and psychological theories, which provide the essential theoretical frameworks for understanding how emotion and opinion are conveyed and interpreted in language (Du, 2022; Li et al., 2012; Taboada, 2016; Zhang, 2024). Together, these domains provide the conceptual basis for building systems that attempt to replicate or approximate the nuanced way in which humans understand sentiment in language.

From a linguistic perspective, sentiment analysis draws heavily on semantics and pragmatics. Semantics focuses on the meaning of words and phrases, which is crucial in determining whether a given text expresses a positive, negative, or neutral sentiment. For example, the word "wonderful" is generally associated with positive sentiment, while "terrible" is negative. However, the true meaning of an utterance often depends on context, which brings in pragmatics, the study of how language is used in context and how meaning is shaped by that context. Pragmatics is essential for interpreting sentiment in more complex linguistic phenomena such as sarcasm, irony, or understatement, where the literal meaning of words may be opposite to the intended sentiment.

Cognitive science also contributes to the understanding of sentiment in language. Cognitive theories emphasize that emotion and cognition are intertwined processes, and that the way we perceive, interpret, and react to linguistic input is shaped by both our mental states and our prior experiences. In sentiment analysis, this manifests in the need to model how humans associate emotional valence with linguistic patterns and how contextual cues, background knowledge, and inference play a role in interpreting sentiment. This needs a model of language understanding that goes beyond the surface level and includes knowledge of discourse, speaker intent, and situational context.

Sentiment analysis is also informed by psychological models of emotion. One foundational framework is Paul Ekman's theory of basic emotions (Ekman, 1992), which identifies a set of universal emotions such as happiness, anger, and fear, that are expressed across cultures and are often reflected in language.

1.4 Applications of Sentiment Analysis in Sensory and Consumer Research

When applied to sensory and consumer research, sentiment analysis allows researchers and practitioners to extract actionable insights from sensory experiences and broader consumer feedback related to food (see Molenaar et al., 2024; Tao et al., 2020; Yan & Li, 2022 for reviews). This approach offers a scalable and efficient way to understand how people feel about products and services, supporting a wide range of applications across product development, quality control, brand management, market segmentation, competitive analysis, and personalized consumer engagement.

Sentiment analysis can be effectively applied to experimental text data collected through open-ended prompts, providing an indirect way for participants to express their affective responses without necessarily being guided toward a specific hedonic judgment (e.g., "*Describe your experience of this product*"). This approach encourages more natural and unconstrained feedback, which helps reduce self-censorship and uncovers more authentic, detailed insights into the factors shaping consumer affective experiences, going beyond simple measures of liking or disliking. Furthermore, participants often feel freer to express hesitation or mild disapproval; for instance, a comment like "*It wasn't really for me, but others might enjoy it*" conveys a nuanced perspective that would be difficult to capture through a standard rating scale. Sentiment analysis can also be applied on transcriptions of focus group discussions or interviews.

The rise of digital platforms has significantly broadened the range of text data sources available for sentiment analysis within the food domain. Consumer reviews on online marketplaces like *Amazon*, *Yelp*, and specialized food apps provide large volumes of user-generated content where customers openly share their opinions and experiences. Forums and community platforms such as *Reddit* offer rich, unsolicited conversations covering food products, recipes, and dietary trends. Additionally, food blogs and influencer content on social media platforms like *Instagram*, *TikTok*, and *X* (Twitter) contribute qualitative insights into consumer attitudes, lifestyle associations, and emerging preferences. Each data source possesses unique characteristics and challenges; for example, social media content tends to be brief, informal, and highly contextualized, while survey responses are generally more focused and descriptive. Consequently, effective sentiment analysis approaches must be tailored to these variations to accurately extract meaningful insights.

This chapter is organized to guide the reader through both theoretical foundations and practical applications of sentiment analysis in the context of sensory and consumer research. The core of the chapter is divided into two main methodological sections. Section 2 focuses on lexicon-based approaches, covering generic and domain-specific sentiment lexicons and how they are used to compute sentiment scores. Section 3 delves into data-driven approaches, including classical machine learning, unsupervised and distantly supervised methods, as well as recent advances in deep learning and large language models. Hybrid approaches that combine rule-based and data-driven methods are also presented. In Sect. 4, the main methods are summarized, with a comparison of their respective advantages and limitations, and practical considerations specific to sensory and consumer research are discussed. The chapter concludes in Sect. 5 by identifying key challenges and future perspectives, emphasizing the potential of sentiment analysis to enrich the understanding of consumer experiences through natural language.

2 Lexicon-Based Approaches to Sentiment Analysis

2.1 *Overview*

Lexicon-based sentiment analysis is among the earliest computational approaches developed to assess the affective content of text. It is grounded in the assumption that words inherently carry affective meaning, which can be systematically evaluated using sentiment lexicons, i.e., precompiled list of words, each annotated with an associated emotional value, or valence. In this context, valence refers to the emotional charge conveyed by a word or expression, indicating whether it expresses a positive, negative, or neutral tone. Words with positive valence communicate pleasant or desirable emotions such as satisfaction, joy, or approval, while negative valence reflects unpleasant or aversive states like frustration, disappointment, or disgust. Neutral valence refers to emotionally neutral or contextually ambiguous terms.

Some sentiment lexicons go beyond simple polarity distinctions by incorporating intensity scores, allowing the assessment not only of whether a sentiment is positive or negative but also of the degree to which it is expressed. In such lexicons, valence is quantified on a numerical scale. A word like "wonderful" might receive a high positive score, whereas "terrible" could be strongly negative. These individual word scores can be aggregated at the level of a single observation (e.g., a consumer's free comment about a product), a paragraph, or an entire text, to estimate its overall affective tone.

Additionally, certain lexicons classify words according to specific emotional categories rather than just valence. This allows for a more nuanced exploration of the emotional content embedded in language, extending sentiment analysis beyond general positivity or negativity toward the mapping of more discrete affective states.

2.2 *Generic Lexicons*

Generic sentiment lexicons are foundational tools in lexicon-based sentiment analysis. They consist of curated lists of words annotated with affective properties, typically valence (positive, negative, or neutral) and, in some cases, emotion categories or intensity scores. These lexicons are designed for general-purpose use across various domains, and while they may not account for context-specific meanings, they offer broad coverage and ease of implementation. Several well-established generic lexicons are commonly used in sentiment analysis.

Bing Liu's lexicon (Ding et al., 2008) is a sentiment resource that categorizes approximately 6800 English words into two classes: positive and negative. This simplicity makes it particularly effective for tasks focused on detecting general sentiment trends, such as determining whether a text is overall positive or negative. Its broad lexical coverage and ease of implementation have contributed to its popularity in both academic research and applied sentiment analysis, especially in domains where fine-grained emotional distinctions are less critical than overall polarity. For example, Bhagat et al. (2023) used this lexicon to extract grocery-specific sentiment terms.

SentiWordNet (Baccianella et al., 2010) is an extension of the *WordNet* (Miller et al., 1990) lexical database, in which each synset (a set of synonym English words representing a concept) is assigned three scores: positivity, negativity, and objectivity, each ranging from 0 to 1 and summing to 1. For example, the word "delightful" may be part of a synset with a high positive score and low negative and objective scores. A single word can correspond to one or more synsets in *WordNet*, as words often carry multiple meanings depending on their usage in context. Each of these synsets may be associated with distinct sentiment profiles in *SentiWordNet*, reflecting different affective interpretations. Ideally, the process of word-sense disambiguation would be applied to determine the most contextually appropriate synset for each word, thereby ensuring greater accuracy in sentiment attribution. However, many practical applications simplify this step by either averaging the sentiment scores across all possible synsets associated with a word or selecting the most common sense as a heuristic. For example, Asani et al. (2020) extracted consumers' food preferences from the TripAdvisor website and computed sentiment scores using *SentiWordNet*, relying on these simplified methods to assign affective values to words without full disambiguation.

AFINN (Nielsen, 2011) is a manually curated lexicon of more than 3300 English words, each assigned a valence score ranging from −5 to +5, with negative values indicating strong negativity and positive values indicating strong positivity. *AFINN* is especially well-suited for short and informal texts, such as social media posts and consumer reviews, due to its emphasis on colloquial and evaluative language. For example, Mostafa (2020) applied *AFINN* to analyse global sentiment toward halal food on Twitter,

The *NRC emotion and sentiment lexicon* (Mohammad & Turney, 2013) includes over 14,000 English words associated with eight basic emotions (anger, fear, anticipation, trust, surprise, sadness, joy, and disgust) and two sentiments (positive or negative). Unlike valence-only lexicons, *NRC* allows for a multidimensional analysis of emotional content, making it especially valuable when the goal is to uncover not just the general tone of a text, but the specific emotions it conveys. For example, Verza et al. (2023) used the *NRC* emotion lexicon to examine consumer sentiments and emotional responses in new seafood product concept development based on analysis of text data collected in focus groups. Singh and Glińska-Neweś (2022) applied the *NRC* lexicon to explore public attitudes toward organic foods by analysing Twitter data.

Linguistic Inquiry and Word Count (LIWC, Pennebaker et al., 1999) is a psychologically grounded text analysis tool that categorizes words into a wide range of psychologically meaningful dimensions, including emotional expression, cognitive processes, social relationships, and behavioural tendencies. While it is not exclusively designed for sentiment analysis, *LIWC* includes dedicated categories for positive and negative emotion words, which are commonly used in affective research. Its structured and theory-driven approach allows researchers to systematically quantify linguistic markers of psychological states, making it particularly valuable in fields such as psychology, health communication, and behavioural science. For example, Tian et al. (2021) used *LIWC* to analyse consumer sentiments expressed in Yelp restaurant online reviews.

The *Syuzhet* lexicon (Jockers, 2020) was developed with a focus on capturing sentiment progression within literary texts. Each word in the lexicon is assigned a numeric sentiment score, similar to the approach used in *AFINN*, allowing for the quantification of emotional dynamics over time. For example, Chen et al. (2021) applied the *Syuzhet* lexicon to identify basic emotions associated with meat alternatives by analysing textual data from scientific articles. Similarly, Pindado and Barrena (2021) used it to examine consumer sentiment of emerging food trends using data collected from Twitter.

Valence Aware Dictionary and Sentiment Reasoner (VADER, Hutto & Gilbert, 2014) is a lexicon and rule-based sentiment analysis tool that includes over 7500 words annotated with valence scores. It was specifically designed to handle the informal, brief, and often nuanced language found in social media texts, such as tweets and online comments. Despite this focus, *VADER* has proven effective across a wide range of consumer-generated content. For example, Kvasnička et al. (2024) employed *VADER* to analyse Twitter data, successfully gauging prevailing attitudes and opinions expressed in tweets related to healthy food.

SentiStrength (Thelwall et al., 2012) is another sentiment analysis tool specifically designed for short, informal text such as social media posts, forums, and online comments. It excels at rapidly assigning both positive and negative sentiment strength scores to each input on two separate scales ranging from 1 (neutral) to 5 (extremely positive/negative). It uses a lexicon-based approach enriched with heuristics to handle informal language features such as emoticons, negations, spelling variants, booster words, and slang patterns commonly found online. For example, Zhou (2024) used *SentiStrength* to perform sentiment analysis on restaurant reviews to assess how sensory synergy or conflict influences ratings.

2.3 Domain-Specific Lexicons

While generic lexicons provide a rapid and interpretable entry point for sentiment analysis, their effectiveness often depends on the domain of application. In the context of sensory and consumer science, these lexicons may fall short in capturing the specific vocabulary used by consumers to describe product experiences. Attributes commonly used in food evaluations may be absent from general-purpose lexicons or misclassified due to context-insensitive sentiment labelling. This is particularly problematic where the affective meaning of a word can vary significantly depending on the product category and individual preferences. For instance, the term "bitter" might carry a negative connotation when describing certain vegetables, suggesting an undesirable flavour. However, the same term may be perceived as neutral or even positively by consumers when used in reference to products like dark chocolate or coffee, where bitterness is an expected and sometimes valued sensory attribute.

While numerous resources exist for measuring emotion through questionnaires (Panagiotou & Gkatzionis, 2022), there is a notable scarcity of studies focused on developing lexicons specifically tailored for sentiment analysis in the context of sensory and consumer science.

Luc et al. (2020) used the *sentimentr* package (Rinker, 2015) to analyse experimental data using a custom-built dictionary tailored to the sensory descriptors relevant to perfumes and car seat leatherettes. By adapting the lexicon to include domain-specific vocabulary, they were able to more accurately capture the affective content expressed by participants in free comment responses. This approach illustrates how customizing sentiment resources to reflect the terminology and evaluative nuances of a particular product category can enhance the sensitivity and interpretability of sentiment analysis in sensory research.

Visalli et al. (2023) developed the concept of "hedonic valence". To create their domain-specific lexicon, they first selected a representative list of attributes frequently used by consumers to describe red wine taste in their verbatim comments. Then, a panel of consumers rated their agreement with statements such as "*I like red wines that are [attribute]*" using a Likert scale ranging from "strongly disagree" (coded −2) to "strongly agree" (coded +2). The hedonic valence of each attribute was calculated as the average score across the panel, providing a direct measure of consumer affective response to these attributes. This scoring system was then reused as a specialized sentiment lexicon tailored specifically for analysing consumer language related to red wines. This method, focusing on generic experiences related to specific attributes, contrasts with that Liapakis et al. (2020), who inferred sentiment scores for food product attributes directly from consumer ratings of "food quality", based on evaluations of specific products described in online reviews. Although their computational approaches differ, the conceptual goal aligns with the work of Mahieu et al. (2022), which focuses on identifying drivers of liking by linking product attributes to consumer preferences. Both methods may enable sentiment model calibration grounded in explicit consumer preferences derived from real tastings. However, because these lexicons are derived from data-driven approaches (see Sect. 3) and based on specific datasets, their generalizability across different consumer groups or product categories may be limited.

The reader can refer to articles not related to sensory and consumer science for alternative methods of construction of domain specific sentiment lexicons (Al-qerem et al., 2024; Hamilton et al., 2016; Labille et al., 2017; Mohamad Beigi & Moattar, 2021; Mohammed & Prasad, 2023).

2.4 Computation of Sentiment Scores

After selecting an appropriate sentiment lexicon, the process of computing sentiment scores involves several key steps. Initially, the text is pre-processed by cleaning, tokenizing into words or phrases, and often converting to lowercase. Depending on the research goals, punctuation, stop words, or domain-specific elements such as brand names or measurement units may be removed or retained. Next, each token is matched against entries in the chosen lexicon; when a match is found, the associated sentiment value or category is retrieved.

The sentiment values for all matched words are then aggregated to produce an overall score or profile at the desired level of analysis: individual comment, aggregated product feedback, or across conditions in an experiment. Common methods for aggregating sentiment scores vary in complexity and the kind of information they capture (see Jamadi Khiabani et al., 2020, for details).

The simplest approach involves summing or averaging the sentiment values of all words identified in the text. This method treats each sentiment-bearing word equally, providing a straightforward overall score that reflects the general emotional tone of the text. However, it assumes that all words contribute equally to the sentiment, which may not always hold true in practice.

To address this, weighted scoring approaches assign different weights to words based on factors such as term frequency, giving more influence to more frequently mentioned terms. Other weighting schemes consider word position, under the assumption that sentiment expressed earlier or later in the text may carry different importance, or syntactic roles, where words with certain grammatical functions (e.g., adjectives or adverbs) may be given greater emphasis due to their stronger affective content. Visalli et al. (2023) first grouped sensory attributes based on their semantic similarity. They then calculated the average sentiment score within each cluster, followed by averaging across all clusters. This approach ensures that semantically related attributes are given equal weight in the overall sentiment analysis.

Another commonly used technique is polarity counting, which involves tallying the number of positive and negative words separately and then calculating a ratio or difference between these counts. This method offers a simple yet effective way to gauge the balance of positive versus negative sentiment in the text, providing insights into overall polarity without relying on numeric sentiment scores.

To improve accuracy, many implementations incorporate linguistic adjustment rules. These may include negation detection, where expressions like "not good" reverse the polarity of the sentiment, or intensifiers such as "very" or "extremely", which amplify the sentiment strength of adjacent words. These refinements enhance the sensitivity and realism of lexicon-based sentiment models. Each of these aggregation strategies has strengths and limitations, and the choice among them often depends on the specific research context, the nature of the text, and the desired granularity of sentiment analysis. For an in-depth review of sentiment computation methods using R, see Naldi (2019).

3 Data-Driven Approaches to Sentiment Analysis

3.1 Overview

While lexicon-based approaches to sentiment analysis rely on predefined sets of words annotated with sentiment values, data-driven approaches represent a fundamental conceptual shift: they learn sentiment patterns directly from data rather than relying on manually crafted linguistic rules. This transition from rule-based to model-based techniques has been pivotal in advancing the field of sentiment analysis, particularly in handling complex, nuanced, or domain-specific language.

Indeed, lexicon-based methods are inherently static: they operate by matching words in a text to entries in a sentiment dictionary. Although this can be effective in controlled or well-understood domains, it assumes that sentiment is a fixed property of individual words. In practice, especially in sensory and consumer science where words and descriptors can have varying sentiment depending on product context, this assumption often breaks down.

In contrast, data-driven approaches are dynamic and adaptive. They treat sentiment as an emergent property of the entire text and learn to infer it from labelled examples. These methods span a spectrum of complexity. Classical machine learning uses handcrafted features extracted from text to train statistical models. Deep learning models learn both the features and the decision rules directly from raw text data, capturing subtle semantic and syntactic patterns. Large language models are pretrained on vast corpora and can be fine-tuned or prompted to perform sentiment analysis with minimal additional data. Hybrid approaches combine elements of both lexicon-based and learned models to improve performance, interpretability, or domain adaptability. This shift allows for more context-sensitive, scalable, and adaptable sentiment analysis, which may be crucial in sensory and consumer science, where affective meaning is often shaped by cultural expectations, product categories, and individual experiences.

The following sections explore the different types of data-driven approaches in detail, evaluating their suitability for various types of consumer feedback and sensory data.

3.2 Machine Learning-Based Approaches

3.2.1 Supervised Approaches

In machine learning sentiment analysis, the problem is generally framed as a supervised classification task. Given a labelled dataset of consumer comments or reviews annotated with sentiment categories (e.g., positive, negative, neutral), numeric scores (e.g., from −5 to +5), or emotion labels (e.g., joy, anger, etc.), an algorithm learns a function that can map new, unseen text to the appropriate sentiment category. This process typically involves three stages: data preprocessing and feature extraction, model training, evaluation and validation.

Data Preprocessing and Feature Extraction

Text data is cleaned and normalized (e.g., lowercasing, tokenization, removal of stopwords or special characters), then converted into numerical representations. Common approaches include Bag-of-Words, TF-IDF (Term Frequency–Inverse Document Frequency), or N-grams (see Chapter 5 for details).

Model Training

A classifier is trained using labelled data to distinguish sentiment patterns. Several well-established machine learning algorithms are used in sentiment analysis, each having unique characteristics and trade-offs (see Ahmad et al., 2017, for a review). Below, some of the most widely used algorithms are introduced.

Naïve Bayes classifiers (Surya & Subbulakshmi, 2019) are based on Bayes' theorem and assume independence between features, a simplification that rarely holds true in natural language. However, despite this assumption, Naïve Bayes remains effective for sentiment analysis, especially when working with large-scale or noisy datasets. Some researchers used *SnowNLP*, a model grounded in a Bayesian algorithm, to compute sentiment scores from product reviews. Xu et al. (2022) focused on comparing green-labelled versus conventional rice, while Zeng et al. (2024) analysed representations of vegetarianism in China.

Support Vector Machines (SVMs, Zainuddin & Selamat, 2014) are robust and highly effective for high-dimensional data, such as text represented via TF-IDF or word embeddings. SVMs work by identifying the optimal hyperplane that maximizes the margin between sentiment classes. This makes them especially suitable for binary classification tasks (e.g., positive vs. negative sentiment), although they can be extended to multiclass settings. SVMs are generally resistant to overfitting and perform well even with relatively small amounts of training data, making them a solid choice for many sentiment applications in consumer research. Gunawan et al., 2022 classified restaurant customers' satisfaction using both SVM and Naïve Bayes algorithms, with both methods achieving comparable accuracy.

Logistic regression (Aliman et al., 2022) is another widely used classifier in sentiment analysis due to its simplicity, interpretability, and solid performance. It estimates the probability that a given input belongs to a particular sentiment class, which allows for flexible use in both binary and multiclass classification tasks. Its probabilistic output is also useful when confidence in predictions is important. Amin et al. (2024) used, among other methods, logistic regression to predict if food were perceived good or poor quality based on consumer reviews. Visalli et al. (2020) performed sentiment analysis on free comment product sensory descriptions through the *Microsoft Azure* machine learning platform based on logistic regression and SVM (Harfoushi et al., 2018).

Decision trees build a hierarchical structure of rules based on feature splits. Although they can be prone to overfitting, especially with sparse text data, they offer clear interpretability and insight into which features (words or phrases) are most influential in determining sentiment. When combined in ensemble methods such as

Random Forests (Bahrawi, 2019) or Gradient Boosting Machines (Alzamzami et al., 2020), decision tree-based models gain much greater predictive power and robustness. These ensemble techniques aggregate the outputs of multiple trees to produce more accurate and stable sentiment predictions. Ensemble models are particularly advantageous when integrating textual features with structured data, such as demographic information, or product categories, making them especially relevant for applied consumer and sensory science contexts. For example, Luc et al. (2022) used both Random Forests and SVMs to classify Free Just-About-Right comments into one of the three categories "too much", "not enough" and "just about right".

Evaluation and Validation

Models are typically assessed using a range of performance metrics, including accuracy, precision, recall, and the F1-score (Dalianis, 2018). Accuracy measures the overall proportion of correctly classified instances, providing a general sense of model performance. Precision evaluates the proportion of true positive predictions among all positive predictions, reflecting the model's ability to avoid false positives. Recall (or sensitivity) measures the proportion of true positives identified out of all actual positive cases, indicating the model's capacity to detect relevant instances. The F1-score combines precision and recall into a single harmonic mean, offering a balanced metric especially useful when dealing with imbalanced datasets. To obtain robust and unbiased estimates of model performance, these metrics are often computed through cross-validation, where the dataset is repeatedly split into training and testing subsets. This process helps detect overfitting, ensures that the model performs well on unseen data, and guides hyperparameter tuning for optimal results.

3.2.2 Distant Supervised Approaches

Distant supervision bridges the gap between fully supervised and unsupervised sentiment analysis by using "weakly labelled" data as a substitute for manually annotated datasets. The central idea is to automatically assign sentiment labels to text using external heuristics, metadata, or proxy signals, such as emoticons, star ratings, or hashtags, which serve as "distant" labels for training models. W. Li et al. (2023a, b) applied this method by leveraging star ratings as weak sentiment labels to guide the learning process.

3.2.3 Unsupervised Approaches

While most machine learning techniques for sentiment analysis rely on supervised learning requiring annotated datasets with sentiment labels, unsupervised learning offers an alternative when such labels are unavailable. Unsupervised approaches attempt to identify sentiment structure or polarity in text without explicit supervision, often using clustering, topic modelling, or representation learning.

Clustering-Based Sentiment Discovery
Methods like k-means, hierarchical clustering, or DBSCAN group texts into clusters based on feature similarity. If clusters show distinct sentiment tendencies (e.g., one with mostly positive words, another with negative phrases), these can be interpreted as sentiment classes. Some of these approaches were used by Gupta & Rattan (2024a, b) to perform aspect-based sentiment analysis of consumer reviews. This is a more fine-grained approach to sentiment analysis that focuses not just on the overall sentiment of a text, but on the sentiment expressed towards specific aspects or features of a product, service, or topic.

Topic Modelling with Sentiment Orientation
Algorithms like Latent Dirichlet Allocation (LDA) can discover latent themes in text. When combined with sentiment lexicons (e.g., identifying topics that co-occur with positive or negative words), topics can be labelled with inferred sentiment orientations. This approach is often termed sentiment-aware topic modelling. For example, S. Zhang et al. (2022) used LDA to extract five topics and aggregate consumer sentiment per topic from a Yelp restaurant reviews dataset.

Word Embeddings, Vector Polarity and Sentiment Lexicon Induction
Using pretrained embeddings (see Sect. 3.3.a), one can compute vector distances between target words and known sentiment anchors (e.g., "pleasant", "disgusting") to infer sentiment orientation of terms or comments not present in sentiment lexicons (Alshaabi et al., 2022).

3.3 *Deep Learning and Large Language Models-Based Approaches*

In recent years, deep learning has significantly advanced the field of sentiment analysis, enabling more sophisticated modelling of linguistic patterns and emotional nuance. Unlike classical machine learning approaches which rely on manually crafted features, deep learning models can automatically learn complex representations of text from raw input. Deep learning models are especially effective at capturing long-range dependencies, negation, sarcasm, and nuanced expressions of preference, challenges that simpler models typically struggle with. Deep learning approaches to sentiment analysis typically involve four stages: text representation, model architecture, fine tuning or transfer learning, prediction and interpretation.

3.3.1 Text Representation

As mentioned earlier, in order to enable machines to process and learn from text, words must first be transformed into numerical representations. Traditional approaches use word embeddings such as Word2Vec or GloVe (Neelima & Mehrotra, 2023). These embeddings map words to fixed-length dense vectors in such a way that semantically similar words lie close together in the vector space. For example, "sweet" and "sugary" may have similar vector representations if they frequently occur in similar contexts. These embeddings are static, meaning that each word has a single representation regardless of how it is used in different sentences.

More advanced approaches use contextual embeddings, generated by models such as *BERT* or *RoBERTa* (Liu et al., 2020). These embeddings dynamically adjust the vector representation of a word depending on its surrounding context. For instance, the word "light" in "light meal" versus "light bulb" would be encoded differently. This context-sensitivity allows for a more accurate and nuanced understanding of meaning, which is especially beneficial in sentiment analysis where the same word can convey different emotions depending on usage.

3.3.2 Model Architecture

In deep learning–based sentiment analysis, neural networks form the backbone of model architecture. Different architectures are selected depending on the nature of the text and the analytical goals.

Recurrent Neural Networks (RNNs) and their more advanced variant, Long Short-Term Memory networks (LSTMs), are commonly used to process text as sequences. These models are well-suited for short- to medium-length documents because they capture the temporal dependencies between words, which is important for understanding context and flow in sentiment-bearing phrases. Convolutional Neural Networks (CNNs), originally developed for image processing, have also proven effective in sentiment analysis by identifying local patterns or n-grams in the text. These models act like filters that can detect key phrases or combinations of words indicative of specific sentiments, making them efficient for extracting position-invariant features from text. For example, Iqbal et al. (2022) implemented RNN and LSTM models for sentiment classification and analysis of consumer feedback gathered from various social media platforms such as Amazon, Facebook, or Twitter.

More recently, transformer-based models have become the state-of-the-art approach. Models such as *BERT* (Bidirectional Encoder Representations from Transformers), *GPT* (Generative Pre-trained Transformer), and *RoBERTa* use self-attention mechanisms to weigh the importance of each word in a sentence relative to the others. These models are pretrained on vast amounts of text, and their ability to understand nuanced context, syntax, and semantics has significantly advanced the accuracy and reliability of sentiment analysis, especially in complex or ambiguous cases. For example, Barbosa and Gomes (2025) used *BERT* to perform sentiment analysis of comments related to food waste on Twitter. Hanić et al. (2024) shown that *BERT* outperformed SVM and random forest for sentiment analysis related to vegan and vegetarian restaurant experiences.

More recent large language models (LLMs), such as *GPT*, are even more powerful, capable of performing sentiment classification with minimal labelled data using prompting, few-shot learning, or zero-shot learning. For example, Torrico (2025) used *ChatGPT* to conduct sentiment analysis by first prompting the chatbot, acting as an experienced taster, to provide detailed descriptions of the sensory characteristics of chocolate brownies with various formulations.

3.3.3 Fine-Tuning or Transfer Learning

Modern sentiment analysis models often leverage transfer learning, where a model is pretrained on large, general-purpose text corpora and then fine-tuned on domain-specific data to enhance performance in specialized applications. Pretrained models like *BERT, RoBERTa*, or *GPT* already capture a broad understanding of language structure and semantics. Fine-tuning adapts these models to the unique vocabulary, tone, and sentiment expressions of a particular domain, often requiring only a small amount of labelled data to achieve high accuracy. In addition to fine-tuning, Retrieval-Augmented Generation (RAG) represents a more recent advancement that combines generative language models with external knowledge retrieval. RAG models consist of two components: a retriever that searches a database or knowledge source (e.g., previous consumer reviews, sensory lexicons, product documentation) for relevant information, and a generator that integrates the retrieved information to generate responses or predictions. Together, fine-tuning and RAG offer powerful tools for adapting general models to domain-specific sentiment analysis tasks, capturing both the depth of language and the specificity of consumer experiences.

3.3.4 Prediction and Interpretation

Once trained, sentiment analysis models output predictions that quantify or categorize the emotional tone of a given text, in a manner similar to other supervised learning approaches. Advanced models, particularly those using attention, can also offer interpretability features, highlighting which words or phrases contributed most to the final sentiment classification.

Prompt-based scoring is a technique that leverages the capabilities of LLMs by providing them with crafted prompts to elicit sentiment judgments directly. Rather than training a dedicated classifier, the model is queried with a natural language prompt that frames the sentiment assessment task. This allows the model to use its pre-trained linguistic and contextual knowledge to generate sentiment predictions. For example, given a consumer review such as "*The texture was smooth, but the flavour was bland and disappointing*", a prompt might be formulated as: "*How did the consumer feel about the product? Respond with positive, negative, or neutral*". *ChatGPT 3.5* might respond with: "*The consumer felt negative about the product. Although the texture was smooth (a positive comment), the flavour being described as bland and disappointing carries stronger negative sentiment, which likely dominates the overall feeling.*" Alternatively, prompts can ask the model to assign a score (e.g., "*Rate the overall sentiment of the review on a scale from -5 to +5*"). *ChatGPT 3.5* might answer: "*I would rate the overall sentiment of the review as -2. The positive aspect ('texture was smooth') slightly offsets the negative feelings about the flavour ('bland and disappointing'), but the overall impression leans toward a mild negative.*"

3.4 Hybrid Approaches

Hybrid approaches to sentiment analysis have gained traction as an effective means to leverage the complementary strengths of lexicon-based and data-driven methods. By combining these techniques, hybrid models aim to address the limitations of relying solely on either approach, achieving greater accuracy, robustness, and flexibility. Hybrid methods are valuable in situations where labelled data is scarce, domain-specific language is critical, interpretability of results is important for stakeholders, and the detection of subtle or ambiguous sentiment expressions requires sophisticated handling.

In practice, hybrid sentiment analysis integrates lexicon-based components for providing domain knowledge and transparent linguistic cues with machine learning or deep learning models that excel at learning complex patterns and contextual information from data. These components can be combined in a variety of ways to optimize performance. For example, lexicons may be used during preprocessing to guide feature selection or to weight words before inputting text into a machine learning or neural network model. Alternatively, ensemble methods might run lexicon-based and learned models in parallel, merging their predictions through majority voting or weighted averaging to enhance overall accuracy. Another common strategy is feature augmentation, where sentiment scores derived from lexicons are incorporated as additional features within supervised learning frameworks, enriching the data representation. Moreover, lexicon-derived rules can serve as guides during model training or as post-processing steps, helping to correct or refine predictions when the model encounters ambiguous or borderline cases. For example, Augustyniak et al. (2015) employed a hybrid approach by first using sentiment lexicons as initial classifiers, followed by a decision tree that acted as a fusion classifier, learning from the outputs generated by the lexicons. Gupta and Rattan (2024b) combined lexicon-based features with LSTM networks to enhance sentiment analysis in restaurant reviews.

4 Summary

4.1 Advantages and Disadvantages of the Different Approaches

Table 8.1 presents a comparison of the key characteristics of lexicon-based and data-driven approaches to sentiment analysis. Despite recent advances in natural language processing and the growing availability of powerful pretrained language models, there is still no definitive or "ultimate" model for sentiment analysis. Machine learning and deep learning models generally outperform lexicon-based methods when sufficient and relevant training data is available, especially after fine-tuning for specific domains. Hybrid approaches that combine lexicon and machine

Table 8.1 Comparison of lexicon-based and data-driven approaches to sentiment analysis

Aspect	Lexicon-based approaches	Data-driven approaches
Core principle	Uses predefined sentiment lexicons	Learns sentiment patterns from data
Sentiment calculation	Aggregates individual word scores	Derives sentiment from context and pattern
Interpretability	Highly interpretable (rule-based logic)	Often opaque ("black-box" models)
Context handling	Weak with context (e.g., sarcasm, negation)	Stronger contextual understanding (especially with deep learning)
Implementation complexity	Simple to implement; low resource needs	Requires significant resources and technical expertise
Speed & efficiency	Fast and lightweight	Slower (especially large models like LLMs)
Data requirements	No labelled data needed.	Labelled data essential (for supervised models).
Adaptability	Needs domain-specific lexicons for accuracy	Adaptable via transfer learning or fine-tuning
Language and domain coverage	Limited by lexicon coverage and language	Broader generalization with multilingual training and transferability
Accuracy	Lower in complex texts	Higher accuracy (supervised or LLMs), but risk of overfitting (supervised)
Bias & fairness	Less bias from data, but lexicons may be outdated or incomplete	Inherits biases from training data; fairness concerns in pretrained models

learning methods often yield improved accuracy by leveraging both interpretability and adaptability. However, performance remains highly dependent on the quality and relevance of the training data, the domain of application, and the nuances of the target language or consumer context.

Outside of sensory and consumer science, numerous studies have compared lexicon-based sentiment analysis methods with machine learning approaches, often using social media data or general text corpora. For example, Kolchyna et al. (2015) demonstrated on *X* (Twitter) data that machine learning models generally outperform lexicon-based methods, while hybrid approaches that combine both techniques achieve even better accuracy. Dhaoui et al. (2017) found that lexicon-based and machine learning methods showed comparable accuracy in identifying positive sentiment, but combining these methods improved classification performance. Sham and Mohamed (2022) evaluated domain-specific tweets and concluded that hybrid models integrating lexicon-based and machine learning approaches yielded the highest F1 scores, reinforcing the advantages of such integration. Ribeiro et al. (2015) benchmarked twenty-four sentiment analysis methods across eighteen datasets and highlighted significant variation in performance depending on the domain, indicating that no single method is universally superior and that tailored approaches are necessary.

In comparisons between deep learning or transformer-based models and lexicon methods, Kotelnikova et al. (2021) found that transformer models like BERT generally outperform lexicon-based approaches, although some lexicons perform better on specific datasets, illustrating the importance of the dataset and domain characteristics. W. Zhang et al. (2024) pointed out limitations in large language models'ability to handle complex sentiment tasks that require nuanced understanding or structured sentiment interpretation, highlighting current challenges in deep learning sentiment analysis.

Domain and dataset dependency are major themes across the literature. Mahajani et al. (2023) showed that individual word sentiment contributions vary depending on context, with no single lexicon entry consistently explaining differences between lexicon and machine learning results. Ribeiro et al. (2015) also emphasized that domain specificity greatly affects model performance, supporting the need for domain-specific training or adaptation.

Limitations and challenges persist in sentiment analysis methodologies. Mara et al. (2025) discussed potential biases inherent in different approaches, noting that lexicon-based methods may oversimplify sentiment by relying on static word lists, while deep learning models can introduce negative biases derived from their training data and architecture. Tan et al. (2023) and Islam et al. (2024) outlined broad challenges in sentiment analyses, including handling poorly structured or sarcastic text, the lack of fine-grained sentiment granularity, dependence on annotated data, limitations of word embeddings, biases present in training datasets, domain dependency, negation handling, and spam detection, all of which complicate accurate sentiment classification.

4.2 Special Considerations in Sensory and Consumer Science

In the context of sensory and consumer science, both lexicon-based and data-driven approaches to sentiment analysis offer distinct advantages depending on the specific research goals, data availability, and resource constraints. However, despite their promising potential, these methods are not yet widely developed or adopted within this field. This is largely due to the dominance of traditional quantitative approaches that rely on direct measurements and quantifications of sentiment and consumer preferences. Furthermore, many sensory and consumer studies are conducted in experimental contexts that generate relatively small amounts of textual data, which limits the applicability and effectiveness of data-driven sentiment analysis methods that typically require larger corpora for training and validation. As a result, the integration of advanced sentiment analysis techniques remains limited, highlighting an opportunity for further methodological development tailored to the unique characteristics of sensory and consumer science data.

Lexicon-based sentiment analysis is well-suited for applications that require rapid, interpretable insights from consumer feedback. For instance, in early-stage product development or small-scale consumer testing, researchers may wish to analyse experimental data such as open-ended responses to free comments without the overhead of training complex models. The transparency of lexicon-based approaches makes them attractive in regulatory or academic settings, where clearly interpretable results are necessary. In multilingual studies or low-resource environments, especially in languages where robust data-driven models are lacking, these methods can offer a viable alternative, provided that appropriate domain-specific lexicons are available or can be developed. However, a significant limitation of lexicon-based approaches in sensory and consumer science is the lack of such resources. Most publicly available sentiment lexicons are generic in nature and have been developed primarily for English-language data. As a result, they may fail to capture the nuances of consumer feedback in specific product categories, and words commonly used in sensory evaluations may be absent from these lexicons or may not carry the appropriate sentiment polarity in a given context, leading to inaccurate interpretations.

In contrast, data-driven approaches are more effective when deeper contextual understanding is needed. These models, which include traditional supervised classifiers, deep learning networks, and more recently, transformer-based language models, are capable of detecting subtle sentiment patterns that go beyond the presence of individual words. This makes them highly valuable in analysing complex consumer expressions, such as mixed sentiments or emotionally nuanced feedback. In large-scale studies, such as those involving thousands of online reviews or social media posts, data-driven sentiment analysis can efficiently uncover emergent trends, consumer preferences, and brand perceptions with greater accuracy than rule-based methods. However, a critical consideration when using pretrained language models is that these models are typically trained on broad, general-purpose text corpora such as news articles, books, or web pages. As a result, their understanding of sentiment is not tailored to the specific vocabulary, context, or emotional tone found in sensory and consumer science. This underscores the importance of domain adaptation, where pretrained models are fine-tuned on relevant, domain-specific corpora to improve their accuracy and relevance for tasks in sensory evaluation, product testing, or consumer research. Without this step, there is a risk that even advanced models will produce sentiment outputs that are misaligned with the subtleties of consumer expression in these fields.

5 Conclusion and Perspectives

This chapter explored the landscape of sentiment analysis methods as applied to sensory and consumer science, a field where understanding the emotional and attitudinal responses of consumers is crucial for product development, quality evaluation, and marketing. Lexicon-based approaches offer interpretable and low-resource sentiment classification by relying on predefined sentiment dictionaries. Classical machine learning models rely on engineered textual features and have proven effective in settings with moderate data availability and clear sentiment distinctions. Deep learning approaches, especially transformer-based models enable nuanced understanding of sentiment in context-rich, unstructured consumer feedback. Hybrid models, which combine rule-based and learning-based elements to achieve a balance between interpretability and adaptability, especially valuable in sensory contexts where subjective language is deeply embedded in cultural and product-specific nuances. Together, these methods represent a toolkit of complementary approaches. The choice of method should depend on the specific application, available resources, and the nature of the consumer data at hand.

To ensure that sentiment analysis provides meaningful and trustworthy insights in sensory and consumer science, it is crucial to tackle potential biases at multiple levels. This involves curating more diverse and representative training datasets, employing models that effectively capture contextual and domain-specific language nuances, and performing regular audits of model outputs to identify and correct systematic errors. Combining automated analysis with expert human review can help reduce risks associated with misinterpretation. Ultimately, bias in sentiment analysis extends beyond a technical challenge to encompass important methodological and ethical considerations. Recognizing and addressing these limitations is essential for establishing sentiment analysis as a reliable tool to understand consumer emotions, inform product development, and support decision-making within sensory science.

Looking ahead, several promising avenues could significantly enhance the effectiveness and applicability of sentiment analysis in sensory and consumer science. One such direction is participatory labelling and feedback loops, where consumers or expert panels contribute to annotation processes through human-in-the-loop or active learning systems. This approach enables dynamic, domain-adaptive training of large language models on specific corpora, thereby improving contextual sensitivity and alignment with the nuanced language used in this field. In parallel, the growing diversity of consumer sentiment expression across multiple modalities, such as text, images, voice, gestures, and direct ratings, calls for integrated analytical approaches. Multimodal, interactive tools like chatbots, paired with inputs from webcams, can facilitate the fusion of these varied data sources. Advances in voice prosody analysis and facial emotion detection further contribute to building unified frameworks that capture a more comprehensive picture of consumer experience. Equally important is the development of transparent and interpretable sentiment analysis models. Explainable artificial intelligence techniques are essential for clarifying the reasoning behind model outputs.

In conclusion, sentiment analysis is emerging as a valuable tool for uncovering the nuanced, affective dimensions of consumer experience in sensory science. As the methods advance, becoming more flexible, interpretable, and attuned to domain-specific language, they present new opportunities to understand, anticipate, and respond to consumer preferences in an increasingly dynamic and complex market environment. Rather than replacing traditional approaches, sentiment analysis serves as a powerful complement, enriching conventional methods with deeper insights into consumer psychology. By integrating these techniques, researchers can achieve a more comprehensive, sensitive, and human-centered understanding of how consumers experience and evaluate products.

References

Ahmad, M., Aftab, S., Muhammad, S. S., & Ahmad, S. (2017). Machine learning techniques for sentiment analysis: A review. *International Journal of Multidisciplinary Sciences and Engineering, 8*(3) www.ijmse.org

Aliman, G., Aliman, G. B., Faye Nivera, T. S., Charmille Olazo, J. A., Jane Ramos, D. P., Danielle Sanchez, C. B., Amado, T. M., Arago, N. M., Jorda, R. L., Jr., Virrey, G. C., & Valenzuela, I. C. (2022). Sentiment Analysis using Logistic Regression. *Journal of Computational Innovations and Engineering Applications*.

Al-qerem, A., Ali, A. M., & Abu Izneid, B. (2024). *Building domain specific sentiment lexicon: Survey studies in systems, decision and control* (Vol. 528, pp. 689–698). Springer. https://doi.org/10.1007/978-3-031-56586-1_50

Alshaabi, T., Van Oort, C. M., Fudolig, M. I., Arnold, M. V., Danforth, C. M., & Dodds, P. S. (2022). Augmenting semantic lexicons using word Embeddings and transfer learning. *Frontiers in Artificial Intelligence, 4*, 783778. https://doi.org/10.3389/FRAI.2021.783778/BIBTEX

Alzamzami, F., Hoda, M., & Saddik, A. E. (2020). Light gradient boosting machine for general sentiment classification on short texts: A comparative evaluation. *IEEE Access, 8*, 101840–101858. https://doi.org/10.1109/ACCESS.2020.2997330

Amin, A., Sarkar, A., Islam, M. M., Miazee, A. A., Islam, M. R., & Hoque, M. M. (2024). Sentiment polarity analysis of Bangla food reviews using machine and deep learning algorithms. https://arxiv.org/pdf/2405.06667v1

Andersen, B. V., Chan, R. C. K., & Byrne, D. V. (2021). A conceptual framework for multi-dimensional measurements of food related pleasure—The food pleasure scale. *Foods, 10*, 2044. https://doi.org/10.3390/FOODS10092044

Asani, E., Nejad, H. V., Hosseinabadi, S., Sadri, J., & Vahdatnejad, H. (2020). Extracting user's food preferences by sentiment analysis. 2020 8th Iranian Joint Congress on Fuzzy and Intelligent Systems (CFIS). https://doi.org/10.1109/CFIS49607.2020.9238725.

Augustyniak, L., Szymánski, P., Kajdanowicz, T., & Tuliglowicz, W. (2015). Comprehensive study on lexicon-based ensemble classification sentiment analysis. *Entropy, 18*, 4. https://doi.org/10.3390/E18010004

Baccianella, S., Esuli, A., & Sebastiani, F. (2010). SentiWordNet 3.0: An enhanced lexical resource for sentiment analysis and opinion mining. http://www.lrec-conf.org/proceedings/lrec2010/pdf/769_Paper.pdf

Bahrawi, N. (2019). Sentiment analysis using random Forest algorithm-online social media based. *Journal of Information Technology and Its Utilization, 2*(2), 29. https://doi.org/10.30818/JITU.2.2.2695

Balahur, A., Hermida, J. M., & Montoyo, A. (2012). Detecting implicit expressions of emotion in text: A comparative analysis. *Decision Support Systems, 53*(4), 742–753. https://doi.org/10.1016/J.DSS.2012.05.024

Barbosa, M. W., & Gomes, A. (2025). Themes and sentiments in conversations about food waste on Twitter: Proposal of a framework using neural topic modeling. *Food Quality and Preference, 122*, 105311. https://doi.org/10.1016/J.FOODQUAL.2024.105311

Betsch, T. (2011). The stability of preferences—A social-cognition view. *Frontiers in Psychology, 2*(NOV), 290. https://doi.org/10.3389/FPSYG.2011.00290

Bhagat, P., Korkankar, P. D., Pawar, J. D., Bhagat, P., Korkankar, P. D., & Pawar, J. D. (2023). Aspect-based sentiment words and their polarities using Chi-Square Test. *Computación y Sistemas, 27*(2), 389–399. https://doi.org/10.13053/CYS-27-2-4397

Boutrolle, I., Delarue, J., Arranz, D., Rogeaux, M., & Köster, E. P. (2007). Central location test vs. home use test: Contrasting results depending on product type. *Food Quality and Preference, 18*(3), 490–499. https://doi.org/10.1016/J.FOODQUAL.2006.06.003

Caballero, R. (2019). The case of wine: Chapter 7. Sensory experiences, meaning and metaphor. *Converging Evidence in Language and Communication Research, 19*, 127–144. https://doi.org/10.1075/CELCR.19.07CAB

Cardello, A. V. (2017). Hedonic scaling: Assumptions, contexts and frames of reference. *Current Opinion in Food Science, 15*, 14–21. https://doi.org/10.1016/J.COFS.2017.05.002

Cardello, A. V., Schutz, H., Snow, C., & Lesher, L. (2000). Predictors of food acceptance, consumption and satisfaction in specific eating situations. *Food Quality and Preference, 11*(3), 201–216. https://doi.org/10.1016/S0950-3293(99)00055-5

Cerri, J., Thøgersen, J., & Testa, F. (2019). Social desirability and sustainable food research: A systematic literature review. *Food Quality and Preference, 71*, 136–140.

Chen, Z., Gurdian, C., Sharma, C., Prinyawiwatkul, W., & Torrico, D. D. (2021). Exploring text mining for recent consumer and sensory studies about alternative proteins. *Foods, 10*, 2537. https://doi.org/10.3390/FOODS10112537

Dalenberg, J. R., Gutjar, S., Ter Horst, G. J., De Graaf, K., Renken, R. J., & Jager, G. (2014). Evoked emotions predict food choice. *PLoS One, 9*(12), e115388. https://doi.org/10.1371/JOURNAL.PONE.0115388

Dalianis, H. (2018). Evaluation metrics and evaluation. *Clinical Text Mining, 45–53*. https://doi.org/10.1007/978-3-319-78503-5_6

Dhaoui, C., Webster, C. M., & Tan, L. P. (2017). Social media sentiment analysis: Lexicon versus machine learning. *Journal of Consumer Marketing, 34*(6), 480–488. https://doi.org/10.1108/JCM-03-2017-2141

Ding, X., Liu, B., & Yu, P. S. (2008). A holistic lexicon-based approach to opinion mining. In *WSDM'08—Proceedings of the 2008 International Conference on Web Search and Data Mining* (pp. 231–239). https://doi.org/10.1145/1341531.1341561

Du, X. (2022). Linguistic features and psychological states: The case of Virginia Woolf. *Frontiers in Psychology, 13*, 823313. https://doi.org/10.3389/FPSYG.2022.823313/BIBTEX

Ekman, P. (1992). An argument for basic emotions. *Cognition & Emotion, 6*(3–4), 169–200. https://doi.org/10.1080/02699939208411068

Fernqvist, F., & Ekelund, L. (2014). Credence and the effect on consumer liking of food—A review. *Food Quality and Preference, 32*(PC), 340–353. https://doi.org/10.1016/J.FOODQUAL.2013.10.005

Giacalone, D., Llobell, F., & Jaeger, S. R. (2022). "Beyond liking" measures in food-related consumer research supplement hedonic responses and improve ability to predict consumption. *Food Quality and Preference, 97*, 104459. https://doi.org/10.1016/J.FOODQUAL.2021.104459

Gunawan, L., Anggreainy, M. S., Wihan, L., Santy, Lesmana, G. Y., & Yusuf, S. (2022). Support vector machine based emotional analysis of restaurant reviews. *Procedia Computer Science, 216*, 479–484. https://doi.org/10.1016/j.procs.2022.12.160

Gupta, V., & Rattan, P. (2024a). Advancing sentiment analysis in restaurant reviews through unsupervised machine learning algorithms. *International Journal of Intelligent Engineering and Systems, 17*(4). https://doi.org/10.22266/ijies2024.0831.82

Gupta, V., & Rattan, P. (2024b). Enhancing sentiment analysis in restaurant reviews: A hybrid approach integrating lexicon-based features and LSTM networks. *International Journal of Intelligent Systems and Applications in Engineering, 12*(3), 185–198. https://www.ijisae.org/index.php/IJISAE/article/view/5240

Gutjar, S., de Graaf, C., Kooijman, V., de Wijk, R. A., Nys, A., Ter Horst, G. J., & Jager, G. (2015). The role of emotions in food choice and liking. *Food Research International, 76*(P2), 216–223. https://doi.org/10.1016/J.FOODRES.2014.12.022

Hallikainen, H., Halinen, M., Tervonen, H., Ruusunen, N., & Wang, Y. (2025). Rating with the senses: How sensory encounters are reflected on online review ratings? *Psychology & Marketing, 42*(1), 159–174. https://doi.org/10.1002/MAR.22119

Hamby, A., Daniloski, K., & Brinberg, D. (2015). How consumer reviews persuade through narratives. *Journal of Business Research, 68*(6), 1242–1250. https://doi.org/10.1016/J.JBUSRES.2014.11.004

Hamilton, W. L., Clark, K., Leskovec, J., & Jurafsky, D. (2016). Inducing domain-specific sentiment lexicons from unlabeled corpora. 595–605. http://nlp.stanford.edu/projects/socialsent

Hanić, S., Bagić Babac, M., Gledec, G., & Horvat, M. (2024). Comparing machine learning models for sentiment analysis and rating prediction of vegan and vegetarian restaurant reviews. *Computers, 13*, 248. https://doi.org/10.3390/COMPUTERS13100248

Harfoushi, O., Hasan, D., & Obiedat, R. (2018). Sentiment analysis algorithms through azure machine learning: Analysis and comparison. *Modern Applied Science, 12*(7), 49. https://doi.org/10.5539/MAS.V12N7P49

Hutto, C. J., & Gilbert, E. (2014). VADER: A parsimonious rule-based model for sentiment analysis of social media text. *Proceedings of the International AAAI Conference on Web and Social Media, 8*(1), 216–225. https://doi.org/10.1609/ICWSM.V8I1.14550

Iqbal, A., Amin, R., Iqbal, J., Alroobaea, R., Binmahfoudh, A., & Hussain, M. (2022). Sentiment analysis of consumer reviews using deep learning. *Sustainability, 14*, 10844. https://doi.org/10.3390/SU141710844

Jamadi Khiabani, P., Basiri, M. E., & Rastegari, H. (2020). An improved evidence-based aggregation method for sentiment analysis. *Journal of Information Science, 46*(3), 340–360. https://doi.org/10.1177/0165551519837187

Jockers, M. (2020). Introduction to the Syuzhet package. https://cran.r-project.org/web/packages/syuzhet/vignettes/syuzhet-vignette.html.

Kaneko, D., Stuldreher, I., Reuten, A. J. C., Toet, A., van Erp, J. B. F., & Brouwer, A. M. (2021). Comparing explicit and implicit measures for assessing cross-cultural food experience. *Frontiers in Neuroergonomics, 2*, 646280. https://doi.org/10.3389/FNRGO.2021.646280/BIBTEX

Kolchyna, O., Souza, T. T. P., Treleaven, P., & Aste, T. (2015). Twitter sentiment analysis: Lexicon method. *Machine Learning Method and Their Combination.*. https://arxiv.org/pdf/1507.00955

Kotelnikova, A., Paschenko, D., Bochenina, K., & Kotelnikov, E. (2021). Lexicon-based methods vs. BERT for text sentiment analysis. Lecture notes in computer science (including subseries lecture notes in artificial intelligence and lecture notes in bioinformatics), 13217 LNCS, 71–83. https://doi.org/10.1007/978-3-031-16500-9_7.

Kvasnička, R., Kvasničková Stanislavská, L., Pilař, L., Kuralová, K., Pilařová, L., & Čejka, M. (2024). Exploring twitter discussions on healthy food: Inspiring, access to healthy food for young, and nutritional economics in focus. *Frontiers in Sustainable Food Systems, 8*, 1357896. https://doi.org/10.3389/FSUFS.2024.1357896/XML/NLM

Labille, K., Gauch, S., & Alfarhood, S. (2017). Creating Domain-Specific Sentiment Lexicons via Text Mining.

Li, H., Chen, Y., Ji, H., Muresan, S., & Zheng, D. (2012). Combining social cognitive theories with linguistic features for multi-genre sentiment. *Analysis, 127–136.* http://www.politicalforum.com/elections-campaigns

Li, W., Chen, Y., Zheng, S., Wang, L., & Lalor, J. P. (2023a). From stars to insights: Exploration and implementation of unified sentiment analysis with distant supervision. W-NUT 2024—9th Workshop on Noisy and User-Generated Text, Proceedings of the Workshop, 89–103. https://arxiv.org/pdf/2305.01710

Li, Y., Guerin, F., & Lin, C. (2023b). The secret of metaphor on expressing stronger emotion. FLP 2022—3rd Workshop on Figurative Language Processing, Proceedings of the Workshop, 39–43. https://doi.org/10.18653/v1/2022.flp-1.6.

Liapakis, A., Tsiligiridis, T., & Yialouris, C. (2020). A sentiment lexicon-based analysis for food and beverage industry reviews. The Greek language paradigm. *International Journal on Natural Language Computing, 9*(2), 21–42. https://doi.org/10.5121/IJNLC.2020.9203

Lim, J. (2011). Hedonic scaling: A review of methods and theory. *Food Quality and Preference, 22*(8), 733–747. https://doi.org/10.1016/J.FOODQUAL.2011.05.008

Liu, Q., Kusner, M. J., & Blunsom, P. (2020). A survey on contextual Embeddings. https://arxiv.org/pdf/2003.07278.

Luc, A., Lê, S., & Philippe, M. (2020). Nudging consumers for relevant data using free JAR profiling: An application to product development. *Food Quality and Preference, 79*, 103751. https://doi.org/10.1016/J.FOODQUAL.2019.103751

Luc, A., Lê, S., Philippe, M., Mostafa Qannari, E., & Vigneau, E. (2022). A machine learning approach for analyzing free JAR data. *Food Quality and Preference, 99*, 104581. https://doi.org/10.1016/J.FOODQUAL.2022.104581

Mahajani, S. J., Srivastava, S., & Smeaton, A. F. (2023). *A comparison of lexicon-based and ML-based sentiment analysis: Are there outlier words?* 2023 31st Irish Conference on Artificial Intelligence and Cognitive Science, AICS. https://doi.org/10.1109/AICS60730.2023.10470734

Mahieu, B., Visalli, M., & Schlich, P. (2022). Identifying drivers of liking and characterizing the ideal product thanks to free comment. *Food Quality and Preference, 96*, 104389. https://doi.org/10.1016/J.FOODQUAL.2021.104389

Mairano, P., Zovato, E., & Quinci, V. (2019). On the relation between lexical and prosodic cues of emotion in English: A preliminary corpus-based analysis. *Http://Journals.Openedition.Org/Anglophonia, 27*. https://doi.org/10.4000/ANGLOPHONIA.2157

Mara, D. A., Culda, L. C., Cristescu, M. P., Nerişanu, R. A., & Moldovan, R.-A. (2025). Exploring the efficacy and bias of sentiment analysis tools in customer reviews. *Journal of Social and Economic Statistics, 14*(1), 19–40. https://doi.org/10.2478/JSES-2025-0002

Miller, G. A., Beckwith, R., Fellbaum, C., Gross, D., & Miller, K. J. (1990). Introduction to wordnet: An on-line lexical database. *International Journal of Lexicography, 3*(4), 235–244. https://doi.org/10.1093/IJL/3.4.235

Mohamad Beigi, O., & Moattar, M. H. (2021). Automatic construction of domain-specific sentiment lexicon for unsupervised domain adaptation and sentiment classification. *Knowledge-Based Systems, 213*, 106423. https://doi.org/10.1016/J.KNOSYS.2020.106423

Mohammad, S. M. (2016). Sentiment analysis: Detecting valence, emotions, and other affectual states from text. *Emotion Measurement, 201–237*. https://doi.org/10.1016/B978-0-08-100508-8.00009-6

Mohammad, S. M., & Turney, P. D. (2013). Crowdsourcing a word-emotion association lexicon. *Computational Intelligence, 29*(3), 436–465. https://doi.org/10.1111/J.1467-8640.2012.00460.X

Mohammed, I., & Prasad, R. (2023). Building lexicon-based sentiment analysis model for low-resource languages. *MethodsX, 11*, 102460. https://doi.org/10.1016/J.MEX.2023.102460

Molenaar, A., Jenkins, E. L., Brennan, L., Lukose, D., & McCaffrey, T. A. (2024). The use of sentiment and emotion analysis and data science to assess the language of nutrition-, food- and cooking-related content on social media: A systematic scoping review. *Nutrition Research Reviews, 37*(1), 43–78. https://doi.org/10.1017/S0954422423000069

Mostafa, M. M. (2020). Global halal food discourse on social media: A text mining approach. *Journal of International Communication, 26*(2), 211–237. https://doi.org/10.1080/13216597.2020.1795702

Naldi, M. (2019). A review of sentiment computation methods with R packages. http://arxiv.org/abs/1901.08319.

Neelima, A., & Mehrotra, S. (2023). A comprehensive review on word embedding techniques. In *Proceedings of the 2023 International Conference on Intelligent Systems for Communication, IoT and Security* (pp. 538–543). ICISCoIS 2023. https://doi.org/10.1109/ICISCOIS56541.2023.10100347

Nielsen, F. Å. (2011). A new ANEW: Evaluation of a word list for sentiment analysis in microblogs. *CEUR Workshop Proceedings, 718*, 93–98. https://arxiv.org/pdf/1103.2903

Norris, C. J., Do, E., Close, E., & Deswert, S. (2019). Ambivalence toward healthy and unhealthy food and moderation by individual differences in restrained eating. *Appetite, 140*, 309–317. https://doi.org/10.1016/J.APPET.2019.05.033

Panagiotou, M., & Gkatzionis, K. (2022). Lexicon development to measure emotions evoked by foods: A review. *Measurement: Food, 7*, 100054. https://doi.org/10.1016/J.MEAFOO.2022.100054

Pennebaker, J., Francis, M., & Booth, R. (1999). Linguistic inquiry and word count (LIWC).

Pindado, E., & Barrena, R. (2021). Using Twitter to explore consumers' sentiments and their social representations towards new food trends. *British Food Journal, 123*(3), 1060–1082. https://doi.org/10.1108/BFJ-03-2020-0192

Ribeiro, F. N., Araújo, M., Gonçalves, P., André Gonçalves, M., & Benevenuto, F. (2015). SentiBench—a benchmark comparison of state-of-the-practice sentiment analysis methods. *EPJ Data Science, 5*(1). https://doi.org/10.1140/epjds/s13688-016-0085-1

Rinker, T. (2015). Dictionary based sentiment analysis that considers valence shifters: Trinker/sentimentr. Tyler. 2018. Dictionary based sentiment analysis that considers valence shifters: Trinkersentimentr. R. https://www.github.comtrinkersentimentr.

Robinson, E., & Higgs, S. (2012). Liking food less: The impact of social influence on food liking evaluations in female students. *PLoS One, 7*(11). https://doi.org/10.1371/JOURNAL.PONE.0048858

Romano, R., Brockhoff, P. B., Hersleth, M., Tomic, O., & Næs, T. (2008). Correcting for different use of the scale and the need for further analysis of individual differences in sensory analysis. *Food Quality and Preference, 19*(2), 197–209. https://doi.org/10.1016/J.FOODQUAL.2007.06.008

Seyeditabari, A., Tabari, N., & Zadrozny, W. (2018). Emotion detection in text: A review.

Sham, N. M., & Mohamed, A. (2022). Climate change sentiment analysis using lexicon, machine learning and hybrid approaches. *Sustainability, 14*, 4723. https://doi.org/10.3390/SU14084723

Singh, A., & Glińska-Neweś, A. (2022). Modeling the public attitude towards organic foods: A big data and text mining approach. *Journal of Big Data, 9*(1). https://doi.org/10.1186/S40537-021-00551-6

Surya, P. P. M., & Subbulakshmi, B. (2019). Sentimental analysis using naive Bayes classifier. *Proceedings—International Conference on Vision Towards Emerging Trends in Communication and Networking, ViTECoN, 2019*. https://doi.org/10.1109/VITECON.2019.8899618

Taboada, M. (2016). Sentiment analysis: An overview from Linguistics. *Annual Review of Linguistics, 2*, 325–347. https://doi.org/10.1146/ANNUREV-LINGUISTICS-011415-040518

Tan, K. L., Lee, C. P., & Lim, K. M. (2023). A survey of sentiment analysis: Approaches, datasets, and future research. *Applied Sciences, 13(7)*, 4550. https://doi.org/10.3390/APP13074550

Tao, D., Yang, P., & Feng, H. (2020). Utilization of text mining as a big data analysis tool for food science and nutrition. *Comprehensive Reviews in Food Science and Food Safety, 19*(2), 875–894. https://doi.org/10.1111/1541-4337.12540

Thelwall, M., Buckley, K., & Paltoglou, G. (2012). Sentiment strength detection for the social web. *Journal of the American Society for Information Science and Technology, 63*(1), 163–173. https://doi.org/10.1002/ASI.21662

Tian, G., Lu, L., & McIntosh, C. (2021). What factors affect consumers' dining sentiments and their ratings: Evidence from restaurant online review data. *Food Quality and Preference, 88*, 104060. https://doi.org/10.1016/J.FOODQUAL.2020.104060

Torrico, D. D. (2025). The potential use of ChatGPT as a sensory evaluator of chocolate brownies: A brief case study. *Food, 14*(3), 464. https://doi.org/10.3390/foods14030464

Tubishat, M., Idris, N., & Abushariah, M. A. M. (2018). Implicit aspect extraction in sentiment analysis: Review, taxonomy, oppportunities, and open challenges. *Information Processing and Management, 54*(4), 545–563. https://doi.org/10.1016/j.ipm.2018.03.008

Ullah, R., Amblee, N., Kim, W., & Lee, H. (2016). From valence to emotions: Exploring the distribution of emotions in online product reviews. *Decision Support Systems, 81*, 41–53. https://doi.org/10.1016/J.DSS.2015.10.007

Verza, M., Camanzi, L., Rota, C., Cerjak, M., Mulazzani, L., & Malorgio, G. (2023). Consumer sentiments and emotions in new seafood product concept development: A co-creation approach using online discussion rooms in Croatia. *Italy and Spain. Foods, 12*(8), 1729. https://doi.org/10.3390/FOODS12081729/S1

Visalli, M., Mahieu, B., Thomas, A., & Schlich, P. (2020). Automated sentiment analysis of free comment: An indirect liking measurement? *Food Quality and Preference, 82*. https://doi.org/10.1016/j.foodqual.2020.103888

Visalli, M., Mahieu, B., Dubois, M., & Schlich, P. (2023). Hedonic valence of descriptive sensory terms as an indirect measure of liking: A preliminary study with red wines. *Food Quality and Preference, 108*, 104861. https://doi.org/10.1016/J.FOODQUAL.2023.104861

Wen, Y., & Laporte, S. (2025). Experiential narratives in marketing: A comparison of generative AI and human content. *Journal of Public Policy and Marketing*. https://doi.org/10.1177/07439156241297973

Wilson, T. D., & Schooler, J. W. (1991). Thinking too much: Introspection can reduce the quality of preferences and decisions. *Journal of Personality and Social Psychology, 60*(2), 181–192. https://doi.org/10.1037/0022-3514.60.2.181

Xu, H., Xiao, M., Zeng, J., & Hao, H. (2022). Green-labelled Rice versus conventional Rice: Perception and emotion of Chinese consumers based on review mining. *Foods, 12(1)*, 87. https://doi.org/10.3390/FOODS12010087

Yan, H. B., & Li, Z. (2022). Review of sentiment analysis: An emotional product development view. *Frontiers of Engineering Management 2022 9:4, 9*(4), 592–609. https://doi.org/10.1007/S42524-022-0227-Z

Zainuddin, N., & Selamat, A. (2014). *Sentiment analysis using Support Vector Machine*. https://doi.org/10.1109/I4CT.2014.6914200

Zeng, G., Chen, Z., & Zhong, S. (2024). "We Chinese just want meat!" an analysis of Chinese netizens' reactions to vegetarian advocacy. *Food Quality and Preference, 115*, 105128. https://doi.org/10.1016/J.FOODQUAL.2024.105128

Zhang, J. (2024). Sentiment and language: A socio-semiotic analysis. *Philosophy Journal, 3*(1), 118–127. https://doi.org/10.23977/PHIJ.2024.030120

Zhang, S., Ly, L., Mach, N., & Amaya, C. (2022). Topic modeling and sentiment analysis of yelp restaurant reviews. *International Journal of Information Systems in the Service Sector, 14*(1). https://doi.org/10.4018/IJISSS.295872

Zhang, W., Deng, Y., Liu, B., Pan, S. J., & Bing, L. (2024). *Sentiment analysis in the era of large language models: A reality check* (pp. 3881–3906). Findings of the Association for Computational Linguistics: NAACL 2024—Findings. https://doi.org/10.18653/V1/2024.FINDINGS-NAACL.246

Zhou, M. (2024). The effects of sensory interaction and sensory conflict on consumer online review rating behavior. *International Journal of Management Science and Engineering Management, 19*(2), 97–105. https://doi.org/10.1080/17509653.2023.2174201

Part V
Reuse of Large Corpus of Food Related Text Data

While previous parts explored theoretical foundations, practical tools for structuring or preprocessing text data, and methods for experimental text data collection and analysis, this final section shifts the focus toward the reuse of large, pre-existing text corpora. It explores both the opportunities and challenges involved in analysing extensive, naturally occurring datasets—such as online reviews and social media content—particularly in the context of food culture, consumer behaviour, and nutrition communication.

Chapter 9 introduces *Wineinformatics*, an emerging field that merges wine expertise with computational text analysis to extract insights from digital wine reviews. Chapter 10 delves into co-occurrence network analysis, a method for uncovering hidden patterns and relationships by mapping associations between terms. Chapter 11 presents netnography, a qualitative approach for investigating online communities and the dynamics of digital food discourse. Finally, Chap. 12 reflects on the socio-cultural implications of food-related narratives on social media, highlighting their influence on dietary norms and the role of media in shaping food-related beliefs and practices.

Chapter 9
Digital Transformation of Wine Opinion: From Critics to Peer Reviews Analysis

Magalie Dubois, Michał Folwarczny, Jing Cao, and Þóranna Bender

1 Introduction

Have you ever spent 30 min debating the subtleties of a carrot or the nuances of fermented milk? Few products inspire such depth of conversation; wine, however, routinely does, as enthusiasts dissect every note of aroma, flavours, mouthfeel, body and balance.

1.1 Earliest Forms of Wine Evaluation

The length of the history of discussion over wine quality bears no comparison with that of any other food product (Amerine et al., 1959). The first traces of wine quality evaluations and recommendations date back to Ancient Greece. Homer praised

M. Dubois (✉)
School of Wine and Spirits Business, Burgundy School of Business, Dijon, France
e-mail: Magalie.dubois@bsb-education.com

M. Folwarczny
Discipline of Marketing, J.E. Cairnes School of Business and Economics, University of Galway, Galway, Ireland
e-mail: michal.folwarczny@universityofgalway.ie

J. Cao
Department of Statistics and Data Science, Southern Methodist University, Dallas, TX, USA
e-mail: jcao@smu.edu

Þ. Bender
Department of Applied Mathematics and Computer Science, Technical University of Denmark, Kongens Lyngby, Denmark

M. Visalli, M. V. Galmarini (eds.), *Collection and Analysis of Text Data in Sensory and Consumer Science*, https://doi.org/10.1007/978-3-032-04851-6_9

Pramnian and Maronean wines, Pliny the Elder commended the wines from Falerne, and Horace the ones from Chios and Lesbos islands.

Before the twentieth century, winemakers were often farmers who suffered significant productivity and quality variations from year to year, making it difficult for traders and retailers to source quality wines (Cawley, 2018; Conca Messina et al., 2019). Wine was mainly sold in bulk to merchants (Lecat et al., 2017). Those latter tasted wines mostly to detect (Simpson, 2004) and consumers relied on the judgment of their wine merchants to avoid fraud and adulterated wines (Shapin, 2012).[1] The question at that time was more "*Is it drinkable?*" rather than "*How good is it?*" Globally, aside from a few exceptions, quality control was left in the hands of traders and retailers rather than in those of producers and few comments were made on the wine quality (Shapin, 2012). There was still a limited direct relationship between wine producers and consumers (Cawley, 2018), and the asymmetry of information prevailed in the wine market, as wine producers and merchants had significantly more information on the quality of the product than wine consumers.

1.2 Formalized Tastings and Sensory Classifications

Following the Second World War, the development of standards by both governments and industries (Carsky et al., 1998) paved the way for the establishment of the International Organization for Standardization (ISO) in 1947. In the middle of the twentieth century, the process of industrialization brought scientific and technical changes, along with an increase in the competitiveness and internationalization of the food and beverage industry (Stone & Sidel, 2003). It fostered the development of sensory science based on standardized sensory evaluation techniques as a systematic way of understanding the human sensory perception of foods and beverages (Lahne, 2016). Organic chemists began to explore the beneficial qualities of wine (Shapin, 2012) and examined the tasting techniques employed by merchants, aiming to establish standardized norms and training methods for wine tasting (Fernandez, 2004).

The 1970s saw the definition of the first standardized terms relating to sensory analysis of food in the framework of *ISO 3972-1979 Sensory analysis—determination of sensitivity of taste* and *ISO 5492-1977 Sensory analysis—vocabulary* (Drake et al., 2009) to allow taster alignment on a shared terminology. The development of standardized tools that ensure the replicability of sensory analysis—most notably the wine-tasting glass defined by the international standard ISO 3591:1977 *Sensory analysis — Apparatus — Wine-tasting glass*—has played a crucial role in facilitating the globalization of wine sensory evaluation.

[1] Today, the same need to verify a bottle's authenticity has led researchers to develop AI-driven methods to detect fraudulent wine. For example, a recent AI driven tool uses machine learning analysis of wine chemical profiles to trace bottles back to their true origin with 99% accuracy (Schartner et al., 2023).

1.3 Emergence of Print Wine Criticism

At a time when most of the population had very limited knowledge of wine, most books about wine were aimed at professionals, dealing with viticulture and oenology, vineyards, regions, and how to serve and drink wine (Cawley, 2018; Fernandez, 2004; Shapin, 2012). Before the 1970s there were no widely circulated magazines offering consumer ratings of wine that would allow consumers to compare relative quality within specific wine categories (James, 2018). Historically, unlike in the continental Europe, both the United States and the United Kingdom have had limited domestic wine production. This meant that local supply was generally insufficient to satisfy consumer demand, creating a strong reliance on imports to supply their markets. The importation of wines also contributed to a broader availability of diverse geographic origins, often influenced by the presence of diaspora communities. While consumers in countries like France typically drink wines from nearby regions, or at least national wines (Alonso Ugaglia et al., 2019), this pattern did not hold in the United Kingdom or the United States. The availability in the market of this greater variety of origins can prove problematic when trying to learn about wine: without local traditions to guide them, where should one begin? In this case, the best solution is to learn by following the advice of connoisseurs.

In the United Kingdom, the exploration of this unfamiliar wine landscape led to the emergence of consumer-focused wine publications, such as *Vine* and *Decanter* in 1975 (Shapin, 2012). Referring to that period, the British wine critic Jancis Robinson wrote in a column published on September 4th, 2015: "*Back in those days, wine used to be one of those subjects about which ordinary people in anglophone countries would hesitate to express an opinion. It used to be left to us experts to tell ordinary tasters what to think and how to describe those thoughts*".[2] The combined emergence of Californian wine publications, cheaper flights to Europe (James, 2018), and the impact of the famous comparative blind tasting between Californian and French wines organized by Steven Spurrier in 1976 (Morrison & Rabellotti, 2017) triggered an interest in wine in the United States. *Wine Spectator* was launched that same year, *Food & Wine* and *Wine Advocate* two years later, and *Wine Enthusiast* in 1979, reflecting the rapid increase in demand for wine experts' opinions in the United States (Storchmann, 2012). "*The emergence of the modern wine critic followed the rise of the United States as a serious wine-consuming country in the early 1960s and then as a serious wine-producing country by the late 1970s. U.S. wine markets lacked the historical and institutional support to offer much guidance to the new consumers who were trying to address the problem of hidden quality information. They concomitantly lacked the ability to provide guidance to the many new producers vying to compete and set prices based on product quality*" (Hsu et al., 2007, p. 11).

[2] https://www.ft.com/content/9de45762-5230-11e5-b029-b9d50a74fd14 (consulted 21.04.2025)

1.4 Impact of Wine Criticism on Consumer Behaviour and Market Performance

This evolution boosted social demand for wine quality assessment and purchasing recommendations among consumers, who wanted to drink high-quality products (Jackson, 2017): "*Around this interest grew the need for describing wines so as to guide consumers in their choice, giving rise to professional wine critics and educators.*" (James, 2018, p. 3). The 1970s marked a pivotal moment for the global wine industry. This period saw a decline in consumption within traditional wine-producing countries and a steady rise in demand from non-producing nations (Morrison & Rabellotti, 2017). Simultaneously, there was a shift toward more premium wine consumption (Wittwer et al., 2003) and the introduction of the Anglo-Saxon tradition of wine tasting and criticism (Fernandez, 2004). These changes contributed to the emergence of a new generation of educated consumers and a growing emphasis on consumer-centered evaluations of wine quality (Carsky et al., 1998).

The evaluative innovation of the 1970s marked a shift toward an increasingly globalized, less technical, and more pleasure-oriented way of speaking about wine in Anglophone settings, especially in the United States (Shapin, 2012). Jackson (2008) also notes that evaluative innovation and academic research impacted both consumption and production during the second half of the twentieth century, as it resulted in a marked improvement in wine quality and brought fine-quality wines to more consumers than ever before. Information about wine quality and quality evaluation has played a major role in current consumption trends: "*In many respects, the resurgence of the U.S. wine industry was predicated upon the existence of a critic-generated classification system that helped new consumers navigate the complexities of products and 'appreciate' wine*" (Jamerson, 2009, p. 385). The democratization of wine purchases and consumption in the 1970s and the transformation of the wine industry from a production-oriented to a market-driven industry increased consumers' dependency on analytical sensory evaluation techniques (Jenster & Jenster, 1993; Langstaff, 2010).

1.5 Digital Revolution and Democratization of Wine Opinion

We live in an era of ubiquitous digital footprints. Consumers generate vast amounts of data—often unintentionally—through activities such as wearing smartwatches, leaving geolocation services enabled on smartphones, jogging with iPods that predict their music preferences, conducting Internet searches, and playing video games online. In the near future, virtual-reality glasses and implanted sensors may further expand this data ecosystem. The volume of digital data is growing exponentially: Rydning et al. (2018) estimated a fourfold increase between 2019 and 2025. Even traditionally in-person experiences like wine tastings transitioned online during the

COVID-19 pandemic. Many of these events adopted virtual formats, which not only broadened access but also enabled the collection of more comprehensive data—unlike physical gatherings, which often go undocumented (Chandra Kruse & Drechsler, 2022). While early research in the 2000s offered limited insight into the use of digital footprints for predicting consumer behaviour, the past decade has demonstrated that tailoring online persuasive appeals to customers' personality traits—derived from digital footprints such as Facebook Likes or Twitter (now X) tweet content—can substantially increase product sales. In three field experiments involving 3.5 million participants, Matz et al. (2017) found that such personalized digital appeals increased both the click-through rates and subsequent purchases by nearly 50%. These findings— that personalized advertising outperforms generic messaging in digital environments—align with industry data reported by EnTribe (2022), indicating that 84% of consumers trust a brand more when it uses user-generated content, and nearly 80% report that their purchase decisions are influenced by such content. The same report also indicates that, compared with user-generated content, influencers are perceived as inauthentic and unrelatable, highlighting the importance of user comments in today's marketing landscape (EnTribe, 2022).

Furthermore, the Internet offers unprecedented access to detailed information about products, including price comparisons (e.g., http://www.wine-searcher.com), community reviews (e.g., www.vivino.com), and expert reviews (e.g., www.robert-parker.com). It results in a subsequent search cost reduction and increased information capability and provides opportunities for better decision-making (Branco et al., 2012; Lynch Jr. & Ariely, 2000; Yuan et al., 2019). Thanks to the internet, the search costs for (i) price, (ii) information, and (iii) expertise have fallen (Anderson, 2004), threatening profits through the search cost for price (price comparison, easy access to other online sellers), but at the same time potentially enhancing them through information and expertise. With the emergence of digital technology, internet content, usage, and the community involved in the creation of content (i.e., prosumers) are evolving at a rapid pace (Choi & Burnes, 2016; Marsh & Rajaram, 2019). The historical development of quality signalling in the wine market could be graphically represented as in Fig. 9.1.

The decline in search costs for price, information, and expertise has posed a challenge to traditional wine experts. In their place, a new and increasingly popular source of quality information has emerged online: review aggregators. *Vivino*, a wine rating app, exemplifies this trend. In 2015, the *Vivino* app had been downloaded by 10 million users and totalled 133 million scanned labels (Kotonya et al., 2018). As of April 2025, *Vivino*—the world's most downloaded wine app and largest online wine marketplace—reports having 72 million users and over 3 billion labels scanned (see Fig. 9.2).

The rapid growth of the number of users, wines, wineries, scanned labels, ratings, and reviews available on the platform reveals a shift occurring in the wine market. *Vivino*'s mission statement illustrates this shift: "*Empower people everywhere to enjoy wine to the fullest. We do this by giving you access to information to*

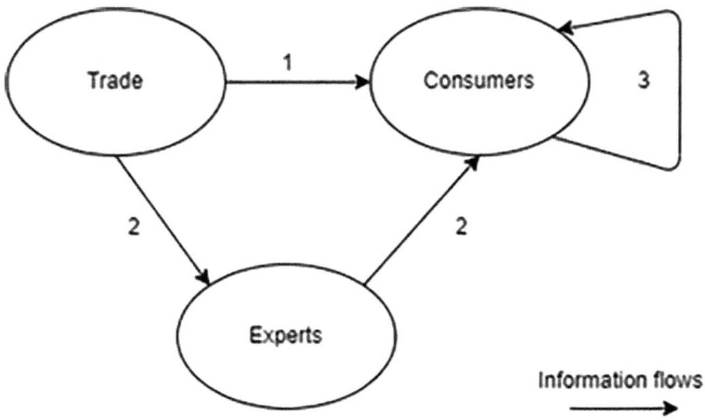

Fig. 9.1 Historical development of quality signalling in the wine market (source: authors)

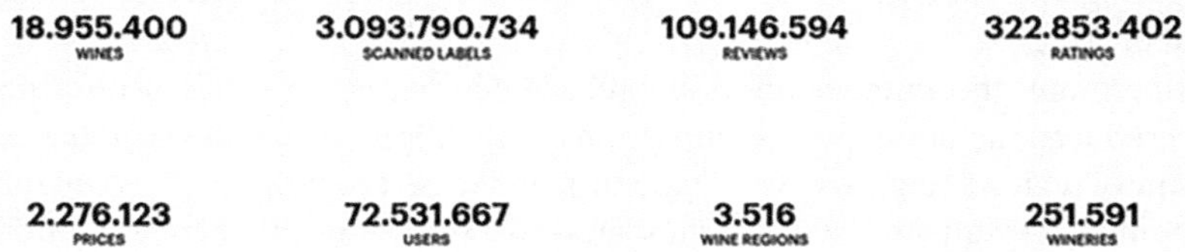

Fig. 9.2 *Vivino* by the numbers (www.vivino.com/about, consulted 21.04.25)

always buy the best wine for you".[3] *Vivino* offers its users to remove barriers to wine enjoyment by providing the information they need to buy the right wine. And this information is experiential information from peers. As mentioned earlier, the Internet lowered search costs for price, information, and expertise. In this context, expertise—now originating from sources beyond traditional authorities—has been commoditized and digitized. Peer reviews are free and easily accessible through apps, with instant access to community ratings by scanning the label of a wine bottle or the wine menu. Despite the popularity of this new type of information source, most marketplaces and online retailers still display the ratings of professional wine experts (e.g., Jeb Dunnuck, James Suckling, Robert Parker, Tim Atkin) to inform consumers' choices. Commercial websites also provide wine awards and expert rating sources as a filtering option (e.g., Antonio Galloni >90). Expert ratings are thus available to online wine customers for free. In an attempt to signal quality to their users, major retailers (e.g., *wine.com* or *Vinatis*) also display aggregated community ratings. They can leverage on the wisdom of the crowd by offering "best sellers" as a filtering option. However, they often fail to offer a significant amount of peer

[3] https://www.vivino.com/wine-news/big-news-about-the-vivino-community (consulted 21.04.2025)

reviews. When the volume of peer reviews is low, online consumers tend to favour expert-generated information (Flanagin & Metzger, 2013). This observation prompts several important questions. Do community ratings and experts' ratings complement or compete with each other? Are they used by wine consumers in different momentum? And should marketers and researchers approach them as distinct forms of influence requiring separate analytical frameworks? The interplay between wine experts and online review aggregators is shaping a new landscape in the wine market (Dubois et al., 2025).

While wine experts have long been a central focus in wine economics, community-based wine ratings have only recently attracted scholarly attention. This new type of information source is bringing heterogeneity to the supply of quality ratings. The emergence of professional wine evaluation in the 1970s has generated evaluation criteria for fine wines, but it has also overlooked lower-priced segments of the market. Some authors capitalized on this gap by writing bestselling guides to inexpensive wines (see Herschkowitsch & Goldstein, 2009). Crowd-based sources of wine evaluation are now covering not only the categories neglected by professional evaluators but also their core market: the fine wine category. In its mission statement, *Vivino* stresses that the experience is "*fun, accessible and effortless*". GQ editors, in an article from December 16, 2016 on "the 20 new rules of drinking (and buying) wine" name its rule #7 "Robert Parker can S.T.F.U[4] now"—stating that the reign of the wine expert dictating what to drink is now over, thanks to new information sources: "*There are now a thousand sources to help us find, understand, order and enjoy more wine*".[5]

1.6 Online Wine Reviews Value and Challenges for Research and Marketing

Online wine reviews represent a significant manifestation of electronic word-of-mouth (eWOM), acting as a critical information source that consumers consult during their decision-making processes (McCannon, 2020). For both academic research and marketing practice, analysing these "free comments" offers immense value by providing large-scale, unsolicited data on consumer perceptions, preferences, and the language used to articulate sensory experiences. Researchers can leverage these data to empirically test and extend theories of consumer behaviour related to online influence, trust dynamics in digital environments, and how information shapes preferences. Marketers can benefit by gaining insights directly relevant to brand-building theories and relationship marketing, understanding consumer engagement with products and brands as expressed in their own voice, informing strategy from

[4] i.e., close the mouthpiece

[5] https://www.gq.com/story/20-new-rules-of-drinking-wine (accessed 21.04.2025)

product development to communication based on authentic market signals derived from this vast big data source.

Despite this potential, analysing online wine reviews poses complex methodological challenges, largely due to the inherently subjective nature of the data. The unstructured open-ended, free-text format of reviewers' comments necessitates specialized computational techniques, as traditional quantitative methods struggle to capture the nuance and context embedded in free text (Yang et al., 2022). Extracting meaningful insights from such data requires the use of advanced computational techniques, including Natural Language Processing (NLP), text mining, and sentiment analysis. These approaches aim to model theoretical constructs from linguistics and sentiment theory in a computationally tractable way. A particularly complex task is bridging the gap between subjective sensory experience—a central concern in sensory science—and its often metaphorical and highly variable verbal expression in user-generated reviews. Addressing this challenge demands rigorous, scalable methodologies capable of interpreting the subtleties of human language (Valente et al., 2018). Furthermore, the inherent characteristics of user-generated content introduce critical data quality issues and necessitate careful consideration of data ethics. Online reviews are prone to various biases, including self-selection bias (aligning with challenges in sampling theory) and strategic manipulation, which impact the data's representativeness and reliability, issues often discussed within eWOM theory regarding source credibility (Verma & Dewani, 2021). The fundamental subjectivity of taste, a core concept in sensory science measurement, means interpreting reviews requires acknowledging the personal nature of the feedback. Moreover, the emerging capability of AI to generate synthetic data introduces a new challenge to data veracity, requiring novel detection methods and posing ethical questions about authenticity and trust in online discourse—areas increasingly relevant to computational social science and data governance frameworks (Petrescu et al., 2023).

2 Wineinformatics: Applying Text Data Analytics Methods to Online Wine Reviews

2.1 Wineinformatics Definition

The advent of the digital age has ushered in a new era for the wine industry, marked by the proliferation of data across its entire value chain, from vineyard management and production logistics to distribution, sales, and, crucially, consumer feedback. This data revolution has given rise to Wineinformatics, an emerging interdisciplinary field that applies principles and techniques from informatics, computer science, statistics, and data science to address complex problems within the wine domain (Tian et al., 2022). Wineinformatics seeks to transform raw data into actionable knowledge, supporting research, decision-making, and innovation across various facets of the wine world (Katumullage et al., 2022).

2.2 Online Wine Reviews as a Core Data Stream for Wineinformatics

Within the expansive scope of Wineinformatics, traditional data sources like weather patterns, soil composition, fermentation parameters, and sales figures have long been central (Dong et al., 2020). However, the rise of Web 2.0 and e-commerce platforms has introduced a novel and particularly insightful data stream: online wine reviews. These reviews, a prime example of User Generated Content (UGC), represent the unsolicited opinions, sensory descriptions, and emotional responses of consumers documented in text format. Unlike controlled sensory panel data, which is structured and gathered under experimental conditions, online reviews are generated organically in real-world contexts and at an unprecedented scale. They provide a direct, voluminous, and often nuanced reflection of consumer perception and experience, making them a crucial component for a holistic understanding of the wine market from a Wineinformatics perspective.

Given that online wine reviews are predominantly composed of unstructured text, text data analytics emerges as a foundational pillar within Wineinformatics for harnessing this specific data stream. Whereas other areas of Wineinformatics may focus on numerical or categorical data, the analysis of online reviews fundamentally relies on methodologies capable of processing, interpreting, and quantifying human language (Chen et al., 2018). Text data analytics acts as a crucial intermediary, transforming the raw, qualitative, and unstructured nature of review text into structured, analysable data capable of generating meaningful insights. It provides the computational means to convert subjective expressions about aroma, flavour, texture, and overall experience into features that can be studied at scale, thus operationalizing this form of big data for scientific and commercial purposes (Calderon-Monge et al., 2025).

2.3 Overview of Key Text Analytics Approaches in Wineinformatics

Unstructured wine-review data presents analysts with significant challenges, including extreme lexical diversity, frequent use of metaphor, and substantial noise. A single database, such as *CellarTracker*, already holds 7.5 million freely written, user-generated tasting notes—each varying in length, orthography, and sensory vocabulary—and submitted without standardized prompts (CellarTracker, 2025). Lexical diversity is so extensive that the 68 million-word corpus analysed by (Hörberg et al., 2025) contained fewer than one in ten words overlapped across the three evaluated domains—food, perfume, and wine—highlighting the richness of such databases. Kwartler (2017) provides a comprehensive overview of the workflow from data acquisition to final analysis. Preprocessing is both extensive and critical: it involves lowercasing text; removing punctuation (e.g., exclamation marks), tabs, extra spaces, numerals, context-irrelevant words, and certain affixes; and spell-checking to correct frequent typos and other user-generated errors (see Chap. 5). Only after this rigorous cleaning can the data be analysed using the chosen analytical technique (Kwartler, 2017).

Analysing online wine reviews within the field of Wineinformatics involves applying several key text analytics approaches. NLP techniques are fundamental; they enable computers to process and understand the linguistic structure of review text, allowing for tasks like identifying key terms or parsing sentences (Yang et al., 2022). Broadly defined, NLP is a quantitative approach to analysing human language, encompassing its understanding, learning, and generation—with recent research placing particular emphasis on language generation (Hirschberg & Manning, 2015). Text mining builds upon NLP to discover patterns, themes, and associations across collections of reviews, revealing recurring ideas or relationships between concepts (Croijmans et al., 2020; Ferreira et al., 2023; Windsor et al., 2023). Sentiment analysis is a crucial method for quantifying the subjective opinions and emotions expressed (see Chap. 8), enabling the assessment of overall sentiment toward a wine or its specific attributes. Furthermore, advanced Artificial Intelligence (AI) is increasingly used to uncover more complex insights, such as predicting wine characteristics from review text or identifying nuanced patterns in consumer language. These approaches collectively provide the computational capability required to extract valuable information from the vast digital discourse on wine quality.

Applying these text analytics methods to online wine reviews generates valuable insights pertinent to various domains. For sensory science, this analysis reveals how consumers verbally describe sensory attributes, offering a complement to controlled panel data and highlighting consumer vocabulary. In consumer science and marketing, insights are gained into the drivers of consumer preference and purchase decisions, often by correlating textual features with ratings or sales data. Understanding market positioning and identifying competitive landscapes also becomes possible by analysing the textual profiles of different wines or brands as perceived by consumers. These analyses contribute data-driven evidence to research questions and provide actionable intelligence for strategic decision-making in the wine industry.

2.4 *The Role of Sentiment Analysis in Wineinformatics*

Sentiment analysis—also called opinion mining—is the branch of Natural Language Processing that automatically detects and quantifies the affective stance (positive, negative, or neutral, and often finer-grained emotions) that a text expresses toward a given target (Taboada, 2016). Although sentiment analysis may seem deceptively simple (it is often reduced to classifying words as positive or negative) it faces significant challenges. These include the cultural and demographic specificity of the data sample, as well as the ambiguity introduced by authors who employ varied emotional frameworks and classifications (Kwartler, 2017). One widely adopted framework is Plutchik's Wheel of Emotions (2001), a psychoevolutionary model that conceptualizes emotions as adaptive responses shaped by natural selection. At its core are eight primary, bipolar emotions arranged on a color-coded "wheel": joy–sadness, anger–fear, trust–disgust, and surprise–anticipation. Each primary emotion has varying intensities (e.g., serenity, joy, and ecstasy), radiating outward from the center to indicate increasing arousal and inward to indicate decreasing arousal. Adjacent emotions on the wheel can blend to form secondary "dyadic" emotions (e.g., joy and trust form love).

However, the sentiment expressed in wine reviews often falls outside the scope of models like Plutchik's Wheel of Emotions (Plutchik, 2001). Rather than encompassing a broad range of emotional polarities such as anger or fear, wine reviews typically reflect varying degrees of positive or negative sentiment. To better capture this domain-specific nuance, Yang and Cao (2025) developed an interpretable sentiment analysis model, which is capable to providing word-level sentiment measure. Their findings revealed that certain words—neutral in everyday language—can carry strong sentiment in the context of wine. For example, words like *"carpet"* and *"fabric"* are associated with a velvety texture, often signalling high-quality wines (e.g., *"...sail across a carpet of superfine tannins lingering on the spicy finish..."*). Conversely, descriptors such as *"quick"* or *"breezy"* (e.g., *"...light with modest citrus and green apple hints, clean quick finish..."*) imply a lack of complexity and are generally viewed negatively. These insights not only enhance our understanding of wine-specific sentiment but also underscore the importance of domain-adapted sentiment analysis—highlighting how different fields may require specialized vocabularies and interpretive frameworks. This knowledge is particularly valuable for generating realistic wine reviews using language models.

3 Research and Marketing Applications

The analysis of this user-generated content yields a diverse range of valuable insights, serving both academic research in sensory and consumer science and informing strategic decisions in wine marketing. These applications collectively demonstrate the significant potential of leveraging online wine review text data to enhance our understanding of consumer behaviour, market dynamics, and product perception (Vanhala et al., 2020).

3.1 Understanding Consumer Perception and Sensory Language

Analysing the text of online wine reviews provides rich data on how consumers perceive and articulate their experiences, offering valuable insights for sensory and consumer science. Text analysis allows researchers to systematically identify and map the terms consumers commonly use to describe sensory attributes, revealing prevalent lexicons that may differ from professional terminology (Bianchi et al., 2021; Croijmans et al., 2020; Ferreira et al., 2023). It also facilitates the comparison of language used by consumers versus professional critics and helps analyses the emotional and experiential language beyond objective descriptors. Understanding the nuances of this language is crucial; as mentioned before, research by Yang and Cao (2025) highlighted that the vocabulary used to express sentiment in this context can have unique meanings distinct from everyday language. This underscores the importance of domain-specific approaches when interpreting consumer sentiment in wine reviews.

3.2 Identifying Drivers of Preference and Purchase Decisions

Online review analysis helps uncover the factors, both sensory and non-sensory, that influence consumer preference and ultimately drive purchase decisions. By correlating textual features—such as specific descriptors, mentioned topics, or overall sentiment—with associated ratings or purchasing behaviour, researchers can identify attributes most strongly linked to consumer satisfaction or dissatisfaction (Fu et al., 2022). Analysis helps pinpoint which sensory descriptors or perceived wine characteristics consistently appear in highly positive or negative reviews, offering insights into critical success or failure factors from a consumer perspective. The discussion of non-sensory factors like price, origin, or brand reputation within reviews can also be analysed to understand their influence on expressed opinion and perceived value.

3.3 Analysing Market Positioning and Competition

Online reviews also offer a valuable source of market intelligence, providing insights into how wines and brands are positioned and perceived relative to their competitors based on their communication and authentic consumer feedback (Fagioli et al., 2022). By comparing the review content and sentiment profiles of different wines or brands in the same category, businesses can benchmark their performance and identify areas of perceived strength or weakness. Analysing review discussions across a market segment can also help identify market niches or emerging consumer interests that are not being fully addressed by existing products. Furthermore, monitoring changes in how a specific brand is discussed over time allows businesses to track shifts in brand perception and manage reputation proactively.

3.4 Informing Product Development and Marketing Strategy

The insights derived from analysing online wine reviews are directly actionable and can inform critical strategic decisions in product development and marketing. Negative feedback or recurring issues mentioned in reviews can guide product improvement efforts, while analysis of desired attributes or unmet needs can inspire new product concepts. Using the language and themes that resonate most positively with consumers in reviews can help craft more effective marketing messages and product descriptions.

Analysing consumer preferences and taste profiles expressed in review data is crucial for powering personalized experiences on e-commerce platforms, including building sophisticated recommendation engines (Müller, 2019). For example, according to Cao (2025), the wine marketplace *Vivino* launched an AI-powered recommendation app *Match for You* in 2021, which reportedly used insights from its extensive database of over 200 million wine ratings and reviews to build unique taste profiles for users after they have rated at least five wines, providing customized match scores across its large wine database. Similarly, *DRINKS*, a company focused on alcohol technology, introduced an AI recommendation system called *DRINKS Recommend*, built on their PAIR (Predictive AI Retailing) models. This model incorporates various factors, including wine prices, regions, varietals, tasting notes, product descriptions, label features, and customer perception derived from data, to generate tailored recommendations for online merchants and their customers. It was reported to significantly increase sales for major online wine retailers where it was deployed. These examples illustrate the substantial commercial value achievable by leveraging online review data for advanced recommendation systems (Liu & Wang, 2025).

The computational analysis of online wine review text offers a wide array of applications for both academic research and marketing practice. From gaining nuanced insights into consumer sensory language and preference drivers to informing competitive strategy and powering sophisticated recommendation systems, the value derived from leveraging this big data source via Wineinformatics methodologies is substantial (Gupta & Katarya, 2024). Achieving these insights reliably, however, depends critically on the robust collection, pre-processing, and analytical techniques employed. It is also important to acknowledge the inherent challenges in working with this type of data, including issues related to data veracity, such as the capability of AI algorithms to generate synthetic wine reviews without requiring actual tasting (Carlson et al., 2023).

4 When AI Writes Reviews: Future Trends and Predictions

4.1 The Problematic of AI-Generated Content

An emerging and significant challenge in analysing online reviews—beyond traditional biases—is the growing use of AI tools to generate content. As people increasingly rely on AI to produce text that closely mimics human writing, serious concerns arise about the authenticity of user-generated content. Studies have shown that AI can now generate convincing wine reviews without ever uncorking a bottle. For example, Carlson et al. (2023) demonstrated this by training a deep learning model on a vast dataset of reviews, including those published by *Wine Enthusiast* between 1996 and 2016. Their study mapped review text to specific wine characteristics—such as winery name and alcohol content—allowing the AI to internalize and reproduce the typical structure and stylistic nuances of human-authored reviews. While such research initially aimed to understand the linguistic patterns of reviews, it highlights AI's capacity to create synthetic content. This capability has profound implications for data veracity and trust, as malicious actors could potentially flood platforms with fabricated reviews, skewing analytical results for researchers and misleading recommendation systems for consumers. Detecting AI-generated content remains a significant challenge,[6] necessitating ongoing research into sophisticated detection methods that can distinguish subtle linguistic and behavioural cues that signal synthetic origins (Ghiurău & Popescu, 2024; Tiwari et al., 2024).

4.2 Future Directions in Text Analysis and Wineinformatics

Despite the challenges, the field of text analysis within Wineinformatics is poised for significant advancements driven by ongoing AI research. Future directions include the development of more sophisticated AI/NLP models capable of better understanding context, identifying subtle emotions, sarcasm, and complex reasoning in wine reviews. The integration of multimodal data, combining review text with analysis of accompanying images or videos (e.g., analysing a photo of a wine label alongside the text of its review), holds potential for deeper, more comprehensive insights into the consumption experience (Rocuant Fuentes et al., 2025). Bender et al. (2024) confirm this potential, showing that fusing free-text reviews, wine-label images, and Napping (a sensory annotation method that asks panellists to place samples on a sheet of paper according to perceived similarities between the samples, providing a quick visual map of the sensory relationships (for a review, see István & Kókai, 2021)) derived sensory annotations yields the strongest performance across tasks such as predicting country of origin, grape variety and region.

---

[6] OpenAI released and then closed down again an AI generated text detector, due to the inaccuracy of it (https://openai.com/index/new-ai-classifier-for-indicating-ai-written-text/)

Furthermore, as AI becomes increasingly integrated into various domains, there is a growing emphasis on ethical AI and model explainability. This shift prioritizes the development of systems that are not only powerful but also transparent in their decision-making processes, incorporating techniques specifically designed to detect and mitigate bias in review analysis.

4.3 Predicting the Future of Online Wine Discourse and the Wine Industry

Looking ahead, the landscape of online wine discourse and its impact on the industry is likely to evolve in several ways. As review platforms evolve, we may see the introduction of features such as verified purchase indicators, more structured feedback mechanisms, and proactive safeguards against AI-generated content. In response, the wine industry is likely to deepen its integration of review analysis into operational workflows while also prioritizing strategies to foster consumer trust amid growing concerns over data authenticity. This could involve increased transparency around review handling processes and the inner workings of recommendation systems. These shifts are poised to reshape market dynamics, amplifying the influence of platforms that effectively uphold review integrity and making consumer trust in online information more critical than ever.

Navigating this evolving future requires proactive strategies from both researchers and industry professionals. For researchers, this involves continuing to develop and refine methodologies that account for data biases, subjectivity, and the threat of synthetic content, potentially focusing on novel validation techniques and interdisciplinary approaches within Wineinformatics. For the industry, it means embracing data-driven insights while exercising critical judgment regarding data quality, investing in ethical AI development, and potentially collaborating with platforms to ensure data integrity. Ultimately, the ability to critically collect, analyse, and interpret the evolving landscape of online wine reviews will remain essential for understanding the digital consumer and succeeding in the future wine market.

4.4 How Can AI Help in Text Analysis?

Traditionally, preprocessing large text corpora and conducting sentiment analysis have been reserved for researchers with extensive programming and analytical skills. However, according to a recent report by the Stanford Institute for Human-Centered Artificial Intelligence (2025), AI now surpasses humans in image classification, visual reasoning, medium-level reading comprehension, English language understanding, multitask language understanding, competition-level mathematics, PhD-level science questions. Humans retain only a slight advantage in multimodal reasoning, and that margin is narrow.

To test AI's feasibility for sentiment analysis, we used R version 4.4.0 (R Core Team, 2024) and the *sentimentr* package version 2.9.0 (Rinker, 2021). We applied these tools to the WineSensed dataset created by Bender et al. (2024), which comprises approximately 897,000 wine-label images paired with 824,000 text reviews harvested from the *Vivino* platform. About 350,000 unique bottlings are annotated with year, region, rating, alcohol percentage, price, and grape composition; a subset of 256 human tastings supplies more than 5000 pairwise flavour-similarity judgments, making the dataset suitable for multimodal and sensory–language modelling (Bender et al., 2024). After filtering out empty rows, the corpus yielded roughly 25,000 reviews. We computed sentiment scores using the *sentimentr* package and appended them as a new column in our data frame. Additionally, we employed the *ChatGPT o3* model (OpenAI, 2024) with the following prompt:

"*In the attached data, please perform sentiment analysis on the 'review' column. Then scale sentiment scores such that −100 corresponds to the most negative affect and +100 to the most positive affect. Add these scores as a new column named 'gpt_o3_sentiment.'*".

We then calculated the Pearson correlation between the two sentiment columns and found a moderately strong association ($r = .47$, $p < .001$). Although AI- and human-generated results via the designated package for R are not perfectly aligned, given the variety of sentiment analysis methods and nuances discussed in this chapter, these techniques remain viable for nonexperts. It is critical to highlight that AI developments often follow an exponential trajectory rather than a linear one; therefore, it will not be surprising if, by the time this chapter is published, results from newer AI models align more closely with those generated by established statistical packages (Stanford Institute for Human-Centered Artificial Intelligence, 2025).

5 Conclusion

This chapter examines the significant transformation in how wine quality is assessed and communicated, tracing its progression from traditional expert-driven evaluations to the modern landscape dominated by vast volumes of user-generated content in the form of online wine reviews. We have positioned the analysis of this unstructured text data as a vital component of Wineinformatics, an interdisciplinary field leveraging computational methods to derive insights for the wine domain. The chapter highlights how core text analysis methods, including Natural Language Processing (NLP), sentiment analysis, and Artificial Intelligence (AI), are employed within this framework to process, interpret, and derive meaning from online reviews.

The application of these methods yields a diverse range of valuable insights with significant research and marketing applications. These range from understanding consumer sensory language and identifying drivers of preference to analysing market positioning and informing strategic decisions in product development and marketing, including powering AI Recommendation Systems. The chapter has demonstrated how analysing online wine reviews provides unprecedented opportunities to study consumer behaviour and market dynamics within sensory and consumer Science.

Whereas the potential of analysing online wine reviews is substantial, this is a dynamic field with inherent challenges, including the complexities of interpreting subjective language, addressing data biases, and navigating the emerging threat posed by AI-generated synthetic content. Nevertheless, the capability to systematically decode the collective digital voice of wine consumers remains crucial. As AI and digital platforms continue to evolve, the critical collection, rigorous analysis, and ethical interpretation of online wine reviews will be increasingly essential for both advancing academic understanding and informing successful strategies in the global wine market. Using one of the wine review datasets, this chapter demonstrates a moderately strong correlation between sentiment scores generated by a newer AI model and those produced by researchers. The future of understanding wine quality is inextricably linked to understanding the digital consumer.

References

Alonso Ugaglia, A., Cardebat, J.-M., & Jiao, L. (2019). The French wine industry. In A. Alonso Ugaglia, J.-M. Cardebat, & A. Corsi (Eds.), *The Palgrave handbook of wine industry economics* (pp. 17–46). Springer International Publishing. https://doi.org/10.1007/978-3-319-98633-3_2

Amerine, M., Roessler, E., & Filipello, F. (1959). Modern sensory methods of evaluating wine. *Hilgardia, 28*(18), 477–567.

Anderson, K. (2004). *The world's wine markets: Globalization at work*. Edward Elgar Publishing.

Bender, T., Sørensen, S. M., Kashani, A., Hjorleifsson, K. E., Hyldig, G., Hauberg, S., Belongie, S., & Warburg, F. (2024). Learning to taste: A multimodal wine dataset (arXiv:2308.16900). *arXiv*. https://doi.org/10.48550/arXiv.2308.16900

Bianchi, I., Branchini, E., Torquati, S., Fermani, A., Capitani, E., Barnaba, V., Savardi, U., & Burro, R. (2021). Non experts' understanding of terms frequently used by experts to describe the sensory properties of wine: An investigation based on opposites. *Food Quality and Preference, 92*, 104215. https://doi.org/10.1016/j.foodqual.2021.104215

Branco, F., Sun, M., & Villas-Boas, J. M. (2012). Optimal search for product information. *Management Science, 58*(11), 2037–2056. https://doi.org/10.1287/mnsc.1120.1535

Calderon-Monge, E., Ripollés-Matallana, V., Baruque-Zanón, B., & Porras Alfonso, S. (2025). Big data analysis of Spanish wine consumers reviews. *International Journal of Wine Business Research, 37*(1), 55–86. https://doi.org/10.1108/IJWBR-12-2023-0084

Cao, J. (2025). AI: A new and impactful player in the quality evaluation of wine. *Harvard Data Science Review, 7*(1). https://doi.org/10.1162/99608f92.be44bdad

Carlson, K., Kopalle, P. K., Riddell, A., Rockmore, D., & Vana, P. (2023). Complementing human effort in online reviews: A deep learning approach to automatic content generation and review synthesis. *International Journal of Research in Marketing, 40*(1), 54–74. https://doi.org/10.1016/j.ijresmar.2022.02.004

Carsky, M. L., Dickinson, R. A., & Canedy, C. R. (1998). The evolution of quality in consumer goods. *Journal of Macromarketing, 18*(2), 132–144. https://doi.org/10.1177/027614679801800205

Cawley, D. (2018). The power of wine language—Critics. *Labels and Sexism*. https://doi.org/10.21427/G5A7-X024

Chandra Kruse, L., & Drechsler, K. (2022). Digitalization of multisensory collective activity: The case of virtual wine tasting. *Journal of Information Technology, 37*(4), 341–358. https://doi.org/10.1177/02683962221096860

Chen, B., Velchev, V., Palmer, J., & Atkison, T. (2018). Wineinformatics: A quantitative analysis of wine reviewers. *Fermentation, 4*(4), 82. https://doi.org/10.3390/fermentation4040082

Choi, H., & Burnes, B. (2016). How consumers contribute to the development and continuity of a cultural market. *Consumption Markets & Culture, 19*(6), 576–596. https://doi.org/10.1080/10253866.2016.1172214

Conca Messina, S. A., Le Bras, S., Tedeschi, P., & Vaquero Piñeiro, M. (2019). *A history of wine in Europe, 19th to 20th centuries*. Volume I.

Croijmans, I., Hendrickx, I., Lefever, E., Majid, A., & Van Den Bosch, A. (2020). Uncovering the language of wine experts. *Natural Language Engineering, 26*(5), 511–530. https://doi.org/10.1017/S1351324919000500

Dong, Z., Guo, X., Rajana, S., & Chen, B. (2020). Understanding 21st century Bordeaux wines from wine reviews using naïve Bayes classifier. *Beverages, 6(1)*, Article 1. https://doi.org/10.3390/beverages6010005

Drake, M. A., Drake, S., Bodyfelt, F., Clark, S., & Costello, M. (2009). History of sensory analysis. In S. Clark, M. Costello, M. Drake, & F. Bodyfelt (Eds.), *The sensory evaluation of dairy products* (pp. 1–6). Springer US. https://doi.org/10.1007/978-0-387-77408-4_1

Dubois, M., Georgantzis, N., & Cardebat, J.-M. (2025). External Evaluations under Quality Uncertainty: The Market for Wine Ratings: Information Dynamics in the Wine Market. *WineEconomics and Policy*. https://doi.org/10.36253/wep-16620

EnTribe. (2022, July 6). New Survey from EnTribe Emphasizes Value of User-Generated Content in Marketing Initiatives. *GlobeNewswire News Room*. https://www.globenewswire.com/news-release/2022/07/06/2475085/0/en/New-Survey-from-EnTribe-Emphasizes-Value-of-User-Generated-Content-in-Marketing-Initiatives.html

Fagioli, F. F., Gallesio, G., & Viganò, E. (2022). Wineries communication strategies. A text mining analysis. *Wine Economics and Policy, 11*(1), 73–87. https://doi.org/10.36253/wep-11305

Fernandez, J.-L. (2004). *La critique vinicole en France: Pouvoir de prescription et construction de la confiance*. Harmattan.

Ferreira, C., Robertson, J., Lam, J., & Vella, J. (2023). Expert reviews uncorked: Contrasting the differences in the language used in online reviews of white and red wine. *Journal of Wine Research, 34*(2), 140–154. https://doi.org/10.1080/09571264.2023.2190086

Flanagin, A. J., & Metzger, M. J. (2013). Trusting expert- versus user-generated ratings online: The role of information volume, valence, and consumer characteristics. *Computers in Human Behavior, 29*(4), 1626–1634. https://doi.org/10.1016/j.chb.2013.02.001

Fu, W., Choi, E.-K., & Kim, H.-S. (2022). Text mining with network analysis of online reviews and consumers' satisfaction: A case study in Busan wine bars. *Information, 13*(3), 127. https://doi.org/10.3390/info13030127

Ghiurău, D., & Popescu, D. E. (2024). Distinguishing reality from AI: Approaches for detecting synthetic content. *Computers, 14*(1), 1. https://doi.org/10.3390/computers14010001

Gupta, G., & Katarya, R. (2024). A computational approach towards food-wine recommendations. *Expert Systems with Applications, 238*, 121766. https://doi.org/10.1016/j.eswa.2023.121766

Herschkowitsch, A., & Goldstein, R. (2009). In T. Walters (Ed.), *The wine trials 2010: The world's bestselling guide to inexpensive wines, with the 150 winning wines under $15 from the latest vintages* (2nd ed.). Fearless Critic Media.

Hirschberg, J., & Manning, C. D. (2015). Advances in natural language processing. *Science, 349*(6245), 261–266. https://doi.org/10.1126/science.aaa8685

Hörberg, T., Kurfalı, M., & Olofsson, J. K. (2025). Chemosensory vocabulary in wine, perfume and food product reviews: Insights from language modeling. *Food Quality and Preference, 124*, 105357. https://doi.org/10.1016/j.foodqual.2024.105357

Hsu, G., Roberts, P. W., & Swaminathan, A. (2007). *Standards for quality and the coordinating role of critics.*, 48.

István, M., & Kókai, Z. (2021). The applicability of napping in the analysis of fermented foods and beverages—A review. *Progress in Agricultural Engineering Sciences, 17*(S1), 119–125. https://doi.org/10.1556/446.2021.30014

Jackson, R. S. (2008). Introduction. In R. S. Jackson (Ed.), *Wine science* (3rd ed., pp. 1–14). Academic. https://doi.org/10.1016/B978-012373646-8.50004-4

Jackson, R. S. (2017). Nature and Origins of Wine Quality (pp. 337–370). *Academic Press*. https://doi.org/10.1016/B978-0-12-801813-2.00008-2

Jamerson, H. (2009). Intoxicators, educators, and gatekeepers: The enactment of symbolic boundaries in Napa Valley wineries. *Poetics, 37*(4), 383–398. https://doi.org/10.1016/j.poetic.2009.06.001

James, A. (2018). How Robert Parker's 90+ and Ann Noble's aroma wheel changed the discourse of wine tasting notes. *ILCEA. Revue de l'Institut Des Langues et Cultures d'Europe, Amérique, Afrique, Asie et Australie, 31*, Article 31. https://doi.org/10.4000/ilcea.4681

Jenster, P. V., & Jenster, L. (1993). The European wine industry. *International Journal of Wine Marketing, 5*(1), 30–73. https://doi.org/10.1108/eb008610

Katumullage, D., Yang, C., Barth, J., & Cao, J. (2022). Using neural network models for wine review classification. *Journal of Wine Economics, 17*(1), 27–41. https://doi.org/10.1017/jwe.2022.2

Kotonya, N., De Cristofaro, P., & De Cristofaro, E. (2018). Of wines and reviews: Measuring and modeling the Vivino wine social network. *IEEE/ACM International Conference on Advances in Social Networks Analysis and Mining (ASONAM), 2018*, 387–392. https://doi.org/10.1109/ASONAM.2018.8508776

Kwartler, T. (2017). *Text mining in practice with R* (1st ed.). Wiley. https://doi.org/10.1002/9781119282105

Lahne, J. (2016). Sensory science, the food industry, and the objectification of taste. *Anthropology of food, 10*, Article 10. https://doi.org/10.4000/aof.7956

Langstaff, S. A. (2010). Sensory quality control in the wine industry. In D. Kilcast (Ed.), *Sensory analysis for food and beverage quality control* (pp. 236–261). Woodhead Publishing. https://doi.org/10.1533/9781845699512.3.236

Lecat, B., Brouard, J., & Chapuis, C. (2017). Fraud and counterfeit wines in France: An overview and perspectives. *British Food Journal, 119*(1), 84–104. https://doi.org/10.1108/BFJ-09-2016-0398

Liu, W., & Wang, Y. (2025). Evaluating trust in recommender systems: A user study on the impacts of explanations, agency attribution, and product types. *International Journal of Human–Computer Interaction, 41*(2), 1280–1292. https://doi.org/10.1080/10447318.2024.2313921

Lynch, J. G., Jr., & Ariely, D. (2000). Wine online: Search costs affect competition on price, quality, and distribution. *Marketing Science, 19(1)*, 83–103. Scopus. https://doi.org/10.1287/mksc.19.1.83.15183

Marsh, E., & Rajaram, S. (2019). The digital expansion of the mind: Implications of internet usage for memory and cognition. *Journal of Applied Research in Memory and Cognition, 8*. https://doi.org/10.1016/j.jarmac.2018.11.001

Matz, S. C., Kosinski, M., Nave, G., & Stillwell, D. J. (2017). Psychological targeting as an effective approach to digital mass persuasion. *Proceedings of the National Academy of Sciences, 114*(48), 12714–12719. https://doi.org/10.1073/pnas.1710966114

McCannon, B. C. (2020). Wine descriptions provide information: A text analysis. *Journal of Wine Economics, 15*(1), 71–94. https://doi.org/10.1017/jwe.2020.3

Morrison, A., & Rabellotti, R. (2017). Gradual catch up and enduring leadership in the global wine industry. *Research Policy, 46*(2), 417–430. https://doi.org/10.1016/j.respol.2016.09.007

Müller, O. (2019). Structuring unstructured data—or: How machine learning can make you a wine sommelier. In K. Bergener, M. Räckers, & A. Stein (Eds.), *The art of structuring* (pp. 307–316). Springer International Publishing. https://doi.org/10.1007/978-3-030-06234-7_29

Petrescu, M., Ajjan, H., & Harrison, D. (2023). The role of AI agents in spreading and detecting fake online reviews: A systematic review: An abstract. In B. Jochims & J. Allen (Eds.), *Optimistic marketing in challenging times: Serving ever-shifting customer needs* (pp. 333–334). Springer Nature Switzerland. https://doi.org/10.1007/978-3-031-24687-6_137

Plutchik, R. (2001). The nature of emotions: Human emotions have deep evolutionary roots, a fact that may explain their complexity and provide tools for clinical practice. *American Scientist, 89*(4), 344–350.

Rocuant Fuentes, C., Rita, P., & Antonio, N. (2025). Assessing the nexus of label content and consumer reviews: Insights from premium wines on Vivino. *International Journal of Wine Business Research.* https://doi.org/10.1108/IJWBR-08-2024-0044

Rydning, D. R. J. G. J., Reinsel, J., & Gantz, J. (2018). The digitization of the world from edge to core. *Framingham: International Data Corporation*, 16, 1–28.

Schartner, M., Beck, J. M., Laboyrie, J., Riquier, L., Marchand, S., & Pouget, A. (2023). Predicting Bordeaux red wine origins and vintages from raw gas chromatograms. *Communications Chemistry, 6*(1), 247. https://doi.org/10.1038/s42004-023-01051-9

Shapin, S. (2012). The tastes of wine: Towards a cultural history. *Rivista Di Estetica, 51*, Article 51. https://doi.org/10.4000/estetica.1395

Simpson, J. (2004). Selling to reluctant drinkers: The British wine market, 1860–1914. *The Economic History Review, 57(1)*, 80–108. JSTOR.

Stone, H., & Sidel, J. L. (2003). Sensory evaluation | descriptive analysis. In B. Caballero (Ed.), *Encyclopedia of food sciences and nutrition* (7th ed., pp. 5152–5161). Academic. https://doi.org/10.1016/B0-12-227055-X/01065-8

Storchmann, K. (2012). Wine economics. *Journal of Wine Economics, 7*(1), 1–33. https://doi.org/10.1017/jwe.2012.8

Taboada, M. (2016). Sentiment analysis: An overview from linguistics. *Annual Review of Linguistics, 2(2)*, 325–347. https://doi.org/10.1146/annurev-linguistics-011415-040518

Tian, Q., Whiting, B., & Chen, B. (2022). Wineinformatics: Comparing and combining SVM models built by wine reviews from Robert Parker and Wine spectator for 95 + point wine prediction. *Fermentation, 8*(4), Article 4. https://doi.org/10.3390/fermentation8040164

Tiwari, S., Sharma, R., Sikarwar, R. S., Dubey, G. P., Bajpai, N., & Singhatiya, S. (2024). Detecting AI generated content: A study of methods and applications. In S. Kumar, S. Hiranwal, R. Garg, & S. D. Purohit (Eds.), *Proceedings of International Conference on Communication and Computational Technologies* (Vol. 1121, pp. 161–176). Springer Nature. https://doi.org/10.1007/978-981-97-7423-4_13

Valente, C. C., Bauer, F. F., Venter, F., Watson, B., & Nieuwoudt, H. H. (2018). Modelling the sensory space of varietal wines: Mining of large, unstructured text data and visualisation of style patterns. *Scientific Reports, 8*(1), 4987. https://doi.org/10.1038/s41598-018-23347-w

Vanhala, M., Lu, C., Peltonen, J., Sundqvist, S., Nummenmaa, J., & Järvelin, K. (2020). The usage of large data sets in online consumer behaviour: A bibliometric and computational text-mining–driven analysis of previous research. *Journal of Business Research, 106*, 46–59. https://doi.org/10.1016/j.jbusres.2019.09.009

Verma, D., & Dewani, P. P. (2021). eWOM credibility: A comprehensive framework and literature review. *Online Information Review, 45*(3), 481–500. https://doi.org/10.1108/OIR-06-2020-0263

Windsor, L. C., Tatara, J. H., Peters, C. B., Kronsted, C., & Windsor, A. (2023). The language of wine reviews. *Journal of Wine Research, 34*(2), 81–100. https://doi.org/10.1080/09571264.2023.2205116

Wittwer, G., Berger, N., & Anderson, K. (2003). A model of the world's wine markets. *Economic Modelling, 20*(3), 487–506. https://doi.org/10.1016/S0264-9993(01)00095-5

Yang, C., & Cao, J. (2025). Interpretable sentiment analysis using the attention-based multiple instance classification model: An application to wine reviews. *Harvard Data Science Review, 7*(2). https://doi.org/10.1162/99608f92.caab9466

Yang, C., Barth, J., Katumullage, D., & Cao, J. (2022). Wine review descriptors as quality predictors: Evidence from language processing techniques. *Journal of Wine Economics, 17*(1), 64–80. https://doi.org/10.1017/jwe.2022.3

Yuan, G., Han, J., Wang, Y., Liang, H., & Li, G. (2019). The product demand model driven by consumer's information perception and quality perception. *Physica A: Statistical Mechanics and its Applications, 535*, 122352. https://doi.org/10.1016/j.physa.2019.122352

Chapter 10
Co-occurrence Network Analysis in Consumer Studies

Amparo Tárrega

1 Introduction

Co-occurrence network analysis is an increasingly popular method for analysing text data, particularly for uncovering hidden patterns, relationships, and structures within large bodies of text. These networks enable researchers to visualize how frequently terms appear together and how they relate to one another within a corpus. By presenting this information in a graphical format, co-occurrence networks help to simplify complex data and support deeper insights into the content and thematic connections embedded in text (Ruiz & Barnett, 2015).

Co-occurrence analysis was initially proposed as a tool in bibliometric studies to examine scientific literature and citation patterns (van Eck & Waltman, 2010). However, its utility extends now beyond to other fields in natural (environment, microbiology, or biomedicine) and social sciences (psychology, education, marketing, and consumer). In marketing and consumer behaviour research, co-occurrence networks are applied to analyse user-generated content to identify patterns about their needs, motivations, and experiences.

In recent years, consumer science has increasingly focused on understanding consumer behaviour from a more ecologically valid perspective. Researchers and practitioners have shifted from the traditional structured methods of measuring consumer responses to exploring their freely generated opinions. Techniques such as open-ended questions, word association, and discussion sessions are now commonly used to capture insights that better reflect genuine consumer experiences and concerns, with fewer researcher-imposed constraints. Additionally, the rise of digital communication platforms has resulted in an unprecedented volume of consumer

A. Tárrega (✉)
Instituto de Agroquímica y Tecnología de los Alimentos (IATA-CSIC), Paterna, Spain
e-mail: atarrega@iata.csic.es

M. Visalli, M. V. Galmarini (eds.), *Collection and Analysis of Text Data in Sensory and Consumer Science*, https://doi.org/10.1007/978-3-032-04851-6_10

data, including online reviews, social media posts, blogs, and more, giving researchers access to spontaneous, naturally occurring consumer expressions (Hutchings et al., 2023). These data sources, typically in the form of unstructured text, require processing to extract meaningful insights (see Chap. 5). For this, text mining techniques (sentiment, topic modelling or co-occurrence networks) which allow researchers to automatically and systematically identify patterns, sentiments, and themes from unstructured data, are gaining increasing interest. In particular, co-occurrence networks have proven valuable in exploring term relationships and thematic trends across various contexts when applied to different types of consumer narratives, such as social media content, interview transcripts, and survey responses.

This chapter offers an exploration of the use of co-occurrence networks in consumer studies, including theoretical insights and practical applications. It aims to provide readers with the knowledge needed to construct and interpret these networks and effectively combine them with other techniques to extract meaningful insights from text consumer-generated data from different sources.

2 Fundamentals of Co-occurrence Network Analysis

The fundamental principles behind co-occurrence network analysis are identifying and visualizing relationships between terms. The different software tools that implement this analysis like *Gephi* (Gephi Consortium, France), *KH Coder* (Higuchi, 2001), and *VOSviewer* (van Eck & Waltman, 2007) follow these same principles but they use different metrics and algorithms for data processing, and visualization. Additionally, R and Python programming environments offer customizable frameworks with packages and libraries for text analysis and constructing co-occurrence networks.

2.1 Identifying Relationships

The first step is the generation of a co-occurrence matrix, where the frequency of co-appearance between each pair of terms is calculated. From this, a similarity or association matrix is derived. In *VOSviewer*, the association strength between two terms (i and j) is computed as the number of times they co-occur divided by the product of their individual total co-occurrences (van Eck & Waltman, 2007). In Gephi, cosine similarity or Jaccard similarity metrics are used to determine the strength of the connection (edge weight) between terms (Bastian et al., 2009). *KH Coder* (Higuchi, 2001, 2016) calculates association between terms using Pointwise Mutual Information (PMI, Church & Hanks, 1990).

2.2 *Visualizing Relationships*

Once the matrix is obtained, the following step is mapping the terms in a two-dimensional space. For this, the nodes, edges and weights are defined based on the similarity or association strength. The position of terms and the distance between them is optimized through different algorithms so that terms that are strongly associated are positioned closer together, and terms with weaker associations are placed farther apart. Algorithms based in Multidimensional Scaling (MDS, van Eck & Waltman, 2007) seek to minimize the weighted sum of squared Euclidean distances between all pairs of items, assigning greater weight to pairs with higher similarity. In the obtained map the physical distance between any two terms reflects their level of co-occurrence: the closer they are, the more frequently they co-occur. This ensures that terms with strong relationships are grouped together. In the so-called force-directed layout, the Fruchterman-Reingold algorithm iteratively adjusts the positions of the nodes (terms) by applying forces. It seeks to minimize the system's energy, pulling related nodes closer together while pushing unrelated nodes further apart (Hansen et al., 2020).

2.3 *Normalization and Final Layout Adjustment*

The final step involves transforming the map through translation, rotation, and mirroring to ensure that the resulting visualization is consistent and reproducible, regardless of variations in the optimization process. This standardization allows researchers to compare maps across different datasets or iterations reliably.

2.4 *Community Detection Algorithms*

Community detection is commonly integrated with co-occurrence network analysis to visualize and identify clusters of terms that are more densely connected between them than to the rest of the network. A wide range of strategies and algorithms are available for clustering text data.

VOSviewer software integrates network layout and clustering simultaneously. The algorithm uses co-occurrence strength between items to both position nodes in the visualization space and assign them to clusters. It minimizes a cost function that considers both the attraction between strongly connected nodes and repulsion between weakly connected or unconnected nodes (Van Eck & Waltman, 2010).

Algorithms based on modularity, such as Louvain (Blondel et al., 2008) are widely used. These aim to maximize the modularity value of a partition, by comparing the observed density of connections within communities to the expected density under a random distribution of links. They are especially suitable for analysing large-scale networks due to their computational efficiency and ability to reveal clear thematic groupings. Louvain is the default algorithm in *Gephi*, an open-source software for network analysis and visualization.

Walktrap is a community detection algorithm based on random walks. A random walk is a mathematical process that consists of taking successive steps from one node to another along the edges of a network, choosing the next step at random from the available neighbours. The algorithm assumes that short random walks are more likely to remain within the same community, and uses this principle to identify groups of nodes that are densely connected. This method is suitable for small to mid-sized networks and is used in tools like *KH Coder* (Pons & Latapy, 2006).

Tools in Python (e.g., *NetworkX*) or R (e.g., *igraph*) are also available to implement these clustering algorithms as well as others such as *Label Propagation*, which identifies communities by spreading labels through the network based on node neighbourhood, and *Infomap*, which uses information theory and simulates the flow of random walks to detect modules that minimize the description length of the paths.

3 Methodology: Constructing Co-occurrence Networks

3.1 Data Sources

Co-occurrence network analyses can be applied to various types of free-text data. The following are common sources from which such textual data can be obtained, each with its own structure and contextual variables.

3.1.1 Surveys

Free comments are written answers to an open-ended question within a structured survey (see Chap. 6 of this book). The data is typically organized with each row corresponding to an individual response. In word association tasks, the text is the list of associations for a given stimuli per participant. These responses are typically collected through online surveys or paper-based forms. In addition to the textual data, surveys often include other data such as respondent code, age, gender, education level, stimulus or product, and numerical ratings.

3.1.2 Focus Groups and Individual Interviews

These data are derived from transcribed spoken responses in moderated discussions or one-on-one interviews. The data is structured by speakers, turns or session segments. Supplementary variables include speaker code, session, demographic background, moderator questions, thematic codes (if annotated), and timestamps. These allow for the nuanced interpretation of term relationships in context, and for cross-participant comparison of discourse.

3.1.3 Social Media

Social media content includes short posts, captions, comments, and hashtags shared by users on public platforms (e.g., X, Instagram, Facebook). This type of data is collected through platform APIs (Application Programming Interfaces), third-party monitoring tools, or web scraping (when permitted). APIs are interfaces provided by platforms like Twitter or Reddit, allowing analysts to access data in a controlled and consistent way for academic or commercial use. Data are organized per post or user entry and typically include variables such as post code, user handle, timestamp, hashtags, mentions, geolocation, and engagement metrics (likes, shares, comments).

3.1.4 Online Reviews

Online reviews are the feedback posted on e-commerce and service platforms such as Amazon, Airbnb or TripAdvisor. These reviews are also collected through public APIs (Yelp, TripAdvisor, or Trustpilot) or web scraping. Each review entry includes the full text and is frequently accompanied by additional variables such as user code, product or service name, numerical ratings, review date, location, language, and indicators like verified purchase status.

3.2 *Text Pre-processing*

As already established in Chap. 5 of this book, pre-processing of text is a critical step and it is also key for enhancing the quality and interpretability of the co-occurrence networks. The specific preprocessing actions applied depend on the nature of the text, the research objectives, and the language involved (which may require language-specific tools).

In consumer research, where textual data is commonly derived from user-generated content the first step is usually text cleaning, including the removal of punctuation, numbers, and special characters, normalization (e.g., converting text to lowercase), and stopword removal to eliminate high-frequency but semantically insignificant words (like "the", "is", "and").

Tokenization is then performed to split the text into meaningful units (usually words), followed by lemmatization or stemming to reduce words to their base or root forms. Specialized tools are required for some languages like Chinese (*Jieba*) or Japanese (*MeCab*) that do not use spaces between words.

Finally, consumer studies usually include grouping synonymous or semantically similar terms (e.g., "tasty" and "delicious") and filtering out low-frequency words that may add noise rather than value to the analysis.

KH Coder or *VOSviewer*, include built-in functionalities for tasks like tokenization, lemmatization, stopword removal, and morphological analysis (especially in multilingual contexts). Alternatively, preprocessing can be performed using programming environments such as R or Python, which provide greater flexibility and control. In R, commonly used packages include *tm*, *quanteda*, and *tidytext* for text cleaning, tokenization, and constructing document-feature matrices. In Python, libraries like *nltk*, *spaCy*, and *TextBlob* allow for efficient and advanced preprocessing of text.

3.3 Building and Visualizing the Network

The cleaned and tokenized terms are used to construct the co-occurrence matrix and to build the co-occurrence network. The analyst must define co-occurrence rules (e.g., sentence-level, paragraph-level) and set thresholds for inclusion. These thresholds typically include the minimum word frequency and the minimum co-occurrence count, which help filter out insignificant or coincidental associations.

In the resulting co-occurrence network or graph different visual elements provide information about the structure and importance of terms:

- Node size (label/circle): the size of each node or label corresponds directly to the frequency with which a term appears in the corpus. Larger nodes and bolder labels represent terms mentioned more frequently, making them visually prominent and indicating key concepts or focal points in the dataset.
- Edge thickness (links): the thickness of the lines connecting nodes represents the strength of the co-occurrence relationship between terms. Thicker edges indicate a higher frequency of simultaneous appearance in the text, reflecting stronger semantic or thematic connections.
- Spatial distance: the layout algorithm positions nodes such that those with stronger or more frequent co-occurrence are placed closer together, while those with weaker or infrequent connections are spaced farther apart. This spatial arrangement provides an intuitive map of relatedness, where proximity implies similarity or association.
- Colour of nodes: nodes are assigned colours based on community detection algorithms, which group terms into clusters of semantically or thematically related concepts. These coloured clusters help identify coherent subtopics or conceptual frames within the corpus, facilitating a deeper understanding of the text's structure.

To illustrate this explanation, a dataset of 40 reviews of hotel experiences was created and presented here below. Table 10.1 shows the occurrences and co-occurrences of ten terms which are represented in Fig. 10.1 where each term is visualized in the network as a circle, with its size proportional to the number of occurrences. For instance, "location" appears as the largest node due to its high frequency, while "good" is the smallest, reflecting its lower mention counts. The links connecting "location" to both "food" and "price" are equally thick, each representing six co-occurrences. Closely related terms, such as "food" and "variety", are positioned near each other and are connected by the strongest link in the network (nine co-occurrences), suggesting a strong thematic pairing.

Table 10.1 Terms, occurrence and co-occurrence with other terms

Term	Occurrence	Co-occurrences
Location	20	Food (6), price (6), clean (5), options (4), service (4), room (3), variety (3), good (3)
Food	14	Variety (9), location (6), price (6), options (6), room (4)
Price	13	Location (6), food (6), room (6), clean (5), good (4), bad (4), options (4)
Room	12	Price (6), clean (6), food (4), service (4)
Service	11	Bad (7), room (4), location (4), food (3)
Variety	10	Food (9), options (5), location (3)
Bad	8	Service (7)
Clean	7	Room (6), location (5), price (5)
Options	7	Food (6), variety (5), location (4), price (4)
Good	5	Price (4), location (3), food (3)

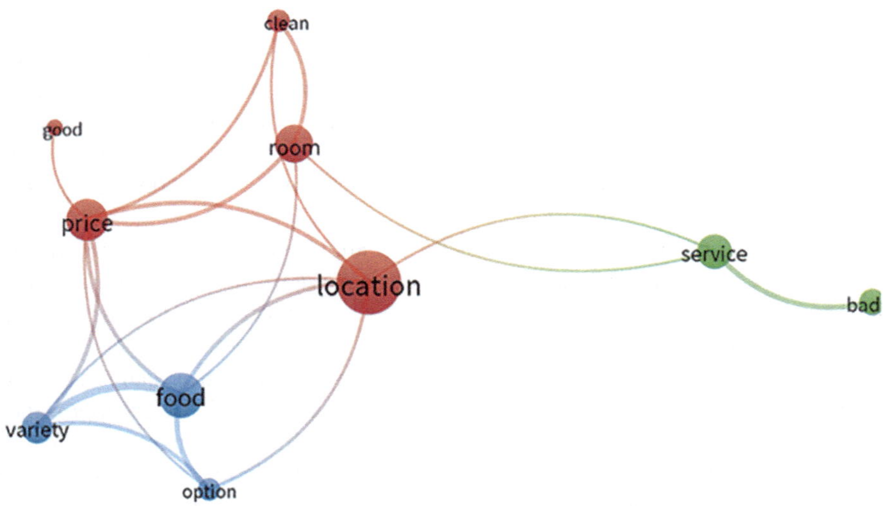

Fig. 10.1 Co-occurrence network obtained from the example data in Table 10.1

The spatial distribution of terms on the map reflects the strength and pattern of their co-occurrences. "Location" occupies a central position, indicating its connections with almost all the other terms. It is the term most mentioned in the reviews and co-occurs with all the different topics. On the contrary, the terms located on the periphery reflect weaker connections to most other terms and only have strong links to the specific terms positioned nearby. For instance, "variety" is only co-mentioned with "food", "location", and "options", and "bad" is only mentioned with "service".

Finally, terms with no direct co-occurrences between them are spatially distant. For example, "service" on the right side, appear on the opposite side from "price", reflecting that these two topics were not mentioned in the same contexts or reviews.

4 Applications in Consumer Research

An overview of the consumer studies using co-occurrence network analysis in the last years is presented in Table 10.2. Co-occurrence network analysis was used in these studies as a visual tool for exploring latent themes, semantic relationships, or discourse structures within textual data. In most of them, co-occurrence networks are integrated with other analytical techniques (sentiment analysis, statistical modelling, topic extraction model) and with other survey analytics or behavioural data to reach a complete or more comprehensive understanding of consumer behaviour.

Sentiment analysis is often used in conjunction with co-occurrence networks to capture the emotional tone associated with specific term clusters, providing insight into how consumers feel about the themes they express. Similarly, topic modelling techniques such as Latent Dirichlet Allocation (LDA) or non-negative matrix factorization (NMF) are used to validate or extend the semantic groupings revealed in network analysis.

When demographic variables are present, network findings can be cross-tabulated by age, gender, location, or cultural background to reveal group-level differences. In studies where quantitative survey data is available, co-occurrence structures are linked to other measures (e.g., acceptance ratings, purchase frequency, or brand trust) through regression analysis, correlation mapping, or probit/logit models (Yano et al., 2019).

In digital contexts, co-occurrence networks are also enriched with engagement metrics—such as likes, shares, or purchase conversions—especially in social media or e-commerce platforms, allowing researchers to connect discourse themes with behavioural outcomes.

The examination of all these studies applying co-occurrence analysis in consumer science reveals a wide diversity on the combined methodological approaches. They are almost unique and different for each study depending on the specific context, research questions, and data characteristics. In this section, the methodologies and key findings of each study are described to help readers identify the most appropriate strategies for their own research needs.

4.1 *Open-Ended Questions and Free Associations Responses in Surveys*

Some examples of co-occurrence network analysis include their use to uncover latent themes and topical structures from survey data to explore motivations and perceptions related to food, sustainability, and bioplastics. Yano et al. (2018) investigated Japanese consumer perceptions of fresh leafy vegetables through an online survey, analysing 897 open-ended responses using word co-occurrence network analysis. The study applied community detection algorithms and centrality measures to identify key motivational clusters in consumption. Six major groups of

Table 10.2 Co-occurrence network analysis in consumer research studies

Source	Authors (Year)	Topic	Purpose	Data type & Size
Survey	Yano et al. (2018)	Leafy vegetables	Understand consumption motivations	Open-ended question, 897 responses
	Huang and Uehara (2023)	Alternative meats	Explore perceptions around alternative meat	Free association, 887 (Japan), 1119 (China)
	Yano et al. (2019)	Plant-based cosmetics	Link themes to consumer traits	Open-ended question, 901 responses
	Huang (2025)	Edible insects	Identify behavioural factors	Open-ended question, 391 responses
	Yano et al. (2021)	Vertical farming	Understand factors influencing acceptance	Free association 289 respondents
	Uehara et al. (2023)	Bioplastics	Explore consumer knowledge of bioplastics	Word association with 12,000 consumers
Online reviews	Li and Zhao (2024)	Fashion & Omnichannel	Identify themes in human-machine interaction	iOS App Store, 5000 reviews
	Mokryn (2024)	Hotels & travel context	Explore how trip purpose influences sentiment	Online hotel reviews, 137,000+ reviews
	Lang et al. (2020)	Fashion rental	Discover motivations/barriers	Consumer reviews, 2398
	La et al. (2021)	Theme parks	Compare sentiment by region	Theme park reviews
	Li and Gao (2024)	Airbnb & culture	Compare satisfaction drivers	Airbnb reviews, 243,215
Social media and blogs	Inoue and Nakajima (2023)	Agriculture exports	Link tweet topics to import behaviour	Tweets
	Puerta et al. (2020)	Gluten-Free & Twitter	Analyse gluten-free conversations	16,386 tweets
	Ghahramani et al. (2025)	Social Media & Public Health	Identify influential hashtags in health campaigns	73,288 tweets (18,143 original, 42,930 retweets, 2519 quotes, 20,846 likes)
	Li and Cao (2022)	Tourism behaviour	Map preferences and evaluations	Travel websites, 536 travel notes
	Lee et al. (2021)	Urban land use	Map retail area transformation	Blogs about Hongdae area. The top 1000 blogs in each year. From 2004 to 2016.
	Taka (2024)	Seafood E-commerce	Identify key service expectations	Japanese websites, 116 seafood e-commerce sites
	Xiong et al. (2024)	Live E-commerce	Detect purchase intent from text	Livestreaming comments, 12,000 barrage comments
	Jin and Li (2021)	Summer tourism	Study tourist decision behaviour	Travel websites (Mafengwo, Ctrip, etc.), 563 travel notes

words emerged, corresponding to health benefits, taste, texture, colour, palatability, and convenience. While health was the most frequently mentioned factor, sensory and practical considerations also played a significant role in shaping consumer attitudes. Yano et al. (2019) analysed consumer perceptions of plant-derived cosmetics by combining co-occurrence network analysis and multivariate probit modelling—a statistical approach used to evaluate multiple binary outcomes simultaneously, accounting for correlations between them. Open-ended responses were used to map how consumers conceptualize natural cosmetics and how these associations relate to demographic and psychographic attributes. Consumers frequently linked plant-based cosmetics with themes of gentleness and safety but also expressed concerns about high prices and slow results. Uehara et al. (2023) explored Japanese consumers' understanding and preferences regarding bio-based and biodegradable plastics, using a discrete choice experiment and co-occurrence network analysis. The researchers surveyed 12,000 respondents, employing a treatment-control design in which the treatment group received information distinguishing bio-based from biodegradable plastics. The co-occurrence network of consumer associations to bioplastics revealed eight subgraphs, reflecting themes such as bio-based properties, environmental expectations, cost concerns, and knowledge gaps.

When accompanied by demographic metadata, these networks allow researchers to segment and compare themes by consumer traits, national groups, or market segments. Yano et al. (2021) used two-mode networks to explore Russian consumer perceptions of vertical farming, finding differences by income and region particularly regarding concerns about naturalness and safety. Huang (2025) applied word association tasks to study Japanese consumers' views on edible insects, distinguishing themes like curiosity and nutrition (motivators) versus disgust and food safety (barriers), with separate networks for consumers willing and unwilling to eat insects.

In cross-cultural studies, co-occurrence networks helped to reveal how conceptual frames vary by language and culture, providing insights into region-specific interpretations. Huang and Uehara (2023) compared Japanese and Chinese consumer perceptions of alternative meats, combining a discrete choice experiment and network analysis. Shared themes included health and plant proteins, but differences emerged in cultural framing, for example, only Chinese consumers referenced insects as protein source, and Japanese consumers linked environmental aspects more to product features than to benefits.

4.2 Online Reviews

The feedback voluntarily provided by users on digital platforms is an interesting data source in consumer research as it reflects authentic experiences, perceptions, and opinions in real-world contexts. These user-generated texts not only influence the purchasing decisions of other potential users but also serve as an input for identifying patterns in consumer satisfaction and dissatisfaction.

To harness the richness of these data, researchers have increasingly combined co-occurrence networks with other analytical techniques such as sentiment analysis (see Chap. 8), topic modelling (a method used to automatically identify themes or topics within a set of texts) and community detection algorithms (used to find clusters or groups of closely related items within a network). These combinations allow for deeper insights into consumer emotions, service expectations, and the attributes that influence satisfaction across different markets.

For example, Lang et al. (2020) explored consumer perceptions in the online fashion rental market by integrating co-occurrence networks with Latent Dirichlet Allocation (LDA), a probabilistic topic modelling technique. Their findings distinguished between experiential and functional themes, and uncovered brand-specific clusters such as "luxury" and "event" for one brand, and "credit check" and "quality" for another. The integration of network structure and probabilistic topic modelling proved to be useful for uncovering brand differentiated consumer narratives.

Building on this approach, La et al. (2021) combined co-occurrence and sentiment analysis to compare theme park reviews across countries. Their results highlighted both universal themes (e.g., joy from "rides" and "family") and shared frustrations (e.g., "queue" and "expensive"), while also revealing regional differences in how consumers express sentiment and prioritize experiences. The cross-national comparison (e.g., U.S. vs. Australia) further showed how regional variations shape sentiment expression and experiential priorities.

A different angle is offered by Mokryn (2024), who applied co-occurrence network analysis to over 137,000 hotel reviews to investigate how traveller priorities shift by trip type. Using the spin glass algorithm and pointwise mutual information (PMI), the study identified clusters around service dimensions like location, food, service, and value, offering a nuanced look at expectations among business, leisure, and couple travellers.

In the context of mobile shopping, Li and Zhao (2025) employed co-occurrence networks to examine user experiences with mobile shopping apps. By segmenting reviews into positive, neutral, and negative sentiment categories, the analysis revealed that app usability and convenience were consistently associated with positive sentiment. Differences also emerged by retailer type, with high-end brands linked to hedonic value and off-price apps to affordability and customer delight.

Finally, Li and Gao (2024) extended this methodology into a cross-lingual framework by analysing Airbnb reviews, using KH Coder to compare consumer discourse across Chinese, Japanese, Korean, and English. Their findings showed how cultural and linguistic differences shaped vocabulary and evaluation criteria consumers use. Korean reviews emphasized family needs and transportation, Japanese reviews focused on soundproofing and cleanliness, while English-language comments centered on location. This multilingual approach underscores the potential of co-occurrence networks to surface nuanced cultural expectations in global consumer platforms.

4.3 Social Media, Blogs and E-commerce

Co-occurrence network analysis of user-generated content on social media—particularly on X (formerly known as Twitter)—has become a valuable tool for tracking public sentiment and uncovering thematic trends in consumer and health-related research. Below are some illustrative examples that highlight the range of its applications.

Ghahramani et al. (2025) analysed over 73,000 tweets from the American Heart Month campaigns from 2019 to 2023 to identify patterns among health-related hashtags. Co-occurrence network analysis and association rules techniques were applied to four subsets of Twitter (X) data: original tweets, retweets, quotes, and likes. The findings revealed that hashtags such as #AmericanHeartMonth, #OurHearts, and #HeartMonth consistently occupied central positions in the networks, with high engagement linked to their co-occurrence with others like #GoRedForWomen and #HeartHealth.

Puerta et al. (2020) examined the text in tweets mentioning gluten-free in Spanish-language. Their findings revealed nine primary themes discussed on Twitter, including products, culinary preparations, places, product characteristics, ingredients, occasions, social context, diet/health and sensory attributes. Moreover, the product-specific networks revealed distinctive topics. The study also concluded that automatic pre-cleaning methods, when applied to large datasets, can yield comparably rich and valid networks to those derived from time-intensive manual coding.

Inoue and Nakajima (2024) examined how social media discourse reflects consumer awareness and influences kiwifruit import trends in Japan. By analysing tweets mentioning Zespri™ using TF-IDF and co-occurrence networks, and applying vector autoregression (VAR) models, they found that mentions of "Kiwi Brothers" significantly predicted future import volumes—explaining nearly 20% of variance. The study demonstrates the effectiveness of integrating text mining with time-series analysis to link online engagement with real-world agri-food trade dynamics.

Other online sources with consumer generated text are e-commerce and blogs, that have been also used as a source of research content. Jin and Li (2021) analysed travel notes from online platforms such as Mafengwo and Ctrip to explore summer tourism behaviour. They combined sentiment analysis, and co-occurrence network mapping. The method revealed key decision-making factors like cost, weather, and transportation in travel narratives and helped to identify behavioural clusters and regional attraction patterns.

Lee et al. (2021) took a spatial and temporal approach, examining co-occurrence networks in blogs focused on Seoul's urban transformation. They used text mining tools and mapped the evolution of retail-related terms (e.g., "café", "brand store", "creative space") over time to detect semantic shifts in consumer descriptions of gentrifying neighbourhoods. This integration of co-occurrence analysis with spatial succession mapping showcased how network methods can support urban and retail landscape studies. Taka (2024) conducted a content analysis of seafood e-commerce websites in Japan, applying morphological analysis (for Japanese text) alongside co-occurrence network construction. The study used linguistic segmentation tools (e.g., *MeCab*) and built frequency-based co-occurrence graphs to identify terms frequently associated with customer expectations, such as "freshness", "delivery", and "packaging." These networks provided actionable insights into service dimensions that mattered most to online seafood consumers.

5 Key Considerations for Applying Co-occurrence Network Analysis

5.1 *When to Apply It*

In consumer research, co-occurrence network analysis is the technique of choice when the objective of the study is one of the following:

- Identify thematic clusters in unstructured data.
- Link themes to demographics and behavioural traits.
- Compare consumer discourse across cultures and languages.
- Uncover emotion-driven or satisfaction-related expressions.
- Support qualitative interpretation with quantitative structure.

5.2 *Strengths*

Co-occurrence network analysis offers several distinctive advantages that make it especially valuable in exploratory consumer research.

- It does not require predefined categories or training data, allowing researchers to analyse large corpora of unstructured text without biasing results.
- It visually represents semantic relationships. This visual dimension aids interpretability and supports intuitive exploration of content by revealing not just what consumers are saying, but how they relate ideas.
- Co-occurrence networks are highly flexible and adaptable across data types and languages. They can be applied to datasets drawn from surveys, interviews, review platforms, social media posts, and blogs, regardless of the language. This versatility makes them especially useful in cross-cultural or multilingual research settings.
- Co-occurrence networks provide a quantifiable structure to qualitative content and are easily integrated with other text analysis techniques such as sentiment analysis, topic modelling and regression analysis.
- Compared with topic modelling techniques that often yield broad thematic categories, co-occurrence analysis can reveal fine-grained associations such as

specific product pairings, emotional co-references, or frequently co-mentioned service attributes.

5.3 Limitations

Co-occurrence networks present several methodological limitations that must be considered when applied to consumer research.

- Co-occurrence networks lack inherent sentiment or contextual understanding. They do not distinguish between positive and negative sentiment, detect sarcasm, or account for emotional nuance. As such, they require integration with sentiment analysis or manual coding to accurately interpret evaluative or affective content.
- Co-occurrence networks identify associations based on proximity rather than true semantic, causal, or syntactic links. This may lead sometimes to misleading or spurious connections between terms that are contextually unrelated.
- Co-occurrence networks are sensitive to preprocessing of the text. Poorly designed cleaning protocols can significantly alter network structure and difficult key themes visualization.
- Large datasets can generate dense and visually complex networks that are difficult to interpret often requiring frequency thresholds that risk excluding rare but meaningful terms.
- The choice and interpretation of network metrics and community structures require a certain level of understanding of network analysis principles. Without this, there is a risk of misinterpreting or overgeneralizing the findings.

6 Conclusions

Co-occurrence network analysis is a versatile and insightful method for analysing textual data in consumer research. Its strength lies in revealing how consumers connect ideas, express needs, and construct meaning around products, services, and experiences.

The method is particularly valuable to explore unstructured or large-scale textual datasets without imposing predefined categories. It excels in uncovering thematic clusters, tracking discourse over time, and comparing consumer narratives across cultures, platforms, and demographic segments. Its flexibility allows it to be applied to diverse data types ranging from short tweets and product reviews to detailed focus group transcripts making it suitable for both exploratory and explanatory research purposes.

While co-occurrence networks offer a valuable entry point into the structure of consumer language, their full potential is realized when used in concert with sentiment analysis, statistical modelling, topic extraction, survey analytics, and behavioural data. This multimethod approach ensures richer interpretation, triangulation of insights, and stronger explanatory power in consumer research.

As future perspectives, providing methodological guidelines or standardized protocols for building and interpreting co-occurrence networks is necessary to ensure comparability and reproducibility across studies, while also reducing the risk of misinterpretation. Co-occurrence network analysis is expected to gain from the integration with Natural Language Processing techniques to provide richer semantic interpretation and greater contextual accuracy.

References

Bastian, M., Heymann, S., & Jacomy, M. (2009). Gephi: An open source software for exploring and manipulating networks. In *Proceedings of the international AAAI conference on web and social media* (Vol. 3, No. 1) (pp. 361–362). AAAI Press.

Blondel, V. D., Guillaume, J. L., Lambiotte, R., & Lefebvre, E. (2008). Fast unfolding of communities in large networks. *Journal of Statistical Mechanics: Theory and Experiment, 2008*(10), P10008.

Church, K., & Hanks, P. (1990). Word association norms, mutual information, and lexicography. *Computational Linguistics, 16*(1), 22–29.

Ghahramani, A., Prokofieva, M., & de Courten, M. P. (2025). Health promotion campaigns using social media: Association rules mining and co-occurrence network analysis of Twitter hashtags. *BMC Public Health, 25*, 67. https://doi.org/10.1186/s12889-024-21255-5

Hansen, D. L., Shneiderman, B., Smith, M. A., & Himelboim, I. (2020). Installation, orientation, and layout. In D. L. Hansen, B. Shneiderman, M. A. Smith, & I. Himelboim (Eds.), *Analyzing social media networks with NodeXL* (2nd ed., pp. 55–66). Morgan Kaufmann. https://doi.org/10.1016/B978-0-12-817756-3.00004-2

Higuchi, K. (2001). Kh Coder: A free software for quantitative content analysis or text mining. In *A free software for quantitative content analysis or text mining*.

Higuchi, K. (2016). *KH Coder 3 reference manual*. Ritsumeikan University.

Huang, S. (2025). Reviving the entomophagy tradition among the younger generation: An application of the theory of planned behavior. *Food Quality and Preference, 124*, 105356.

Huang, S., & Uehara, T. (2023). Young consumers' perceptions of and preferences for alternative meats: An empirical study in Japan and China. *Frontiers in Sustainable Food Systems, 7*, 1290131. https://doi.org/10.3389/fsufs.2023.1290131

Hutchings, S. C., Dixit, Y., Al-Sarayreh, M., Torrico, D. D., Realini, C. E., Jaeger, S. R., & Reis, M. M. (2023). A critical review of social media research in sensory-consumer science. *Food Research International, 165*, 112515. https://doi.org/10.1016/j.foodres.2023.112515

Inoue, Y., & Nakajima, S. (2024). The role of tweets in agricultural export: An approach from text-mining and timeseries analyses. *British Food Journal*, 126(4), 1597–1616.

Jin, Y., & Li, X. (2021). Research on the development of Changbai mountain summer tourism from the perspective of consumer behavior. *E3S Web of Conferences, 292*, 03038. https://doi.org/10.1051/e3sconf/202129203038

La, L., Xu, F., Ren, Q., Zhen, F., & Lobsang, T. (2021). Understanding consumers' sentiment expressions in online reviews: A hybrid approach. *Journal of International Consumer Marketing, 34*(4), 465–479. https://doi.org/10.1080/08961530.2021.1982807

Lang, C., Li, M., & Zhao, L. (2020). Understanding consumers' online fashion renting experiences: A text-mining approach. *Sustainable Production and Consumption, 21*, 132–144. https://doi.org/10.1016/j.spc.2019.12.003

Lee, Y., Lee, Y., Seong, J., Stanescu, A., & Hwang, C. S. (2020). A comparison of network clustering algorithms in keyword network analysis: A case study with geography conference presentations. *International Journal of Geospatial and Environmental Research, 7*(3), Article 1. https://dc.uwm.edu/ijger/vol7/iss3/1

Lee, M., Kim, H., & Cheon, S. (2021). A network approach to revealing dynamic succession processes of urban land use and user experience. *Sustainability*, 13, 11955. https://doi.org/10.3390/su132111955

Li, Y., & Gao, J. (2024). Satisfaction of customers in Airbnb: A cross-culture text mining of online reviews. *Anatolia, 1–15*. https://doi.org/10.1080/13032917.2024.2363814

Li, M., & Zhao, L. (2025). Understanding fashion omnichannel experience through mobile application customer reviews. *Journal of Global Fashion Marketing, 16*(1), 90–108. https://doi.org/10.1080/20932685.2024.241730

Li, M., & Zhao, L. (2024). Understanding fashion omnichannel experience through mobile application customerreviews. *Journal of Global Fashion Marketing*, 16, 90–108.

Mokryn, O. (2024). Consumer sentiment and hotel aspect preferences across trip modes and purposes. *Journal of Theoretical and Applied Electronic Commerce Research, 19*, 3017–3034. https://doi.org/10.3390/jtaer19040145

Pons, P., & Latapy, M. (2006). Computing communities in large networks using random walks. *Journal of Graph Algorithms and Applications*, 10, 191–218.

Puerta, P., Laguna, L., Vidal, L., Ares, G., Fiszman, S., & Tárrega, A. (2020). Co-occurrence networks of Twitter content after manual or automatic processing: A case-study on "gluten-free". *Food Quality and Preference, 86*, 103993.

Ruiz, J. B., & Barnett, G. A. (2015). Exploring the presentation of HPV information online: A semantic network analysis of websites. *Vaccine, 33*(29), 3354–3359. https://doi.org/10.1016/j.vaccine.2015.05.017

Taka, T. (2024). Content analysis of seafood e-commerce sites using a text mining approach: A case study of Japan. *Journal of International Food & Agribusiness Marketing, 36*(2), 316–327. https://doi.org/10.1080/08974438.2022.2104416

Uehara, T., Nakatani, J., Tsuge, T., & Asari, M. (2023). Consumer preferences and understanding of bio-based and biodegradable plastics. *Journal of Cleaner Production, 417*, 137979. https://doi.org/10.1016/j.jclepro.2023.137979

van Eck, N. J., & Waltman, L. (2007). VOS: A new method for visualizing similarities between objects. In *Advances in data analysis* (pp. 299–306). Springer. https://doi.org/10.1007/978-3-540-70981-7

van Eck, N. J., & Waltman, L. (2010). Software survey: VOSviewer, a computer program for bibliometric mapping. *Scientometrics, 84*(2), 523–538. https://doi.org/10.1007/s11192-009-0146-3

Xiong, Y., Wei, N., Qiao, K., Li, Z., & Li, Z. (2024). Exploring consumption intent in live e-commerce barrage: A text feature-based approach using BERT-BiLSTM model. *IEEE Access, 12*, 69288–69298. https://doi.org/10.1109/ACCESS.2024.3399095

Yano, Y., Blandford, D., Maruyama, A., & Nakamura, T. (2018). Consumer perceptions of fresh leafy vegetables in Japan: An application of word co-occurrence network analysis. *British Food Journal*. https://doi.org/10.1108/BFJ-09-2017-0500

Yano, Y., Kato, E., Ohe, Y., & Blandford, D. (2019). Examining the opinions of potential consumers about plant-derived cosmetics: An approach combining word association, co-occurrence network, and multivariate probit analysis. *Journal of Sensory Studies, 34*, e12484. https://doi.org/10.1111/joss.12484

Yano, Y., Nakamura, T., Ishitsuka, S., & Maruyama, A. (2021). Consumer attitudes toward vertically farmed produce in Russia: A study using ordered logit and co-occurrence network analysis. *Food, 10*, 638. https://doi.org/10.3390/foods10030638

Chapter 11
Netnographic Approach to Analyse Social Media Content

Carlos Gómez-Corona

1 Introduction on Qualitative Studies

One of the most important distinctions between qualitative and quantitative studies lies in the type of data. In the basic sciences, researchers often seek to collect numerical (quantitative) data that can be subjected to statistical analysis. The concept of data itself can be traced back to its Latin origin—data is the plural of *datum*, meaning a piece of information or "something given" (Wilkinson, 2005). In qualitative research, data are not numbers but stories, narratives, and lived experiences shared by individuals. The results of a qualitative study are far from being a mean score where we can calculate the variance of responses, instead, qualitative studies are experiences that enable us to frame and explore a set of different variables or theories that come from the dialogue of the participants of the study.

Contrary to popular belief, qualitative studies do not need to be preceded or followed by quantitative research. Qualitative research can stand on its own as a rigorous and insightful method, capable of explaining patterns of behaviour, generating hypotheses, or confirming the existence of trends. Nevertheless, mixed-method approaches (combining qualitative and quantitative techniques) have become increasingly popular to gain a more comprehensive understanding of given phenomena. Traditionally, qualitative studies were often conducted first, followed by quantitative research to validate initial findings. However, over the past decade, there has been a growing trend toward reversing this sequence—qualitative studies are now frequently used after quantitative surveys, to provide richer context and deeper explanation of statistical results. And in some cases, qualitative studies are

C. Gómez-Corona (✉)
Human Insight Department - TTH, dsm-firmenich, Neuilly-sur-Seine, France
e-mail: carlos.gomez@dsm-firmenich.com

M. Visalli, M. V. Galmarini (eds.), *Collection and Analysis of Text Data in Sensory and Consumer Science*, https://doi.org/10.1007/978-3-032-04851-6_11

the only and primary source of information in disciplines like anthropology, sociology, and even some psychological research.

Qualitative methods have always been at the core of anthropology, a discipline with one of the longest-standing traditions in the study of human habits and attitudes. The origins of anthropological inquiry can be traced back to the sixteenth century, when information arriving in the West from distant cultures began to be systematically recorded and studied. While early accounts were often unsystematic, ethnocentric, and of limited quality (Reygadas, 2014), the emergence of scientific societies such as the *Société Ethnologique de Paris* (1839–1848) and the Ethnological Society of London (1843–1871) helped foster the academic study of cultural diversity (Urry, 1984). Among the most widely used anthropological methods is ethnography, which, according to Pink (2009), is a process for creating and representing knowledge—about societies, cultures, or individuals—based on the researcher's immersive experience (the ethnographer). The origins of modern ethnography are still debated. Some scholars attribute its methodological foundations to Bronislaw Malinowski, particularly after the publication of Argonauts of the Western Pacific in 1922, which was based on several years of immersive fieldwork in the Melanesian islands. Today, the iconic image of Malinowski—a solitary Western male anthropologist spending extended periods in distant locations using rigorous observational methods—has become the archetype of the ethnographer (Fig. 11.1). Other scholars highlight Margaret Mead's influential ethnographic work on Adolescence, Sex, and Culture in Samoa (1928), which, while also impactful, was shaped by a more psychological perspective.

Contemporary ethnography continues to offer valuable cultural insights, particularly in consumer research, by involving researchers in everyday consumers' activities such as cooking, shopping, and eating. However, the objective is not simply to observe these behaviours but to understand the meanings that individuals attach to them. While the ethnographies of Malinowski and Mead lasted for years, modern

Fig. 11.1 Original image of the 1922 publication of Malinowski, and Fig. XXXV from the book, with the caption "Communal Cooking of Mona (Taro Dumplings)". (Image from: The Project Gutenberg eBook of Argonauts of the Western Pacific (Guthenberg Project, 2024))

ethnographic research is typically limited to a few hours or days of fieldwork. This is largely due to the increasing specificity of research questions as well as the practical constraints of time and funding that make long-term immersion less feasible (Valentin & Gómez-Corona, 2018).

2 Netnography as a New Method from Anthropology

2.1 *What Is Netnography*

Just as digital technologies have enabled the development of mobile ethnographies (Novoa, 2015), they have also fostered the rise of netnography—sometimes referred to as virtual, digital, or online ethnography. Coined from the fusion of "network" and "ethnography," netnography is a methodological adaptation of traditional ethnographic approaches, specifically designed to investigate cultures and communities that emerge in online environments. Introduced by Robert Kozinets, netnography seeks to interpret the communicative practices of individuals and groups operating within digital spaces (Kozinets, 2015). Netnographic studies can be conducted across a wide range of online platforms, including microblogging social media sites (e.g., X, formerly Twitter), image and video-sharing platforms (e.g., Instagram, YouTube), hybrid social networks (e.g., Facebook), discussion forums (e.g., TripAdvisor), and personal or professional blogs (e.g., HuffPost, Business Insider). Beyond social media, netnography can also be applied to mainstream media content such as news outlets, broadcast media, and corporate websites. Regardless of the digital setting, the core ethnographic principles still apply: sufficient immersion in the field and rigorous qualitative data analysis are essential.

The role of the researcher in netnography may vary—from passive observation to active participation—depending on the platform and research objectives. Importantly, participants may or may not be aware that they are part of a research study, raising unique ethical considerations. The use of netnography has expanded rapidly in recent years, with applications across diverse consumer domains, including tourism (Mkono & Markwell, 2014), or the identification of lead users in food and beverage innovation. As digital interactions increasingly shape consumer behaviour, netnography is poised to become an essential tool in the researcher's methodological toolkit. Much like mobile ethnography, it offers a practical and timely response to the realities of a hyper-connected, dynamic, and mobile world—transforming not only how we study consumers, but how ethnographic inquiry itself is evolving.

2.2 *Application of Netnography in Social Media Studies*

The exponential rise of social media in recent years has had a significant impact on consumer research. Platforms such as Facebook, X, and Instagram are increasingly leveraged to explore consumer perceptions of various products, including wine

(Mellish et al., 2023), beer (Arellano-Covarrubias et al., 2022), sustainable consumption behaviours (Choudhary et al., 2019), and attitudes toward natural food products (Fatemi et al., 2023). Although the systematic use of social media is relatively recent in the field of consumer research, its roots can be traced back to studies of social networks in the 1990s (Lewis et al., 2008). Several factors contribute to the growing appeal of social media as a data source for consumer researchers. The first is the accessibility and richness of digital content. Social media enables individuals to publish content with minimal effort, low cost, and an unprecedented reach—potentially addressing millions of people worldwide (Sibilia, 2017). This leads to the second factor: the vast quantity of user-generated data available for analysis. With billions of active users globally, social media offers researchers a wealth of information that can be systematically examined to understand consumer behaviours, attitudes, and food choices.

Most consumer research using social media data—particularly within sensory and food domains—has relied on quantitative text analysis methods. Data is typically extracted in bulk and processed through automated or semi-automated techniques such as natural language processing or sentiment analysis. However, qualitative approaches remain underutilized, despite their potential to provide richer contextual insights. The aim of this chapter is to highlight and advocate for the use of qualitative methods in the analysis of social media text data. By doing so, we explore how such approaches can deepen our understanding of consumer meaning-making and enhance the interpretive dimensions of consumer research in digital environments.

Netnography is a relatively recent methodological approach that adapts traditional ethnographic techniques to the study of cultures and communities within computer-mediated environments (Anne-Marie et al., 2017). First introduced by Kozinets (1998) in the context of marketing research, netnography—also known, though less commonly, as cyber-ethnography, digital ethnography, or online ethnography—offers a framework for understanding digital social behaviour through immersive observation and qualitative analysis. As with classical ethnography, netnographic research requires sufficient time in the online field to observe and interpret the phenomena under study. It also adheres to established qualitative methods for data collection and analysis (Valentin & Gómez-Corona, 2018). When applied to social media, netnography enables the longitudinal observation of user behaviour. For example, studies tracking interactions on platforms like X have demonstrated the method's capacity to uncover nuanced consumer dynamics over time (Arellano-Covarrubias et al., 2019).

In the domain of food and beverage research, netnographic studies have emerged primarily in tourism-related contexts. Examples include investigations into wine tourism (Thanh & Kirova, 2018), shared meal experiences (Atsız et al., 2022), and local food consumption practices (Kodaş & Aksoy, 2022). These studies highlight netnography's potential to offer rich cultural and behavioural insights that go beyond what traditional methods can reveal. Given the growing digitalization of consumer experiences, netnography presents a valuable opportunity for researchers in the food and beverage sector. Its ability to capture authentic, naturally occurring interactions makes it a powerful tool for uncovering the meanings consumers attach to food-related practices in online environments.

2.3 Methodological Implications

Translating established ethical principles into the realm of social media research presents significant challenges, particularly for consumer researchers. As McKee (2013) notes, it is often difficult to clearly delineate the responsibilities and ethical obligations of researchers working with data generated in digital environments. Numerous early studies in this area highlighted ethical ambiguities and missteps, underscoring the need for greater clarity and sensitivity in digital research practices (Gómez-Corona & Nguyen, 2023; Casar, 2025; Zeller & Woods, 2024). One methodological approach that raises particular concern is of course netnography, where, as we have mentioned before, we use non participatory observations of a specific online community or group of people (or institutions such as associations, restaurants, etc.) without directly interacting with participants. The person being observed may never know that they are being observed for research purposes. While this approach involves collecting publicly accessible data, it nonetheless raises ethical concerns about privacy and informed consent. Zimmer (2010) provocatively stated in his article fifteen years ago, "*But the data is already public*", however, even publicly available social media content warrants careful handling when used in research. The best source of ethical guideline should be found in anthropological associations, as non-observational methods are a key part of behavioural studies in anthropology and sociology. One example can be the "Anthropological Ethics" from the American Anthropological Association (2025), which has a statement on ethics for anthropological studies, codes, handbooks, and even a public mail for queries. Other associations with ethical guidelines are: Association of Social Anthropologists of the UK, international organizations like UNESCO (2016), or the European Commission (2021).

While ethical concerns around digital research are increasingly recognized, there remains a lack of dedicated regulations or standardized ethical guidelines specifically addressing consumer research in social media environments. In the absence of such tailored frameworks, researchers are often left to interpret and apply broad data protection principles—most notably those set forth in the General Data Protection Regulation (GDPR)—to complex and evolving digital contexts. Given the growing relevance of digital platforms in consumer research, the authors argue that it is urgent to address the ethical vulnerabilities that emerge in these settings. To effectively navigate these challenges, researchers must rely on comprehensive and context-sensitive ethical frameworks. Figure 11.2 presents five frequent types of privacy violations encountered in social media research. For additional ethical guidance, researchers are advised to consult the Internet Research: Ethical Guidelines developed by the Association of Internet Researchers (AOIR, 2019), alongside the GDPR, which has been in force across the European Union since May 25, 2018, to harmonize and reinforce data privacy standards.

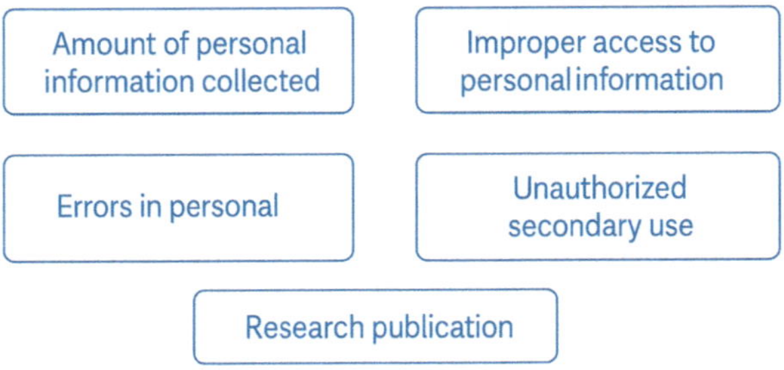

Fig. 11.2 Five frequent types of privacy violations encountered in social media research (applied in consumer science)

3 Qualitative Text Data Analysis in Netnography

3.1 Methodological Considerations of Qualitative Analysis

Qualitative approaches in consumer research enable exploration and, in some cases, the corroboration of hypotheses. However, it is important to avoid terms such as "validation" or "confirmation", which are typically reserved for quantitative studies involving representative samples and statistical margin errors. Even within a qualitative paradigm, a clear research objective, well-defined research questions, and hypotheses to be explored remain essential. In the same way as with quantitative data, qualitative data must demonstrate reliability, understood as the consistency or stability of the findings (Cozby, 1997). While qualitative research does not rely on standardized instruments, researchers should aim to obtain similar findings if the study were to be repeated under comparable conditions. One strategy to strengthen the robustness of qualitative insights is triangulation, which involves the use of multiple sources or methods. Álvarez-Gayou Jurgenson (2012) identifies four types of triangulations:

- Data triangulation, which involves drawing from multiple data sources.
- Investigator triangulation, which entails having more than one researcher involved in data collection or analysis, enhancing interpretive consistency.
- Theory triangulation, which brings in different theoretical perspectives to interpret the same data set.
- Methodological triangulation, which applies various research methods to examine the same phenomenon.

In qualitative research, various observational approaches are used to access participant data, including non-participant observation, participant observation, and controlled observation. In the case of netnography, observation is typically non-participant in nature as already mentioned. This form of observation can only occur in online naturalistic settings where the researcher does not intervene but rather observes behaviours as they unfold. In the case of netnographies, the naturalistic context can be social media or online platforms, and the qualitative analysis performed prioritized (at least for this chapter) is text analysis of social media conversations, in other words, social media posts. Besides text, the images can also be analysed (e.g., Instagram images) but the scope of the analysis of the images falls outside the objective of this chapter. For more information on image analysis see Gómez-Corona and Salazar Lama (2023).

There are several approaches in the qualitative text analysis in netnographies such as content analysis, discourse analysis, or interpretative phenomenological analysis (Forrester, 2010). For this chapter, we are going to focus on two different approaches: discourse analysis, and interpretative phenomenological analysis. It is important to mention that these analyses may use some qualitative data visualization also, for example, simple graphs with number of mentions of certain variables or sensory descriptors, correspondence analysis of text data, word clouds, or any other type of data visualization that relies in descriptive methods, rather than statistical approaches.

3.2 Discourse Analysis

In the context of sensory and consumer research, discourse analysis (often aligned with thematic analysis) goes beyond identifying what consumers say about products; it explores how language shapes and reflects shared experiences, preferences, and cultural meanings related to food and sensory perception. Language in this setting does not simply report facts; rather, it constructs social realities that influence how consumers evaluate taste, texture, aroma, or product identity. Various types of discourse analysis exist, including discursive psychology, Foucauldian discourse analysis, and general interpretive approaches. These methods are not bound by rigid analytical steps, which gives them the flexibility to be applied in diverse contexts—from analysing online discussions about plant-based foods to exploring how consumers articulate trust in food technologies or quality cues (Wiggins & Riley, 2010).

The primary objective of discourse analysis in consumer and sensory research is to explore how consumer language constructs distinct versions of food-related reality. For example, discourse can shape perceptions of a product as "natural", "authentic", or "artificial". This approach recognizes that language is socially and culturally embedded, and therefore never neutral. The meanings consumers attach to sensory attributes or product claims are influenced by broader social norms, values, and cultural representations. Discourse analysis allows researchers to investigate how language functions in consumer narratives, what is emphasized, challenged, resisted, or normalized, and to assess the implications of these discursive patterns for sensory perception and product acceptance.

Table 11.1 outlines the typical steps involved in conducting discourse analysis on qualitative textual data within the context of consumer and sensory studies. This approach reflects a traditional, manual method of analysis and does not incorporate the use of artificial intelligence (AI)-based language models. While AI tools—such as those integrated into qualitative research platforms like *Indeemo*, which specializes in digital ethnographies (https://indeemo.com)—are increasingly available, a manual approach is presented here for clarity and pedagogical purposes.

Table 11.1 Text analysis steps in discourse analysis in qualitative studies

Step	Description	Example
Transcription	Produce a verbatim transcript (e.g., indicating pauses, tone, emphasis). Transcripts are stored in editable formats.	*"Honestly, I used to drink beer mostly for social reasons, it wasn't really my thing..."*
Coding	Assign interpretive labels to segments of text. This can be done manually (e.g., highlighting) or using qualitative analysis software. Coding is iterative.	Highlighting expressions related to social behaviour, taste preferences, or identity shifts.
Initial impressions	Leave space for notes in the transcript margins. On the left a shit, note potential key themes; on the right, summarize the participant's main point.	Left margin: Social motivationright margin: Shift in personal preference
Key themes	Identify recurring ideas or meanings across participants. Focus on statements that represent deeper meanings or patterns.	Social drinking, exploration of new flavours, individual differentiation
Verbatim collection	Select illustrative participant quotes for each theme. Review all transcripts before choosing final quotes to ensure representativeness and diversity.	*"I used to drink beer socially, but now I want to try different things..."*
Discourse identification	Organize selected quotes into coherent narratives. Reorder and refine quotes to reveal underlying discourses (e.g., identity, pleasure, authenticity).	Two key discourses: 1. Sensory pleasure2. Search for personal identity

Adapted from Gómez-Corona (2019)

3.3 *Interpretative Phenomenological Analysis*

Interpretative Phenomenological Analysis (IPA) comes from phenomenology, which serves as a philosophical school, a psychology current and a qualitative research approach, particularly influential in psychology and human sciences. At its core, it aims to describe and interpret how individuals perceive and make sense of their lived experiences (Forrester, 2010). Originating from the work of philosopher Edmund Husserl (1859–1938), who placed subjective experience at the center of philosophical reflection (Ashworth, 2008), phenomenology was later expanded by Martin Heidegger (1889–1976) to incorporate existential and contextual dimensions of being.

In consumer research, phenomenological perspectives have been adopted to explore "product experiences"in-depth, with an approach of user experience, and perhaps more with a design or marketing approach (Gómez-Corona & Valentin, 2019). In recent years, the concept of "product experience"has attracted growing attention among consumer scientists, not only as a holistic framework for understanding the dynamic interaction between individuals and products (Desmet & Hekkert, 2007; Schifferstein, 2009), but also as a lens through which to interpret the realities of modern consumer society. In this context, the use of IPA in consumer research with a sensory approach for food and beauty products can be a perfect fit and several studies have been published with this approach, especially in studies focusing on how individuals engage with food, drinks, and everyday sensory environments (Eatough & Smith, 2006; Valtonen et al., 2010; Gómez-Corona et al., 2017). Phenomenology is particularly relevant to netnographic studies and sensory ethnographies (Pink, 2009), as it centers on the interpretation of meaning derived from personal and contextualized experiences. Rather than seeking causal explanations, phenomenology seeks to understand how meaning is constructed from experience—how consumers perceive and make sense of flavour, texture, ambiance, or ritual in consumption practices (Gallagher & Zahavi, 2008). This interpretative stance is essential when the aim is to understand *how a product or a situation is experienced*, not just *what is said* about it.

Typically, phenomenological research relies on semi-structured interviews to collect rich, first-person accounts of specific events, products, or sensory interactions. Participants or the people observed share detailed narratives of their experiences. According to Langdridge (2007), four guiding principles define the phenomenological approach:

- A focus on *human experience* as the central object of study.
- A concern for *meaning-making*, and how meaning shapes perception and behaviour.
- A preference for *descriptive and relational analysis*, rather than explanatory or causal models.
- An awareness of the *researcher's role* in co-constructing meaning, influenced by cultural, historical, and personal contexts.

In its more interpretative forms, IPA provides a nuanced framework to explore how people make sense of their emotional, social, and sensory worlds. Researchers using IPA acknowledge that access to participants' experiences is always partial, filtered through both the participant's narration and the researcher's interpretive lens. It is akin to attempting to walk in someone else's shoes—while fully recognizing that this understanding will always be an informed approximation. This reflexivity is especially important when conducting netnographic research, where researchers analyse digital traces of consumer narratives and must remain vigilant not to impose their own meanings onto participants' voices. Sensory experiences discussed in online communities, for instance, descriptions of a wine's aroma or a nostalgic memory linked to a traditional dish, can be rich entry points for phenomenological analysis, provided researchers attend to context, tone, and subjective framing. Table 11.2 outlines the main steps involved in conducting a phenomenological analysis in consumer and sensory research, adapted to both interview-based and netnographic data sources, again, this example is a "manual" analysis of the text data.

4 Case Study: Mint Uses and Habits in Latin America

The objective of this study was to understand the perception, uses and habits around mint consumption in Latin America. It was an exploratory study performed in twenty countries in Latin America. The data for this study was collected using

Table 11.2 Text analysis steps in IPA in qualitative studies

Step	Description	Example (mezcal consumption)
1. Reflection & Familiarization	Begin by reflecting deeply on the data: Its quality, context, and origin. Keep a reflexive journal throughout the analysis. Then, become familiar with the data by listening to audio recordings and reading transcripts multiple times.	Reflexive insight: Consumers of mezcal often speak about taste and origin. A strong emotional connection and sense of pride seem to emerge when discussing mezcal.
2. Identifying emerging themes	Re-read the transcripts analytically to identify preliminary themes.	Verbatim example: "*I started drinking mezcal with a friend... we made it a habit to try new types. Now I can't drink just any mezcal, I'm quite selective.*"
3. Descriptive summaries	Create short summaries of key phrases that can later be interpreted more deeply.	- Mezcal consumption as a social activity among friends - exploring new flavours - developing refined taste preferences
4. Initial interpretation	Reflect on the summaries and explore what they reveal about the participant's experience with the product and broader meanings. Consider emotions, values, and identity.	- friendship as shared experience - Mezcal as a personal journey - from casual to symbolic consumption
5. Grouping themes	Organize themes that share experiential similarities. Group them under broader thematic clusters to clarify different experiential domains.	Group 1: Friendship, shared memories, travel stories group 2: Identity formation, differentiation group 3: Emotional, social, symbolic consumption
6. Final themes	Select the most meaningful themes that capture the essence of the consumer experience and align with the research objective.	Mezcal as an identity-forming product—The symbolic nature of mezcal consumption.

Adapted from Gómez-Corona (2019)

Synthesio®, a commercial social media listening platform (https://www.synthesio.com/). *Synthesio*® enables access to both social and mainstream media content across 197 countries and in more than 80 languages. It allows researchers to conduct keyword-based searches within specified timeframes and geographic areas, down to the regional level. One notable advantage of this platform is the capacity to analyse text without character limitations, a constraint found in many other platforms. Moreover, *Synthesio*® can automatically retrieve public demographic information—such as age and gender—when made publicly available by users on their profiles.

For the data analysis, the approach diverged from that of Vidal et al., (2015), who recommended excluding retweets in Twitter-based research to avoid redundancy. In contrast, this study retained retweets and repeated posts, based on the rationale that the recurrence of specific content, especially regarding mint in beverages, sweet goods and savoury food products. To obtain the social media data, three sets of Boolean searches were performed in *Synthesio* platform (Table 11.3). For the Boolean search the connector used were OR, NEAR/7 (meaning that mint OR

Table 11.3 Boolean search used in Synthesio platform to retrieve the information of mint per product segment

Beverages	((mint OR minty OR menthol OR spearmint OR peppermint) NEAR/7 (wine OR spirits OR beer OR water OR nectar OR "hot chocolate" OR cocktail OR cocktails OR mocktail OR mocktails OR drinks OR drink OR drinkable OR beverages OR beverage OR juice OR juices OR coffee OR smoothie OR smoothies OR tea OR milk OR milky OR shaked OR shaking OR shake OR blended OR blend)) NOT (amazon OR buy OR amzn OR offer OR order OR voucher OR deal OR deals OR lenovo OR nike OR google OR iphone OR android OR candles OR candle OR adidas OR gin OR "deal" OR "deals" OR "offer" OR "coupon" OR "shipping" OR "set" OR "price" OR "pack" OR "hot deal" OR "buy" OR "coupons" OR "packs" OR "set" OR "job" OR "cell phone" OR "ebay" OR "pope" OR "marriage" OR "blue tooth" OR "1950" OR "waste water" OR "low carb" OR "ovarian" OR "jowl" OR "Rio" OR "pet" OR "dog" OR voucher OR vouchers OR offer OR offers OR todayDealsSteal OR mydealdotcom OR 2mydeal)
Sweet goods	((mint OR minty OR menthol OR spearmint OR peppermint) NEAR/7 (baked OR baking OR bar OR Barley OR Biscuit OR Bran OR Bread OR Brioche OR Brownies OR Bun OR Butter OR Cake OR camembert OR candies OR candy OR Cereals OR Cheese OR Cheesecake OR chewingum OR chocolate OR confit OR Cookies OR Corn OR Cracker OR Cream OR Croissant OR Dairy OR dessert OR diet OR dietary OR Dough OR Doughnut OR Gingerbread OR gum OR gummy OR Jam OR jellies OR jelly OR lassi OR latte OR licorice OR lollipop OR Macaroon OR marshmallow OR Milk OR milkshake OR Muesli OR nutrition OR nutritional OR Oats OR Panettone OR "panna cotta" OR Pie OR Rice OR Shortbread OR sorbet OR soy OR Speculatius OR sugar OR Tart OR teddy OR Toast OR Wheat OR Yogourt OR Yoghurt)) NOT (amazon OR buy OR amzn OR offer OR order OR voucher OR deal OR deals OR lenovo OR nike OR google OR iphone OR android OR candles OR candle OR adidas OR gin OR "deal" OR "deals" OR "offer" OR "coupon" OR "shipping" OR "set" OR "price" OR "pack" OR "hot deal" OR "buy" OR "coupons" OR "packs" OR "set" OR "job" OR "cell phone" OR "ebay" OR "pope" OR "marriage" OR "blue tooth" OR "1950" OR "waste water" OR "low carb" OR "ovarian" OR "jowl" OR "Rio" OR "pet" OR "dog" OR voucher OR vouchers OR offer OR offers OR todayDealsSteal OR mydealdotcom OR 2mydeal)
Savory	((mint OR minty OR menthol OR spearmint OR peppermint) NEAR/7 (toast OR salad OR cheese OR camembert OR meat OR Focaccia OR pizza OR bread OR beef OR pork OR chicken OR turkey OR fish OR pepperoni OR salami OR burger OR sandwich OR appetizers OR pasta OR Spaghetti OR bruschetta OR ((tartine OR snack OR snacks OR snacking OR cracker OR crackers OR chips OR appetizer OR "finger food" OR spread OR dip OR aperitif OR munchies OR "hors oeuvre" OR canape OR zakouski OR tapas) NEAR/5 (salted OR salt OR salty)))) NOT (amazon OR buy OR amzn OR offer OR order OR voucher OR deal OR deals OR lenovo OR nike OR google OR iphone OR android OR candles OR candle OR adidas OR gin OR "deal" OR "deals" OR "offer" OR "coupon" OR "shipping" OR "set" OR "price" OR "pack" OR "hot deal" OR "buy" OR "coupons" OR "packs" OR "set" OR "job" OR "cell phone" OR "ebay" OR "pope" OR "marriage" OR "blue tooth" OR "1950" OR "waste water" OR "low carb" OR "ovarian" OR "jowl" OR "Rio" OR "pet" OR "dog" OR voucher OR vouchers OR offer OR offers OR todayDealsSteal OR mydealdotcom OR 2mydeal)

minty, etc.; were at least closed in 7 words to wine OR spirits, etc.), and NOT (to exclude inappropriate content that is not related to the objective of the study). The Boolean search was translated into Spanish and Portuguese.

For the data analysis the data was coded into its main context. The social media information contained multiple phrases, tweets, Facebook publications, and extracts from forums or blogs, where the keywords (flavour-related words) were mentioned by users. The data retrieved included an ID number for each mention, the country and date when the mention was published, the website name where it was posted, the URL from which it was extracted, and the username and gender, if available. The information was downloaded in complete format (e.g., all the tweets by country) in a summary table of frequencies. Word counting was applied to all data (social media and mainstream) to obtain the more popular flavour-related word associations in each country, as it has been a common method for analysing information about food studies involving X (Twitter).

After a brief discourse analysis, all the social media mentions were categorized in frequencies of flavour-related words per country through an automatic count using *Synthesio*®, some flavour-related word frequencies were grouped due to their similar nature: the group "alcoholic drinks" included mojito, sparkling wine, cocktail, and drinks. After the grouping, percentages of each category name per country were calculated and flavour-related words with a lower occurrence value than 1% for all countries were discarded to avoid low frequency data. Once an initial list of words was obtained, the same list was searched through Google Trends to identify the number of searches and be able to cross the social media mentions with the Google searches engine (Fig. 11.3). From the scatterplot, a mean score was projected in the X and Y value in a qualitative way to try to split the data into four different categories. (1) Prioritize zone, where there is a high number of social media mentions (sharing), and high number of Google (searches); (2) Watch, where there is a high number of Google searches (consumer interest), but low number of

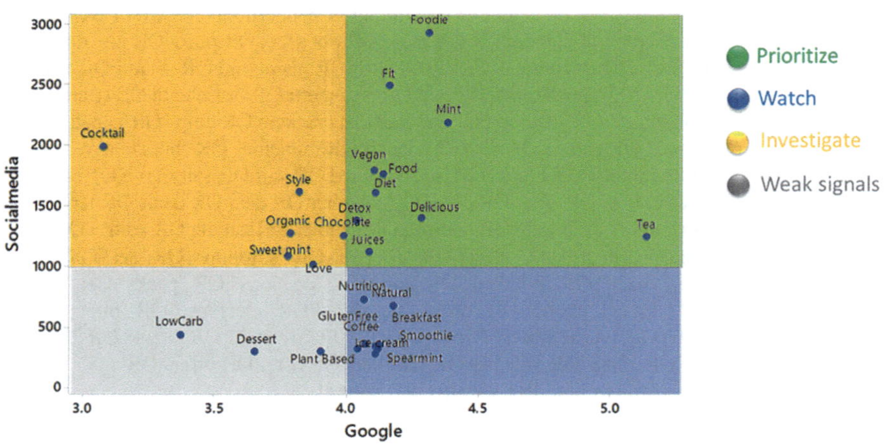

Fig. 11.3 Cross of the social media mentions in the Y axes, and the number of searches in Google (in thousands)

sharing; (3) Investigate zone, which has a bigger number of social media sharing (perhaps denoting knowledge), but low Google searches; and finally, (4) Weak signals zone, where there is a low number of mentions in social media and low Google searches.

For this period of three years, 68,000 total number of social media conversations (posts) were identified. The biological sex declared in the accounts corresponded to 39% of men, and 61% of women. The main countries with the biggest volume of social media conversations where Brazil (40.9%), Argentina (17.9%), Mexico (15.8%), Chile (5.9%), and Colombia (4.4%). With the cross of the sensory descriptors and the product categories and benefits (Fig. 11.3), the author decided to focus on social media mentions that where in the Investigate and Prioritize zone. This zone was qualitatively divided into five dimensions: *alcoholic drinks*, *organic & natural*, *healthy-fit-detox-vegan*, *foodie*, and *tea*. We believe that this can be five dimensions of use of mint in social media (Fig. 11.4).

Figure 11.5 shows an extract of the text and images of the different social media dimensions of mint use. The text of the images is kept in their original language and the key content to identify its dimension is highlighted. The social media channel (e.g., Instagram, X, etc.) was deleted for this publication, together with the data and number of followers and likes, to avoid any possible tracking to the original source. The original images that the consumers used were replaced with royalty free images that were close enough to the original picture. The key content or category, as well as the country of origin, is marked below in the green box. These simple analyses enable the researcher to better illustrate the content and characteristics of each dimension. The discourse analysis in this example is illustrated in its most simple form.

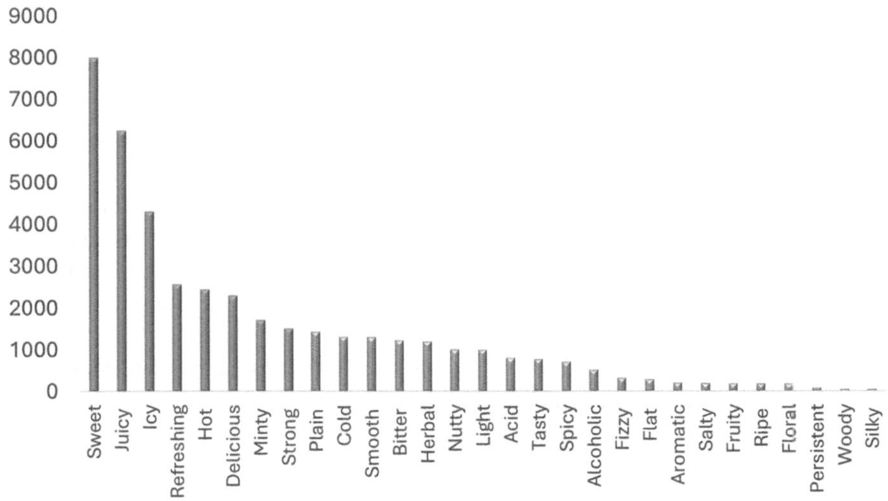

Fig. 11.4 Sensory descriptors identified in social media conversations. The green bars are above average mentions (1400 mentions), and the grey bars are below average

Fig. 11.5 Five qualitative dimensions of mint use in social media (combining the social media conversations with Google search engine). (Images from pixabay.com, Royalty free)

Fig. 11.6 Alcoholic drinks dimension in four different posts showing mint & cocktails, sparkling cocktails, and mixes using peppermint and spearmint. (Images from pixabay.com, Royalty free)

Figure 11.6 illustrates four different posts about mints and alcoholic drinks. The main qualitative categories identified where the use of mint in cocktails, sparkling beverages, and the use of spearmint and peppermint. The original text is at the top part of each image and the original language is also used, either Spanish or Portuguese.

On the other hand, Fig. 11.7 shows the dimension of organic and natural, organic was the common word used in all the social media posts, and here the mint was used in a wide variety of product categories such as ice-cream, salads, roasted vegetables, and desserts.

In Fig. 11.8, the dimension is *healthy, fit, detox, and vegan*. This dimension was harder to cluster, and therefore a set of descriptors are used, contrary to the organic and natural dimension that was very clear. In this dimension, healthy and vegan

Fig. 11.7 Organic & natural dimension; showing organic in different product categories like ice-cream, salad & veggies, and desserts. (Images from pixabay.com, Royalty free)

Fig. 11.8 Healthy, fit, detox, vegan qualitative dimensions of mint use in social media in four different posts. (Images from pixabay.com, Royalty free)

were frequently paired together, even though they do not denote the same benefits, in the social media posts they co-existed. Additionally, cereals and fruits were also present in this dimension. In the case of the Foodie dimension (Fig. 11.9) they included both sweet and savoury products. And finally, in Fig. 11.10 there is the tea dimension, where tea is the dominant, unique product category, with a combination of green tea, fruits and spices.

5 Conclusions and Recommendations

This chapter has demonstrated how netnography and qualitative text analysis have become essential tools in contemporary consumer research, particularly within digital environments. Building on the methodological heritage of classical ethnography, netnography offers researchers the opportunity to explore online communities and

Fig. 11.9 Foodie qualitative dimensions in four different posts: from dessert to meat & proteins, and sea food. (Images from pixabay.com, Royalty free)

Fig. 11.10 Tea qualitative dimensions used in combination with green tea, fruits & spices, and matcha. (Images from pixabay.com, Royalty free)

cultural practices, enabling the investigation of not only consumer behaviours but also the meanings and symbolic constructions attached to consumption. Qualitative text analysis, through methods such as discourse analysis and interpretative phenomenological analysis, provides the depth needed to capture the subjective and contextual dimensions of consumer experiences—insights that purely quantitative approaches often fail to reveal on their own. However, this type of research is not without its challenges. Among the most significant is the ethical dimension, especially concerning the handling of personal data or content published in digital spaces that, while publicly accessible, still warrant respectful and responsible treatment. The absence of specific ethical guidelines for consumer research in social media contexts underscores the need for researchers to rely on robust ethical frameworks, such as those developed in anthropology or by organizations specializing in internet research ethics.

The case study on mint consumption in Latin America clearly illustrates the potential of these methodologies. By collecting and analysing social media data, the research identified cultural patterns and specific dimensions of mint usage, ranging from applications in alcoholic beverages to associations with health and wellness trends. The integration of social media data with tools like Google Trends proved effective for triangulation and qualitative validation, enabling the differentiation between well-established trends and emerging weak signals in the market.

For future research, the author recommends to further develop mixed-method approaches, leveraging the strengths of automated techniques without losing sight of the interpretative and contextual analysis that characterizes qualitative research. Additionally, there is a need to strengthen researchers' training in ethical considerations and to promote the development of specific protocols for netnographic studies within consumer research. This will help ensure that investigations remain rigorous, relevant, and ethically responsible. In sum, netnography and qualitative text analysis offer a fertile path for understanding, in all its complexity, the cultural and sensory dynamics shaping consumer behaviour in the digital age.

References

Álvarez-Gayou Jurgenson, J. L. (2012). *Cómo hacer investigación cualitativa. Fundamentos y metodología*. PAIDOS.

American Anthropological Association. (2025). *Anthropological ethics*. Retrieved from [June, 2025]: https://americananthro.org/about/anthropological-ethics/

Anne-Marie, T., Chau, N., & Kai, K. K. (2017). Ethical questions related to using netnography as research method. *The ORBIT Journal, 1*(2), 1–11.

AOIR–Association of Internet Researchers. (2019). *Internetresearch: ethical guidelines 3.0*. Retrieved from: https://aoir.org/reports/ethics3.pdf.March2021

Arellano-Covarrubias, A., Gómez-Corona, C., Varela, P., & Escalona-Buendía, H. B. (2019). Connecting flavors in social media: A cross-cultural study with beer pairing. *Food Research International, 115*, 303–310.

Arellano-Covarrubias, A., Escalona-Buendía, H. B., Gómez-Corona, C., & Varela, P. (2022). Pairing beer and food in social media: Is it an image worth more than a thousand words? *International Journal of Gastronomy and Food Science, 27*, 100483.

Ashworth, P. (2008). Conceptual foundations of qualitative psychology. En Smith, J. A. (Ed.). Qualitative psychology: A practical guide to research methods. : SAGE.

Atsız, O., Cifci, I., & Rasoolimanesh, S. M. (2022). Exploring the components of meal-sharing experiences with local foods: A netnography approach. *Current Issues in Tourism, 25*(6), 919–936.

Casar, A. (2025). When "anonymous" is not enough: Methodological issues and the safety of human subjects in social media research. *Proceedings of the Linguistic Society of America, 10*(1), 5907–5907.

Choudhary, S., Nayak, R., Kumari, S., & Choudhury, H. (2019). Analysing acculturation to sustainable food consumption behaviour in the social media through the lens of information diffusion. *Technological Forecasting and Social Change, 145*, 481–492.

Cozby, P. C. (1997). *Methods in behavioral research* (6th ed.). Mayfield.

Data, G. (2018). *GDPR: General Data Protection Regulation*. Retrieved from [June, 2025]: https://gdpr-info.eu/

Desmet, P., & Hekkert, P. (2007). Framework of product experience. *International Journal of Design, 1*(1), 57–66.

Eatough, V., & Smith, J. A. (2006). I feel like a scrambled egg in my head: An idiographic phenomenological analysis. Case study of meaning making and anger using interpretative. *The British Psychological Society, 79*, 115e135.

European Commission. (2021). *Research ethics in ethnography/anthropology*. Retrieved from: [June, 2025]: https://ec.europa.eu/info/funding-tenders/opportunities/docs/2021-2027/horizon/guidance/research-ethics-in-ethnography-anthropology_he_en.pdf

Fatemi, H., Kao, E., Schillo, R. S., Li, W., Du, P., Jian-Yun, N., & Dube, L. (2023). Using social media to analyze consumers' attitude toward natural food products. *British Food Journal, 125*(9), 3145–3159.

Forrester, M. (2010). *Doing qualitative research in psychology | A practical guide*. SAGE.

Gallagher, S., & Zahavi, D. (2008). *The phenomenological mind. An introduction to philosophy of mind and cognitive science*. Routledge.

Gómez-Corona, C. (2019). Estudios cualitativos tradicionales. In C. Gómez-Corona & M. Lelièvre-Desmas (Eds.), *Introducción al Analisis Sensorial y Estudios con Consumidores: Nuevas Perspectivas*. XOC ESTUDIO.

Gómez-Corona, C., & Nguyen, H. (2023). Consumer research in social media: Guidelines and recommendations. In C. Gómez-Corona & H. Rodrigues (Eds.), *Consumer research methods in food science*. Humana Press.

Gómez-Corona, C., & Salazar Lama, D. (2023). The analysis of images in consumer research. In C. Gómez-Corona & H. Rodrigues (Eds.), *Consumer research methods in food science*. Humana Press.

Gómez-Corona, C., & Valentin, D. (2019). An experiential culture: A review on user, product, drinking and eating experiences in consumer research. *Food Research International, 115*, 328–337.

Gómez-Corona, C., Escalona-Buendía, H. B., Chollet, S., & Valentin, D. (2017). The building blocks of drinking experience across men and women: A case study with craft and industrial beers. *Appetite, 116*, 345–356.

Guthenberg Project. (2024). *The Project Gutenberg eBook of Argonauts of the Western Pacific*. Retrieved from [June, 2025]: https://www.gutenberg.org/cache/epub/55822/pg55822-images.html

Kodaş, B., & Aksoy, Z. (2022). Understanding of tourists' memorable local food experiences: A netnography study. *Journal of Economy Culture and Society, 66*, 379–396.

Kozinets, R. V. (1998). On netnography: Initial reflections on consumer research investigations of cyberculture. *Advances in Consumer Research, 25*(1), 366–371.

Kozinets, R. V. (2015). *Netnography: Redefined*. SAGE.

Langdridge, D. (2007). *Phenomenological psychology. Theory, research and method*. PEARSON.

Lewis, K., Kaufman, J., Gonzalez, M., Wimmer, A., & Christakis, N. (2008). Tastes, ties, and time: A new social network dataset using Facebook.com. *Social Networks, 30*(4), 330–342.

Malinowski, B. (1979 [1922]). *Los argonautas del Pacífico Occidental*. Peninsula.

McKee, R. (2013). Ethical issues in using social media for health and health care research. *Health Policy, 110*(2–3), 298–301.

Mead, M. (1990 [1928]). *Adolescencia y Cultura en Samoa*. Paidos.

Mellish, E., Gómez-Corona, C., Mariutti, L. R., & Rodrigues, H. (2023). How do people communicate about sensory descriptors in social media? An investigation comprising floral descriptors, four beverages, and thirteen English-speaking countries. *Journal of Sensory Studies, 38*(4), e12837.

Mkono, M., & Markwell, K. (2014). The application of netnography in tourism studies. *Research Notes and Reports/Annals of Tourism Research, 48*, 289–291.

Novoa, A. (2015). Mobile ethnography: Emergence, techniques and its importance to geography. *Human Geographies: Journal of Studies & Research in Human Geography, 9*(1).

Pink, S. (2009). *Doing sensory ethnography*. Sage.

Reygadas, L. (2014). Todos somos etnógrafos. Igualdad y poder en la construcción del conocimiento antropológico. En Oehmichen C. (Ed.), *La etnografía y el trabajo de campo en las ciencias sociales*. Universidad Nacional de México.

Schifferstein, H. N. J. (2009). The drinking experience: Cup or content? *Food Quality and Preference, 20*(3), 268–276.

Sibilia, P. (2017). *La intimidad como espectáculo*. Fondo de Cultura Economica.

Thanh, T. V., & Kirova, V. (2018). Wine tourism experience: A netnography study. *Journal of Business Research, 83*, 30–37.

UNESCO. (2016). *ASA ethics. Ethical guidelines for good research practices*. Retrieved from [June, 2025]: https://ich.unesco.org/doc/src/Example_code_of_ethics-Association_of_Social_Anthropologists_(UK-Commonwealth).pdf

Urry, J. (1984). A history of fields methods. En Ellen, R. F. (Ed.), *Ethnographic research: A guide to general conduct*. Academic.

Valentin, D., & Gómez-Corona, C. (2018). Using ethnography in consumer research. In *Methods in consumer research* (Vol. 1, pp. 103–123). Woodhead Publishing.

Valtonen, A., Markuksela, V., & Moisander, J. (2010). Doing sensory ethnography in consumer research. *International Journal of Consumer Studies, 34*, 375–380.

Vidal, L., Ares, G., Machín, L., & Jaeger, S. R. (2015). Using twitter data for food-related consumer research: A case study on "what people say when tweeting about different eating situations". *Food Quality and Preference, 45*, 58–69.

Wiggins, S., & Riley, S. (2010). Discourse analysis. In M. Forrester (Ed.), *Doing qualitative research in psychology | A practical guide*. Londres.

Wilkinson, L. (2005). *The grammar of graphics. Statistics and computing*. Springer.

Zeller, F., & Woods, J. (2024). *19 Qualitative research quality in social media research*. The Sage Handbook of Qualitative Research Quality.

Zimmer, M. (2010). "But the data is already public": On the ethics of research in Facebook. In *The ethics of information technologies* (Vol. 12, pp. 229–241). Routledge.

Chapter 12
Renewal of Dietary and Nutritional Norms in the Age of Social Media and Educational Challenges

Clémentine Hugol-Gential

1 Introduction

Eating practices are at the core of societal, cultural, and economic transformations. They reflect changes in social norms, aesthetic values, and health concerns. Today, these transformations are accelerating due to the increasing mediatization of food and the influence of social media. Platforms such as Instagram, TikTok, and YouTube shape new ways of perceiving and representing eating, no longer as a mere biological necessity but as an aesthetic, social, and identity-driven experience (Hugol-Gential, 2024). Hashtags like #foodporn and #healthyfood, associated with carefully staged food images, contribute to the establishment of aesthetic and nutritional norms. The "nutritionalization" of food discourses (Poulain, 2009; Romeyer, 2015) is central to this dynamic. This phenomenon, initiated by public initiatives such as the National Nutrition and Health Program (PNNS), has now been amplified on social media (Hugol-Gential, 2022; Munro et al., 2024). Images of balanced meals and before/after body transformations circulating widely on digital platforms establish new bodily norms. Thin and sculpted bodies become symbols of personal success as well as health (Hubert, 2004; Adamiec, 2016). This dynamic, exacerbated by algorithmic logic, encourages individual responsibility for health.

Alongside dominant discourses, resistance narratives emerge. Movements such as fat activism or advocacy for a food transition (Hugol-Gential, 2023) contribute to the revalorization of non-normative bodies and the promotion of more sustainable and inclusive eating practices. However, these resistance narratives clash with algorithmic logic, which favors engaging and visually appealing content, thereby limiting the visibility of counter-discourses on digital platforms.

C. Hugol-Gential (✉)
CIMEOS Laboratory, Université Bourgogne Europe, Dijon, France
e-mail: clementine.hugol-gential@u-bourgogne.fr

M. Visalli, M. V. Galmarini (eds.), *Collection and Analysis of Text Data in Sensory and Consumer Science*, https://doi.org/10.1007/978-3-032-04851-6_12

These transformations in food and body norms call for an educational response. Media and Information Literacy (MIL) appears as a strategic lever to help "eating citizens" critically engage with digital content. By training adolescents to decode algorithmic manipulation mechanisms and produce alternative narratives, MIL can enhance their critical thinking and autonomy. Educational institutions, families, health professionals, and digital platforms all have a collective role to play in promoting more inclusive and responsible practices. This need for collective regulation suggests a shift from individual accountability to shared public responsibility.

The findings presented here are based on an analysis of a selection of social media posts collected through netnographic research (Kozinets, 2009). This approach combines hashtags and keywords related to the body and food while also considering audience engagement, interactions, content, and media extensions (books, blogs, etc.) to capture circulations and interconnections. The two investigated platforms (Instagram and TikTok) are not Open Source, preventing automatic data retrieval. Thus, the research process remains time-consuming since data collection must be done manually. One of the only tools facilitating data retrieval is hashtag-based search. Moreover, the "filter bubble" effect of platforms requires the creation of multiple profiles with different avatars (gender, age, followed content) to maximize thematic coverage and bypass existing algorithmic biases. The manual collection process allows for the aggregation of diverse content and profiles.

The examined corpus, based on several hundred posts, reveals a diversity of profiles: activists, professional influencers openly endorsing products and services, bloggers who sometimes started before Instagram's emergence, professionals (dietitians, nutritionists), and others. The selected profiles vary in positioning, follower count, and thematic scope, extending beyond food issues to topics such as fashion, lifestyle, health prevention, and fitness. Consequently, typology creation proves complex due to the diversity of activities, content production beyond social networks, thematic variations, and influencers who clearly announce their professional status, thematic focus, or promotional links in their bio. While not exhaustive, the selected corpus aims to reflect the plurality of existing situations.

Furthermore, studying body transformation narratives and dietary injunctions requires a dual analysis—textual and iconographic—to understand how bodies and practices are represented, normalized, or contested. By analysing discourse and visual materials, this study explores the discursive and visual mechanisms through which body representations are (re)constructed and showcased in relation to food, as well as their impact on social practices.

This chapter examines food practice transformations through three key dimensions: the impact of social media on dietary habits, the mediatization of food and its methodological challenges, and the perspectives offered by Media and Information Literacy (MIL). This analytical framework seeks to better understand how food and body narratives are constructed, disseminated, and contested in digital spaces while identifying actionable levers to promote more inclusive, diverse, and respectful representations of eating practices and bodies.

2 Social Media and Eating Practices

Social media platforms serve as an echo chamber for the transformation of contemporary eating practices. Far from being mere showcases for sharing experiences, these platforms actively contribute to the dissemination of norms regarding what should be eaten (David, 2021). Through images, short videos, and personal narratives, social media stages food consumption in content where injunctions of pleasure and self-control intersect. Bodies themselves become vehicles for this narrative, particularly through the valorisation of physical transformations and dietary routines.

Eating practices on social media emerge at the intersection of three major dynamics. First, the "nutritionalization" of food discourse transforms eating into a matter of health and performance. Second, social media acts as a prescriber of norms, imposing aesthetic and behavioural models through influencers and algorithms. Finally, the influencer economy reinforces the commercialization of these practices, encouraging consumers to follow diets, purchase food products, or enrol in coaching programs.

2.1 Nutritionalization of Food Discourse

Nutritionalization has led to the massive dissemination of nutritional and health information in public and media food discourse. This process, initiated in the 2000s by public initiatives such as the National Nutrition and Health Program (PNNS), has profoundly altered perceptions of eating (Adamiec, 2017). Health recommendations have turned into normative prescriptions based almost entirely on individual responsibility (Carof, 2021; Hugol-Gential, 2022).

Digital platforms, particularly Instagram, TikTok, and YouTube, play a central role in the dissemination of nutritional injunctions. Hashtags like #healthyfood aggregate millions of posts promoting food perceived as healthy and balanced (Hugol-Gential, 2022). These posts often highlight colourful and aesthetically pleasing dishes, which, while visually attractive, do not always reflect the realities of a balanced diet (David, 2021). Health becomes a visual performance, where the value of a meal is judged based on its Instagram-worthy aesthetics or its association with visible bodily transformations.

One consequence of this nutritionalization of discourse is the individualization of health responsibility. Users are encouraged to see their diet as a personal project to be controlled and optimized. The body becomes the visible marker of dietary success and, by extension, social success (Hubert, 2004), often equated with thinness (Carof, 2021). However, this individualistic logic obscures the social, economic, and cultural determinants of eating practices. The injunctions to eat well are inaccessible to many low-income households, which may lack both financial resources and logistical access to healthy foods (Cavaillet et al., 2021). Additionally, producing these Instagram-worthy meals is often time-consuming and requires culinary skills. The mental and cognitive load associated with "eating well" adds an extra burden, particularly for women, who often remain the primary organizers of meals within families (Lavrilloux & Masson, 2018).

2.2 Social Media Influencers as Prescribers of Norms

Food-related content on social media does not merely reflect eating practices; it actively contributes to their construction. Food influencers serve as both relays and prescribers of dietary and aesthetic norms.

Social media content presents two major trends:

- The enjoyment of food, promoted by the #foodporn movement, which highlights rich, calorie-dense dishes. This type of content celebrates indulgence and the transgression of dietary rules (Marti, 2022).
- Healthy and controlled eating, represented by the #healthyfood hashtag, which emphasizes portion control, ingredient selection, and nutritional balance.

Although seemingly contradictory, these two trends coexist on the same platforms, reinforcing conflicting dietary norms. Users face a dual injunction: they must both indulge (hedonistic logic) and control their eating habits (hygienist logic).

Food influencers play a key role in spreading these norms. Through posts, short videos, and stories, these content creators showcase their meals, eating routines, and body transformations. This process of personal storytelling lends an authentic and emotional dimension to their publications, enhancing their credibility. The perceived closeness between influencers and their audience strengthens their prescriptive power. This symbolic authority is based on personal experience rather than scientific or professional expertise, which explains the widespread acceptance of influencer advice (Staii, 2022).

2.3 The Influencer Economy and Its Implications

The commercialization of eating practices via social media has led to the emergence of new markets dominated by influencers. In 2023, the influencer economy reached a global turnover of €19.5 billion, with a projected 20% growth in 2024 (Hugol-Gential, 2024). This influencer economy operates through several key mechanisms related to food and other industries:

- Brand partnerships: companies producing dietary supplements, protein bars, and various food products collaborate with influencers to promote their items via sponsored posts or discount codes.

- Personal branding: many influencers launch their own food product lines or coaching programs.
- Content monetization: on platforms like TikTok and YouTube, monetization often relies heavily on transformation storytelling. For instance, influencers may showcase their own body transformation, reinforcing their credibility as dietary advisors. This personal narrative, perceived as visual proof of the sincerity of their advice, increases their influence and ability to mobilize their community. These posts fully integrate into influencers'economic models by boosting their visibility and potential commercial partnerships.

However, this dynamic gives rise to ethical concerns. The 2023 French law regulating influencers (Law No. 2023-451) now requires content creators to disclose their commercial partnerships and indicate when their posts are advertisements. This legislative framework aims to protect consumers from "disguised advertising" that previously proliferated on social media, though enforcement mechanisms remain limited, and issues related to public health and dietary content remain largely unregulated.

In terms of eating practices, the influencer economy is transforming consumption norms. Commercial nutritionalization becomes omnipresent, with meals no longer perceived solely as sources of pleasure or social interaction but as activities to be optimized. Influencers actively promote diets or "superfoods" such as high-protein products marketed as enhancing health or physical performance. This logic exacerbates bodily performance injunctions and fosters individual guilt over dietary choices.

3 Issues of Food Mediatization and Methodological Challenges in Analysing Circulated Messages

The mediatization of food on social networks goes beyond the mere presentation of dietary practices. It actively contributes to the reconfiguration of norms that govern the relationship with food and the body. Social networks, through images, videos, and personal narratives, transform individual practices into collective experiences and identity symbols (Fischler, 1988).

The narratives unfolding on these platforms oscillate between hygienist logics, which emphasize health and self-control, and hedonistic logics, embodied by culinary and food-related pleasure. At the same time, social networks also offer spaces of resistance. Counter-discourses emerge, driven by activists advocating for body diversity and sustainable food practices. These narratives, such as those of Fat Activism and Food Activism, contribute to rebalancing dominant norms by valuing bodies and practices that are often made invisible (Carof, 2021). These critical narratives aim to redefine standards of food and body representation.

The study of online dietary practices presents significant methodological challenges. Adopting a combinatory approach, based on manual data collection and a detailed analysis of algorithmic dynamics, helps to capture the tensions between normative injunctions and alternative narratives. However, this approach remains dependent on platform evolutions, highlighting the need to develop adaptive methodologies capable of grasping the complexity of digital practices. Furthermore, analysing these dynamics requires a methodological framework suited to the specificities of digital media. Observing content disseminated on Instagram and TikTok has necessitated an adaptation of traditional social science methods. Following Noortje Marres' (2012) reflections on the redistribution of methods, research has employed manual data collection to closely examine the modes of staging and interaction characteristic of food influencers. Since data access is constrained by platform policies, it was crucial to adopt a flexible methodological approach, combining post analysis with a reflection on the inherent limitations of platform structures.

3.1 The Narrativization of Bodies and Meals

The mediatization of food relies on narrative devices that transform dietary practices into identity symbols. On social networks, as in traditional media, images, immersive videos, and personal narratives no longer simply describe what people eat. They construct a visual and emotional narrative around meals, embodied bodies (Giddens, 1991), and dietary practices.

In this research, Instagram and TikTok were identified as key spaces for staging food and the body, particularly due to their highly visual nature and central role in structuring digital trends. The selection of influencer profiles studied was based on several criteria to ensure a diversity of discursive and aesthetic positions, while avoiding an analysis limited to the most visible figures of influencer marketing.

The analysis of posts highlighted how bodies are transformed into objects of visual performance. Before/after photos, ubiquitous on Instagram and TikTok, reflect a valorisation of self-control where food becomes a tool for bodily transformation (Creuzet, 2021). This narrative is rooted in an ideal of body optimization that obscures the material and social constraints that make these transformations inaccessible to everyone. The body becomes visible proof of success, observable by all and subject to public judgment (Lupton, 2017). These narratives contribute to the stigmatization of non-transformed or non-optimized bodies. The discourse of bodily performance assumes that transformation is possible for anyone who follows the right diet or exercise program (Régnier, 2017). This rhetoric of personal transformation is based on a neoliberal vision of individual autonomy, where body control becomes a prerequisite for personal success.

Simultaneously, the aestheticization of meals on these platforms transforms eating into a primarily visual experience. The staging of dishes follows precise codes where presentation takes precedence over the nutritional or gustatory dimension of food. This research has highlighted the structuring effect of algorithmic logics, which prioritize content generating high engagement and, consequently, favour the most aesthetically standardized publications. This staging reinforces the imperative of perfection and food performance (Munro et al., 2024). The narrativization of meals and bodily transformations leads to the construction of implicit social norms (Contois & Kish, 2022). The ideas of control, mastery, and performance thus become imperatives to achieve.

3.2 *Spaces of Resistance and Empowerment*

Social networks do not merely disseminate dominant food and body norms; they also provide spaces of resistance. These counter-discourses, led by social movements and activists, challenge aesthetic and dietary standards. Social networks become collective empowerment spaces where new capabilities emerge (Cefaï, 2016), allowing users to assert their agency against normative injunctions. Two emblematic movements of this dynamic are Fat Activism and Food Activism, which reshape the relationship with bodies and food.

Fat Activism embodies resistance against fatphobia by making non-normative bodies visible and denouncing the stigmatization of larger bodies. Activist accounts contribute to the deconstruction of aesthetic norms and the dissemination of critical knowledge (Marx, 2024). These narratives do not merely recount individual experiences; they carry a collective demand for visibility. However, this visibility is accompanied by symbolic violence and silencing, as highlighted by Roxane Gay (2017), who denounces how larger bodies are forced into discretion to avoid disturbing normative order. This paradox of stigmatized visibility underscores the limitations of empowerment on platforms where algorithms favour standardized content. Despite these constraints, these resistance narratives contribute to redefining body norms, particularly by opening recognition spaces for users seeking new models of identification.

Similarly, Food Activism introduces a critique of dominant dietary practices and promotes sustainable and frugal eating habits. Unlike #healthyfood and #foodporn narratives, which often focus on visual performance, Food Activism advocates for food transition. These narratives aim to repoliticize eating by highlighting the social and environmental implications of food choices.

However, these resistance dynamics encounter the algorithmic logics of platforms. Narratives promoting transition and body diversity often receive less engagement than spectacular or normative content. Additionally, brands sometimes appropriate these activist narratives (Sedda & Botero, 2002), particularly by marketing sustainably labelled products or endorsing a body-positive approach, neutralizing the initial critique. Nevertheless, these movements demonstrate that social networks can be collective empowerment spaces where the redefinition of norms and the politicization of narratives contribute to transforming collective imaginaries.

Observing discursive dynamics around Fat Activism and Food Activism highlights a fundamental paradox of social networks: while they serve as spaces of empowerment, they also impose constraints that limit the reach of contestations. These tensions emphasize the importance of combining content analysis with reflections on digital architectures to better grasp the underlying logics that govern the circulation of food norms and counter-norms.

4 Media Education: Perspectives and Challenges

Given the powerful influence of social networks, Media and Information Literacy (MIL) is emerging as a crucial tool to help younger generations navigate food and body norms. Digital platforms do not merely disseminate content; they shape collective imaginaries and contribute to the construction of social norms. Algorithms, influencers, and engaging visual content impose beauty standards and dietary practices that can have harmful effects on physical and mental health, particularly among adolescents, who are highly active on social media and in the midst of identity formation.

To counter these dynamics, it is no longer enough to place the burden of responsibility solely on families and teachers. A broad coalition of actors—including platforms, public institutions, health professionals, and young people themselves—must be mobilized. MIL should help individuals understand algorithmic mechanisms, decode aesthetic and commercial narratives, and promote alternative discourses. The goal is to equip young people with critical thinking skills and the ability to produce counter-discourses that highlight the diversity of body types and eating practices.

Beyond the messages circulating on social networks, it is also crucial to examine how these messages are received. This can be achieved through participatory and qualitative methodologies as well as survey-based research. Participatory research methods (including collaborative research and action research) allow for various levels of engagement from participants. While some approaches involve participants in defining research objectives and methodology, others limit their role to providing data (Audoux & Gillet, 2015).

In this study, in addition to analysing the discourse and its circulation, we sought to understand the reception of these messages and the social representations of food on social networks. This was done through workshops conducted in middle school classes with 222 students aged 12–13 and a questionnaire-based survey involving 357 young people aged 13–18. The workshops aimed to capture representations through implicit activities, such as sorting tasks where students categorized about fifteen influencers based on their own criteria and explained their choices. Another method included free association tasks, asking students: *"When you hear 'food influence', what comes to mind?"* This approach avoids direct confrontation and instead focuses on identifying internalized social representations.

4.1 What Happens on Social Media Does Not Stay on Social Media: Concrete Impacts on Youth Food and Body Norms

Social networks are not just virtual spaces for sharing and interacting; they are powerful tools shaping behaviours and perceptions of the body among (young) people. What happens on social media does not remain confined to the digital sphere. One of the most concerning consequences of this influence is the rise of eating disorders

(EDs). In France, an estimated one million people suffer from EDs, with a particularly high prevalence among adolescents. Disorders such as anorexia and bulimia often emerge during adolescence and represent the second leading cause of mortality among 15–24-year-olds, after road accidents.

Social networks are believed to amplify this phenomenon. A 2016 study (Sidani et al., 2016) found that young adults who spend the most time on digital platforms are 2.2–2.6 times more likely to develop an ED. This trend is not solely explained by the amount of time spent online, but also by how the platforms are used. A 2021 literature review (Padin et al., 2021) identified five key factors influencing the development of EDs among young people: time spent on social networks, types of interactions generated, sociocultural influences, self-esteem, and body image satisfaction. These findings indicate that platforms are not neutral; they amplify body norms that weigh heavily on young people.

Research conducted through a questionnaire-based survey of 357 young people aged 13–18 revealed clear trends in the types of content adolescents follow (Hugol-Gential, 2024). Entertainment, sports, food, and beauty are among the most popular themes, highlighting the central role of aesthetic norms in their daily lives. Popular influencers in France like *Squeezie* and *Léna Situations* attract young audiences with entertaining content, but many also follow figures in the food and fitness industries. Notably, 40% of surveyed youth reported following influencers in these fields.

Social networks exert significant normative pressure on young people, particularly regarding body image and consumption practices. This aesthetic pressure leads some to alter their appearance using beauty filters or image cropping. Algorithms, by prioritizing the most engaging content, promote slim, muscular, and photogenic bodies, reinforcing these aesthetic imperatives. Meanwhile, influencers act as trendsetters, endorsing fitness and nutrition products, legitimizing their recommendations through personal body transformation narratives. This model of legitimacy is reinforced by key popularity indicators such as follower count, which grants influencers authority in health and nutrition matters. This dynamic fuels a mimetic consumption pattern, with 15.9% of surveyed youth stating that they had purchased fitness products recommended by influencers, further contributing to the commodification of body transformation.

Additionally, it is important to note that 80% of 13-year-olds are active on social media, and 50% of children join platforms as early as age 11 (Digimind Study). This early access has led the French legislature to pass a law on July 7, 2023, prohibiting minors under 15 from registering on social media platforms without parental authorization (Law No. 2023-566). This measure represents an attempt to regulate digital usage and limit young people's early exposure to potentially harmful content.

4.2 Media Education: Addressing Food and Body Norms

Media and Information Literacy (MIL) is not limited to technical training on digital tools; it is about fostering adolescents' critical thinking skills so they can analyse and deconstruct the normative narratives circulating online (Cordier, 2021; Jehel,

2022). This is especially important as digital platforms amplify body performance narratives through algorithms that favour "perfect" bodies and highly aestheticized meals, at the expense of more inclusive and diverse representations (Hugol-Gential, 2024).

A broader concept of digital literacy is necessary (Bosler et al., 2019). This goes beyond understanding digital tools and interfaces; it encompasses the ability to analyse, critique, and produce content. It involves understanding algorithmic logic, recognizing persuasive techniques used by influencers and platforms, and distinguishing commercial narratives from critical discourse. Young people need to learn how to identify body transformation narratives and dietary injunctions promoted by influencers.

A key aspect of MIL is the critical assessment of algorithms. Social media content feeds are not neutral; they are driven by algorithms designed to maximize user engagement. The most visually appealing content—sculpted bodies and vibrant meals—are prioritized, while alternative and critical narratives are pushed to the margins. Young users must become aware of how their interactions (such as likes, follows, comments) influence algorithms and shape their personalized content feeds. This awareness is essential to mitigate algorithmic reinforcement loops that trap adolescents within normative narratives on body and food.

MIL also extends to parents and teachers, who play a crucial role in guiding young people. Acknowledging this need, some platforms have begun implementing initiatives to better protect young users. For instance, in 2023, Meta launched the *"Parents' Campus"* program to educate families on parental control tools and digital usage regulations. This initiative was expanded in 2024 with the launch of a youth-specific version of Instagram, designed to provide a safer environment with enhanced parental controls and age-appropriate content.

However, it is important to remain critical of these initiatives. While they may be useful, they do not challenge the fundamental logic of algorithms or the dominance of commercial narratives. Instead, they shift the responsibility for regulation onto parents and children while preserving the economic interests of platforms. Families, often ill-equipped to counter the persuasive power of visual narratives, find themselves on the front lines. Strengthening collective digital literacy is therefore essential to help young people, parents, educators, and health professionals better understand the mechanisms of digital platforms and the media representation of bodies and food.

4.3 From Individual to Public and Collective Responsibility

Historically, digital regulation has relied on individual responsibility, assuming that adolescents and their parents should develop enough critical thinking to recognize social media risks and self-regulate their behaviour. However, this approach has proven inadequate against the persuasive power of platforms and their algorithms. Today, a shift from individual to public and collective responsibility is necessary.

Digital platforms play a central role in this transformation. Engagement-driven algorithms amplify the most engaging content. While Instagram has introduced content warnings, these are reactive measures that do not alter algorithmic operations or challenge commercial narratives. To redistribute regulatory responsibility, platforms must be held accountable. Potential regulatory measures include algorithmic transparency and independent audits to ensure that inclusive narratives are not marginalized in favour of commercial ones. Additionally, allowing users to personalize their content feeds could promote a more diverse range of narratives about food and body image (Hugol-Gential, 2024).

Public institutions play a major role in this process. They must not only regulate platforms and influencers but also strengthen Media and Information Literacy. Shifting from individual responsibility to collective responsibility is essential to counter the overrepresentation of commercial and normative narratives on social media.

5 Conclusion

The analysis of food and body-related practices on social media raises several methodological challenges, requiring an approach adapted to the specificities of the digital environment. The collection and analysis of data from platforms like Instagram and TikTok face limitations imposed by algorithmic architectures, which restrict data access and influence the visibility of studied content. The netnographic method used here (Kozinets, 2009) relies on manual content harvesting, made necessary due to the lack of open-source access on these platforms. Content identification is based on hashtag and keyword searches, complemented by the creation of profiles with diverse characteristics (gender, age, interests) to circumvent the “filter bubble” effect (Pariser, 2011) and access a broader range of narratives. This approach allows for the observation of the circularity of food and body-related narratives, but it remains time-consuming and subject to the evolving nature of digital platforms.

Building the corpus also presents challenges related to the hybrid nature of the analysed profiles. The influencers studied vary in their positioning (activists, nutrition professionals, bloggers, entrepreneurs), the number of followers they have, and the topics they address, often extending beyond the food domain. This diversity complicates the creation of homogeneous typologies and requires a contextualized analysis of content and self-presentation strategies. A qualitative discourse analysis is essential to understanding the symbolic and emotional dimensions embedded in these narratives, complementing a quantitative approach that identifies trends and patterns in the disseminated content.

Finally, integrating audience reception into the analysis constitutes a key methodological axis for assessing the impact of food and body-related narratives on audiences, particularly adolescents. The implementation of participatory workshops and survey-based questionnaires with young people aged 12–18 allows for a cross-examination of content analysis with audience perceptions. These participatory methodologies, inspired by action research and collaborative approaches (Audoux & Gillet, 2015), enable a deeper understanding of internalized social representations and the influence of social media on food and body norms. However, they face inherent limitations related to self-reporting and biases linked to the digital identity construction of respondents. The articulation of these different methodological approaches is therefore essential for comprehending the complex dynamics of food and body-related practices in the digital age.

Social media platforms are transforming food practices and body perception by amplifying logics of prescription, resistance, and education. Sculpted bodies and aesthetically curated meals are upheld as models of social and health success, reinforcing injunctions of performance and self-control. This dynamic, driven by the influencer economy, fosters the commodification of food practices, the promotion of diet-related products, and the individual responsibility of food consumers. In response to these pressures, counter-narratives are emerging. Movements such as Fat Activism and Food Activism pave the way for new collective imaginaries. These narratives promote body diversity and call for a more sustainable and inclusive food transition. However, these counter-discourses struggle to gain visibility within digital ecosystems structured by recommendation algorithms that prioritize engaging and aesthetically standardized content. The commercial appropriation of these movements- particularly by brands that advocate body positivity or sustainability—also weakens the initial critical scope of these narratives.

In this context, Media and Information Literacy (MIL) appears as a strategic lever to equip young people, as well as families, educators, and health professionals, with critical skills to navigate the norms imposed by social media. Understanding algorithmic mechanisms, recognizing influencers'persuasion strategies, and distinguishing commercial narratives from critical ones are becoming essential skills. MIL not only strengthens individual reflexivity but also creates spaces for the reappropriation of digital narratives. By encouraging the production of alternative counter-discourses, MIL contributes to redefining social norms related to food and the body, paving the way for greater diversity in online narratives. Finally, it is necessary to shift from a logic of individual responsibility to one of collective and shared responsibility. This means fostering a collective dynamic in which food and body norms are no longer unilaterally dictated by digital platforms and influencers but are redefined through inclusive and diverse narratives. Reclaiming these narratives requires redistributing responsibilities among users, platforms, influencers, and public institutions. Only under these conditions can more inclusive, equitable, and sustainable food practices develop, enabling collective emancipation from normative injunctions.

References

Adamiec, C. (2016). *Devenir sain. Des morales alimentaires aux écologies de soi*. Presses Universitaires de Rennes.

Adamiec, C. (2017). *Manger sain n'est pas si sain. Raisonner l'alimentation santé*. Hachette.

Audoux, C. et Gillet, A. (2015). Recherches participatives, collaboratives, recherches-actions. Mais de quoi parle-t-on ? Dans Les chercheurs ignorants (dirs.), *Les recherches-actions collaboratives. Une résolution de la connaissance*. Presses de l'EHESP.

Bosler, S., Pascau, J., Pleau, J., & Fastrez, P. (2019). Des concepts au terrain : questionnements relatifs à la culture numérique en éducation aux médias et par les médias. *Communication et langages, 201*(3), 41–66.

Carof, S. (2021). *Grossophobie : sociologie d'une discrimination invisible*. Maison des Sciences de l'Homme.

Cavaillet, F., Darmon, N., Dubois, C., Gomy, C., Paturel, D., & Pérignon, M. (2021). Vers une sécurité alimentaire durable : enjeux, initiatives et principes directeurs. Rapport de synthèse. *Terra Nova*.

Cefaï, D. (2016). Publics, problèmes publics, arènes publiques… Que nous apprend le pragmatisme ? *Questions de communication, 30*, 25–64.

Contois, E. J. H., & Kish, Z. (2022). *Food Instagram: Identity, influence, and negotiation*. University of Illinois Press.

Cordier, A. (2021). Exercer un esprit critique en confiance : le défi de l'éducation à l'information et aux médias. *Hermès, La revue, 88*, 168–174.

Creuzet, M. (2021). *Corps sous influence. Les réseaux sociaux, entre carcan et émancipation*. L'Harmattan.

David, M. (2021). *Contribution des médias sociaux aux représentations et aux pratiques d'une alimentation saine chez les jeunes*. Thèse de doctorat en gestion, Normandie Université.

Fischler, C. (1988). Food, self and identity. *Social Science Information., 27*, 275–292.

Gay, R. (2017). *Hunger. A memoir of (my) body*. HarperCollins.

Giddens, A. (1991). *Modernity and self-identity: Self and society in the late modern age*. Cambridge University Press.

Hubert, A. (Ed.). (2004). *Corps des femmes sous influence : questionner les normes*. Les cahiers de l'Ocha.

Hugol-Gential, C. (2022). Des corps et des assiettes sur Instagram : diffusion des normes de santé et diététiques. *Etudes de communication, 58*, 105–124.

Hugol-Gential, C. (2023). L'alimentation durable en France : enjeux de politisation et de communication. *Communiqué*, 37.

Hugol-Gential, C. (2024). *Corps, Alimentation et Réseaux Sociaux*. Editions le murmure.

Jehel, S. (2022). *L'adolescence au cœur de l'économie numérique. Travail émotionnel et risques sociaux*. Etudes et controverses, Institut National de l'Audiovisuel.

Kozinets, R. (2009). *Nethnography: Doing ethnographic research online*. Sage.

Lavrilloux, M., & Masson, E. (2018). Le régime : une affaire de femme et non d'homme ? Influence des représentations sociales sur la description et la nomination des pratiques de contrôle alimentaire. *Les cahiers internationaux de psychologie sociale, 117–118*, 71–95.

Lupton, D. (2017). Digital media and body weight, shape and size: An introduction and review. *Fat Studies, 6*(2), 119–134.

Marres, N. (2012). The redistribution of methods: on intervention in digital social research, broadly conceived. *The Sociological Review, 60*, 139–165.

Marti, C. (2022). #Foodporn : nourrir le dispositif des images. *Communication et Langages, 213*, 107–122.

Marx, D. (2024). *Dix questions sur la grossophobie*. Editions Libertalia.

Munro, E., Wells, G., Paciente, R., Wickens, N., Ta, D., Mandzufas, J., Lombardi, K., & Woolard, A. (2024). Diet culture on TikTok: A descriptive content analysis. *Public Health Nutrition, 27*(1).

Padin, P. F., González-Rodríguez, R., Verde-Diego, C., & Vázquez-Pérez, R. (2021). Social media and eating disorder psychopathology: A systematic review. *Cyberpsychology: Journal of Psychosocial Research on Cyberspace, 15*(3), Article 6.

Poulain, J.-P. (2009). *Sociologie de l'obésité*. Presses Universitaires de France.

Régnier, F. (2017). Vers un corps féminin sur mesure : l'alimentation et les techniques de la corpulence en France et aux États-Unis (1934–2010). *L'année de la sociologie, 37*, 131–162.
Romeyer, H. (2015). Le bien-être en normes : les programmes nationaux nutrition santé. *Questions de communication, 27*, 41–61.
Sedda, P., & Botero, N. (2002). La voix inaudible de la différence. Présence numérique des personnes porteuses d'anomalies dermatologiques. In *Donner la parole aux "sans voix"? Construction sociale et mise en discours d'un problème public, Ferron B, Née E, Oher C.* Presses Universitaires de Rennes.
Sidani, J. E., Shensa, A., Hoffman, B., Hanmer, J., & Primack, B. A. (2016). The association between social media use and eating concerns among US young adults. *Journal of the Academy of Nutrition and Dietetics, 116*(9), 1465–1472.
Pariser, I. (2011). The filter bubble. What the internet is hiding from you. The Penguin Press.
Staii, A. (2022). La participation des "influenceurs" non spécialisés aux communications publiques de santé : enjeux de légitimation. *Études de communication, 58*, 125–145.

Index

M. Visalli, M. V. Galmarini (eds.), *Collection and Analysis of Text Data in Sensory and Consumer Science*, https://doi.org/10.1007/978-3-032-04851-6

GPSR Compliance

The European Union's (EU) General Product Safety Regulation (GPSR) is a set of rules that requires consumer products to be safe and our obligations to ensure this.

If you have any concerns about our products, you can contact us on ProductSafety@springernature.com

In case Publisher is established outside the EU, the EU authorized representative is:

Springer Nature Customer Service Center GmbH
Europaplatz 3
69115 Heidelberg, Germany

Batch number: 09457520

Printed by Printforce, the Netherlands